Bui's Maths Book

A Compendium of
Mathematical Invention

Volume 1

R H S White

T0222816

Order this book online at www.trafford.com
or email orders@trafford.com

Most Trafford titles are also available at major online book retailers.

Note for Librarians: A cataloguing record for this book is available from Library
and Archives Canada at www.collectionscanada.ca/amicus/index-e.html

Printed in Victoria, BC, Canada.

ISBN: 978-1-4251-7868-0 (sc)
ISBN: 978-1-4269-1413-3 (eBook)

Library of Congress Control Number:

*We at Trafford believe that it is the responsibility of us all, as both individuals and corporations,
to make choices that are environmentally and socially sound. You, in turn, are supporting this
responsible conduct each time you purchase a Trafford book, or make use of our publishing services.
To find out how you are helping, please visit www.trafford.com/responsiblepublishing.html*

*Our mission is to efficiently provide the world's finest, most comprehensive book publishing
service, enabling every author to experience success. To find out how to publish your book, your
way, and have it available worldwide, visit us online at www.trafford.com*

Trafford rev. 08/06/09

 www.trafford.com

North America & international
toll-free: 1 888 232 4444 (USA & Canada)
phone: 250 383 6864 ♦ fax: 250 383 6804 ♦ email: info@trafford.com

Preface

On contemplating his life as a mathematician, Sir Isaac Newton said that at times he felt like a small boy playing on the seashore and occasionally finding an interesting pebble while the vast ocean of mathematics lay before him.

We all enjoy the tales of Shakespeare or Charles Dickens and feel good if we manage to solve a crossword. We can also enjoy the subtlties of logic, symmetries and patterns, the rediscovery of a theorem that we had learnt at school but had since forgotten or uncovering the reasoning in a proof that we had not seen before.

Mathematics is a neglected part of British culture. In England, if you watch the quiz shows that are a popular diversion on British TV you will find plenty of evidence. On 'The Weakest Link' the host, Ann Robinson, thinks that the height of mathematical achievement is to be able to add together a pair of two digit numbers. Jeremy Paxman, the quizmaster for 'University Challenge' will sneer at the student who thinks that 'The Jew of Malta' was written by Shakespeare and not Christopher Marlowe while happily professing an ignorance of mathematics. When interviewing Professor Brian Cox, John Humphrys freely admitted to 'knowing nothing about science'. It seems that the celebrities of today may reasonably be expected to have some knowledge of poetry, literature, history, the cinema and pop music but to be asked a question concerning mathematics would be deemed to be unfair (unless it was adding two numbers together).

If the ideas that we learn at school are not continually used or refreshed then they will recede into the dusty corners of our minds and slowly fade away. Children study Shakespeare and Dickens at school and enjoy the stories on film or TV in later life but most children will study mathematics to the standard of O level and from that point onwards lose what interest they had. TV and radio are very good at refreshing our love of the Arts but very poor at shoring up the mathematical confidence of their audiences. Without such stimulation one's mathematical confidence will diminish and not grow.

Part of the function of this book is to show how a set of simple axioms and definitions can quickly build up into a solid theory by means of a few theorems and that the results we obtain are often evident in or relevant to the real world around us.

Chapter 1 investigates how the Babylonians, the Egyptians, the Greeks and the Maya represented numbers and how Archimedes calculated the number of grains of sand needed to fill the Universe. Chapter 2 illustrates how quickly Euclid's axioms build up a theory of plane geometry. Chapter 3 discusses Pythagoras, his music and his Theorem. We give eight proofs of the Pythagoras theorem and discuss Pythagorean triples and Fermat's Last theorem. Chapter 4 proves some properties of the Greek Pythagoras diagram. Chapters 5,6,7 and 8 are related in that chapter 5 introduces the binary, octal and Hex number systems and codes that are used in computers. Chapter 6 introduces Boolean algebra and chapter 7 shows how Boolean algebra can be applied to the design of computer circuits. Chapter 8 is concerned with clock arithmetic and how it is used to represent negative numbers in the computer. Chapter 9 is a collection of diverse problems including the wobbly table, the design of a football, how to cross a river and how to make a snooker ball return to its starting point. Chapter 10 presents some theorems from topology including Euler's Theorem for solid shapes, the Squiggles Theorem for plane networks, Pick's Theorem and Plato's solids (all with proofs). Chapters 11, 12 and 13 are also related. Chapter 11 introduces coordinate geometry with applications to the straight line and circle, chapter 12 shows how transformations can be represented using coordinates and matrices and chapter 13 introduces the theory of groups and shows how transformations and symmetries have the structure of a group. Chapter 14 is devoted to the important method of mathematical proof called induction and chapter 15 concerns probability, arrangements and selections.

Volume 2 continues the story with chapters on Sequences and Series, Trigonometry, Special Relativity, Complex numbers, Vectors, Calculus and Conics.

This is not a mathematics textbook although there are many solved examples and exercises, all with answers, and it is hoped that it will appeal both to the general reader and to the mathematics specialist.

Bui's Maths Book is also an appeal to TV and radio programme makers to bring mathematics into mainstream programming and not to hide it away at 5 o'clock in the morning, but the treatment needs to be more than that. We need programmes that are there for the general public interest and not just for those studying for an examination in mathematics. It is not only those studying for an examination in English Literature who would watch "The Merchant of Venice" on TV. In the age of the Internet and space travel people want to see the equations and how they are solved. The theorems that we learned at school should be recalled and built upon, not forgotten. The cynical publishers dictum that "each equation will halve the number of readers" should not hold any more than an equation on the TV screen would halve the number of viewers.

However, the main motivation behind Bui's Maths Book is simply a passion for a subject that I feel should be enjoyed and better understood by many more people.

"Now we can see what makes mathematics unique. Once the Greeks had developed the deductive method they were correct in what they did, correct for all time. Euclid was incomplete and his work has been extended enormously but it has not had to be corrected. His theorems are, every one of them, valid to this day" — Isaac Asimov

"It isn't right to admit that you know nothing about the Arts so why should it be acceptable for science?"
— Professor Brian Cox commenting after John Humphrys said he knew nothing about science.

I would like to express my appreciation and gratitude to Dr Guy Vine, Jeff Terrell and Gerry Craddock for their encouragement and for reading (and criticizing) large chunks of the text.

Any errors belong to me and if you find any I would be grateful if you would tell me where they can be located.

Lastly, of course, I express my gratitude to my wife Phillipa for her encouragement, her penetrating criticisms of some of my verbiage and grammar and for her help with the design of the cover.

Introduction

Friday afternoon in mid-summer is not the most favourable time to teach a class of electronics students in their late teens but that was the situation in which I found myself when I took up a temporary teaching post at Croydon College in the third term of the 1998-99 academic year. Most of the students were anxious to get away early and only a handful stayed on till 5pm to catch up with assignments. Two students, however, known to me as Miah and Bui would always wait till the end of the session when all the other students had left and then start plying me with all kinds of questions on mathematics.

Friday afternoons always meant an extra half an hour or so at the board "doing maths".

Realistically, as long as you can read the gas meter and check the speedo on your car, that's about all the maths the average citizen needs to know to survive in the modern world. Once you can press the right buttons on the calculator, the arithmetic skills needed today are minimal. So, why do maths? But we could also ask why read poetry? Why listen to music? Why look at paintings? Why do anything that is not really necessary for your survival as an intelligent animal in a competitive world? Perhaps the answer is that you are not just an animal trying to survive. You have in your head, the most intricate and complicated computing machine known to man. A dog has a brain but a dog cannot read poetry or gain pleasure from hearing music or looking at paintings. You have an enquiring mind that appreciates patterns, symmetries, logical argument and paradoxes. This is why we 'do maths', not to survive (unless you happen to be an engineer or a maths teacher) but because you can gain some pleasure and satisfaction from recognizing patterns, symmetries and subtle arguments. Mathematics is not just another subject on the school syllabus, it is a way of looking at the world and a way of thinking about things. I recall sixty maths students packed into the Mill Lane lecture theatre, in Cambridge, many years ago. The lecture on mathematical topology was coming to a close. The 20 foot blackboard was covered wall to wall with symbols, without a word of English and I, like most of those present, had drifted into a bemused soporific torpor having been completely lost after the first ten minutes. But the lecturer concluded with the words "and there you have it ladies and gentlemen – the Jam Sandwich Theorem" – there was a rustle and stirring among the students – here at last was something we could recognize. The lecturer proceeded to explain that he had proved that any three point sets in three dimensional space could "each and severally" be divided into equal halves by a single flat plane. In other words, given a hunk of bread (two slices if you like), a lump of butter and a blob of jam, you can always cut the jam sandwich exactly in half so that each half of the sandwich has half the bread, half the butter and half the jam. It was most refreshing to realize that such a mass of mathematical symbols and notation could produce such an easily understood result. Maybe this is the problem that many people have with mathematics, being able to cut through the mass of notation and equations that are so often encountered, to see some relevance to the real world or simply stated ideas.

This is a book for mathematics enthusiasts. Not necessarily competent mathematicians but those who have the curiosity to want to know how it works and what it is all about. I hope that it will appeal to both beginners, engineers, scientists, teachers and those of a less scientific persuasion. No attempt has been made to avoid equations and symbols, after all, that is the language of mathematics, but I hope that each chapter will impart some of the wonder and thrill that we experienced on that day in Mill Lane. It is dedicated to Bui Van Dong who once playfully suggested that I should write a book on "all that stuff" and to those students who have provided me with many of those golden moments when the spirit of youthful enquiry awakens the interest of a teacher to realize that his passion for mathematics is something worth sharing.

R.H.S.White
Surrey, England
2008

Volume 1

Contents

Chapter 1

The Invention of Numbers 1

Chapter 2

Rotations, triangles and Circles 61

Chapter 3

Pythagoras 99

Chapter 4

Squares   136

Chapter 5

Number bases, Codes and Factors   148

Chapter 6

Boolean Algebra   182

Chapter 7

Logic Circuits   208

Chapter 8

Complements and Clocks   234

Chapter 9

Mathematical Thinking   255

Chapter 10

Topology   313

Chapter 11

Chapter 12

Transformations and Matrices   400

Chapter 13

Chapter 14

Induction   483

Chapter 15

===============================

CHAPTER 1

The Invention of Numbers

The Lubombo Bone

The Lubombo mountains rise in the east of Swaziland, sandwiched between the Nyetane river and the border with Mozambique. The southern end of the range borders on South Africa.

200 000 years ago, a small community of hunter gatherers set up home in the caves of the Lubombo mountains and the remains of their charcoal fires, stone axes, knives, painted shells and cave paintings have been discovered during excavations carried out in the 1970's. In one particular cave, called Border Cave, the fossilised leg bone of a baboon was found. It seems that it was carved by a woman who wanted to keep a record of the passing of the days of the month. It has 29 notches carved into it. It has been estimated to be 37000 years old and it may be the oldest fossil to be discovered which has a mathematical significance. It is the earliest evidence we have of one of our ancestors taking an interest in counting.

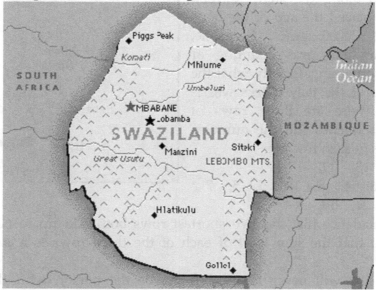

Web pages from South Africa seem to write Lebombo as in the map above, whereas pages from Swaziland write Lubombo. It is not clear where the border lies, some pages put the caves in South Africa and some place it in Swaziland. There is now an interest in tourism the area and it makes one wonder if there are rival claims for the ownership of Border Cave but, be that as it may, some 37 000 years ago an African lady patiently recorded the passing of the days of the month on the leg bone of a baboon.

Ishango

Lake Edward straddles the border between Uganda and the Congo. It is one of the sources of the river Nile. The Congo is now called Zaire and lake Edward appears on some maps as Lake Idi Amin Dada, but I do not think this name will last. For a few hundred years, about 20 000 years ago, the Ishango peoples settled the shores of Lake Edward, hunting, gathering and fishing. The settlement survived for only a few hundred years before being destroyed by a volcanic eruption.

In 1962, Jean de Heinzelin de Braucourt, a geologist and field worker for the Royal Belgian institute of Natural sciences made the find that brought him international fame. It is a bone that has a quartz writing tip fixed to one end and a number of groups of tally marks along the shaft. Carbon dating has placed the age of the bone to around 20000 years. Someone, 20,000 years ago, was counting or recording something. Exactly what the significance of the tally marks was is not sure. There are three rows of groups of scratches and the numbers of scratches in each row are:

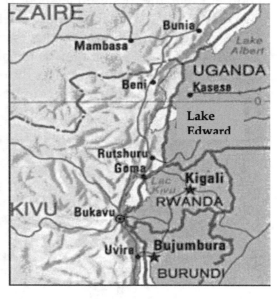

    11, 13, 17, 19

    3, 6, 4, 8, 10, 5, 5, 7

    11, 21, 19, 9

Is it a coincidence that the two shortest rows both add up to 60? Is it a coincidence that the sum total of each of the three rows is a multiple of twelve?

Of course, we will never know, but maybe the person with the quartz writing stick twenty thousand years ago was an Archimedes, a Newton or an Einstein of the ancient world puzzling over the patterns of the stars in the sky or the cycles of the sun and the moon or simply fascinated by the abstract concept of number.

We may never know, but here seems to be the earliest evidence of the germ of an interest in abstract numbers.

The Ishango Bone

In 1937, Dr Karl Absolon was searching for relics of stone age man in excavations at Vestonice in Czechoslovakia and he was fortunate to come across what is known as the wolf's bone. It is the bone of a wolf that has been dated to about 30,000 years ago. It is believed that a stone age hunter was keeping a record of his hunting prowess. The bone has 57 scratches along its length in groups of five.

But scratches on a bone would not do for larger numbers. Each scratch represented just one thing, but it is clear from these tally marks that these ancient peoples had realised that it was useful to group the tally marks together into bundles. Mankind would soon decide that it was easier to deal with larger numbers if they invented symbols for the number of objects in a group. Inventing symbols for the number of objects in a group would lead to the invention of a number system and the beginnings of arithmetic.

The Sumerians

The nomadic tribes of the Middle East at around 4000 B.C. could not stay in one area for long. The food they gathered and the animals they hunted would quickly be used up if they did not move on. However they were beginning to gain expertise in domesticating the wild sheep and goats and planting food crops and as a result their way of life would take on significant change.

It is not known where they came from, but around 3500 B.C. the Sumerians, one of these wandering tribes, found themselves in the fertile valley between the river Tigris and the Euphrates in what is now modern Iraq. Here they learnt how to grow crops and tend livestock so that they could abandon their nomadic ways and adopt a new way of life, living in permanent settlements, towns and villages and develop a new civilization and culture. By 3000 BC, the Sumerians had invented various symbols for recording numbers and lengths so that they could count sheep, jars of oil and quantities of grain or fish and measure cloth. At first they used a different symbol for each different kind of article because the idea of an abstract number that exists on its own was missing. We are not sure of all the symbols they used, but three sheep may have been recorded 🐑🐑🐑 or five jars of oil ♥♥♥♥♥ and so on. By 2000 BC they discovered that it was not necessary to used different symbols for different things and they invented the cuneiform (Latin: cuneus=wedge) system using just two symbols that could easily be pressed into the surface of a clay tablet. The clay could be baked hard if a permanent record was needed.

Cunieform Symbols

The symbol for one   T

The symbol for ten   ◄

The number system exhibits the beginnings of a base ten place value system.

The numbers 1 to nine were:

1	2	3	4	5	6	7	8	9

The numbers ten to fifty were

10   20   30   40   50

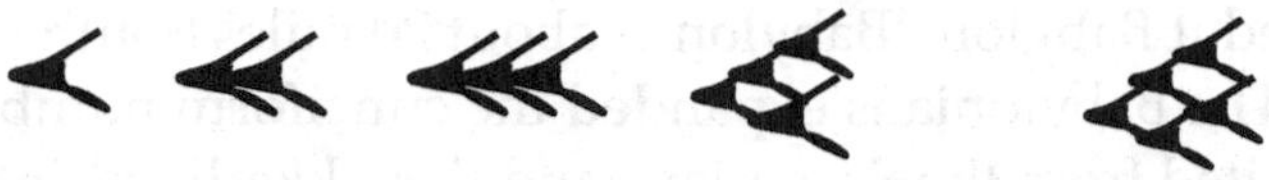

Combining these symbols, they developed a positional number system for example:

31 =

42 =

57 =

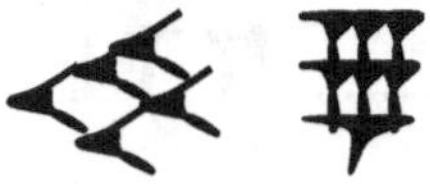

Exercise 1

Perform these additions in cuneiform numbers

1.

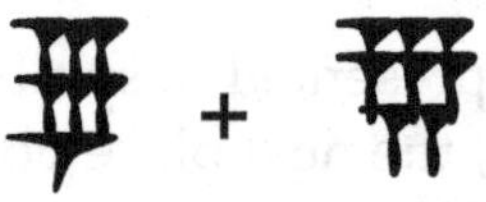

=

2.

=

3.

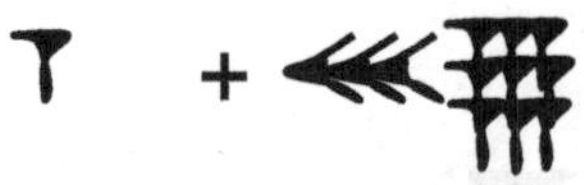

=

4.

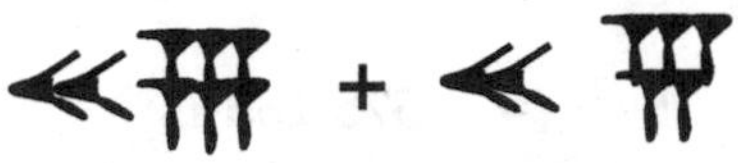

=

But peace and tranquillity for the Sumerians was not to last. In 2300 BC the area was invaded by a tribe called the Akkadians or Assyrians. The Akkadians named their capital the city of Akkad. In 1900 BC Mesopotamia, the land between the rivers, was invaded by the Babylonians who renamed the city of Akkad and called it Babylon. Babylon is about 20 miles South of Baghdad in modern Iraq. The Babylonians expanded the cuneiform number system that they had inherited from the Sumerians and the Akkadians, into a base 60 positional number system. Sixty was probably used because the Babylonian scholars and priests took a year to be 360 days. Today we still use 360 degrees in a turn, 60 seconds in a minute and 60 minutes in one hour, thanks to the scholars of ancient Babylon.

Numbers greater than sixty were represented by showing how groups of sixty there were thus we have:

61	62	63	64	....	121

Notice the important gaps e.g. in 61, to distinguish 61= ⟙ ⟙ from 2= ⟙⟙

However there are still problems: Does ⟙ represent **1** or **60** ?

Sometimes the numerals had to be context dependent! There was no special symbol for zero so you had to recognise from the context whether ⟙ was meant to be a **1** or a **60**.

However, numbers of any size could now be represented in symbols. The first place position represented how many ones, the next place position would represent how many **60s**, then **60x60 = 3600** and the next place position is **60x60x60** ...and so on.

Thus, the Babylonian place values were

60x60x60 =216000	60x60 =3600	60 =60	ones =1

for example

= 21x216000 + 44x3600 + 13x60 + 5 = 375 185

Exercise 2

1. What numbers do the following represent?

   (i)   𝗬𝗬 𝗬

   (ii)   𝗬 𝗬𝗬

   (iii)   𝗬𝗬𝗬

   (iv)   𝗬 𝗬 𝗬

2. What is one million in cuneiform?

Answers:

Exercise 1

1.   2.   3.   4.

Exercise 2

1.   121,   62,   3,   3661

2.

The Egyptians

While the Sumerians were developing their culture in Mesopotamia between 5000 BC and 3000 BC, another civilization was emerging from pre-history along the banks of the river Nile. Agriculture and the domestication of livestock had allowed the Egyptians to enjoy a settled way of life resulting in the development of one of the most impressive early cultures in the history of mankind.

From 3000 BC, along the banks of the Nile, they developed stone architecture, building expertise and a unique hieroglyphic writing style. (Greek hieros=sacred, gluphe=carving)

About 15 miles south of Cairo, at Sakkara we can see the first step pyramids, build for Pharaoh Zosser (2667-2648 BC) but the Egyptian builders soon switched from constructing step pyramids to building pyramids with

smooth surfaces and twelve miles south west of Cairo we find the ruins of Memphis, the capital of the Egyptian state from 2575 to 2130 BC. At Giza, near Memphis, we see the Great Pyramid of Khufu (2589-2566) and the slightly smaller pyramid of his son Khafre (2558-2532). The third pyramid and the Sphinx, with the face of the king and the body of a lion, the largest statue of ancient Egypt, was built for Menkaura son of Khafre.

http://whc.unesco.org/en/list/86

Memphis and the pyramids of Giza are a world heritage site.

Thebes which is now called Luxor, developed from a settlement on East bank of the Nile in 2650 BC and was the capital city of Egypt from 2100 to 1400 BC. The temples at Luxor were built during this period and added to by a succession of 30 different Pharaohs. The complex of temples at Luxor is now regarded as the largest religious site in the world.
Karnak is one and a half miles North of Luxor.
A carved stone dating from 1500 BC, which is now displayed in the Museum of the Lourve in Paris, was found at Karnak.

The stone has the following symbols carved on its surface:

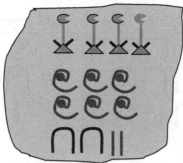

This represents the number **4622** in the Egyptians hieroglyphic system:
The hieroglyphics used for their numbers were these:

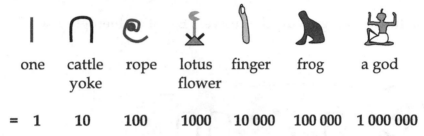

one	cattle yoke	rope	lotus flower	finger	frog	a god
= 1	10	100	1000	10 000	100 000	1 000 000

The Egyptians used a simple base ten system but had no concept of place value and did not have a symbol for zero. The number on the temple at Karnak would give the same value, wherever the symbols happened to be, for example, the following hieroglyphics represent the same number:

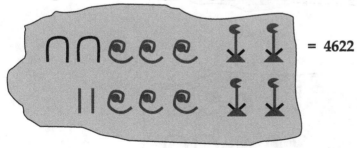

= 4622

Addition using Egyptian hieroglyphs is rather clumsy.

Example

Using the hieroglyph 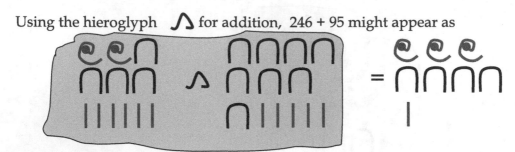 for addition, 246 + 95 might appear as

Exercise 1

Perform the following giving the answers Egyptian hieroglyphics.

1

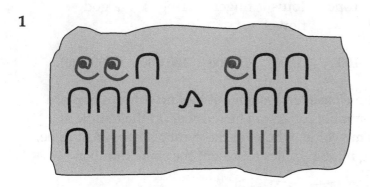

2

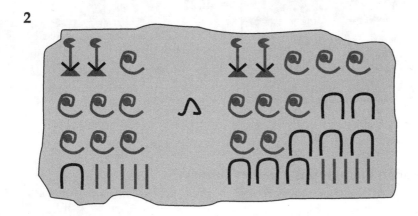

3

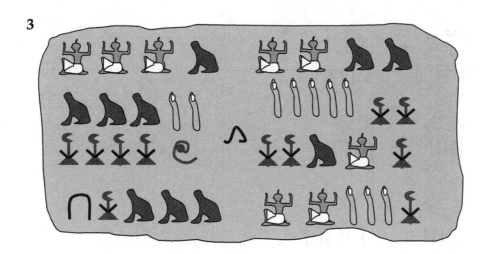

Egyptian Fractions

The Egyptians used the symbol for a mouth, ⊖ to indicate division only used fractions with numerator equal to one, except in these three special cases:

$$\rightleftharpoons = 1/2 \qquad ⊖ = 2/3 \qquad ⊖ = 3/4$$

otherwise, the mouth symbol always indicated "one over", for example,

$$⊖_{|||} = \frac{1}{3}.$$

$$⊖∩∩∩ee∩∩||| = \frac{1}{253}.$$

Exercise 2

Use the Egyptian mouth to find

[1]   ⊖_{|||} + ⊖_{|||}

[2]

⊂⊃ + ⊂⊃
IIII   ∩II

[3]

⊂⊃ + ⊂⊃
IIIII   ∩∩

[4]

⊂⊃ + ⊂⊃
IIIIII   ∩∩∩

[5]

⊂⊃   +   ⊂⊃∩∩∩∩II
IIIIIII

Hieratic Numerals

When the religious leaders of ancient Egypt started to use papyrus they developed a different number system that was more suitable for writing rather than carving on stone and which was used alongside the hieroglyphics. It is called the Hieratic script (Greek: hieratikos=priest). The hieratic script is more concise and easier to write than hieroglyphic but it is still not a place value system.

Alexander Rhind, a Scotsman, purchased a papyrus from a dealer in Luxor in 1858. It is the best surviving manuscript of Egyptian mathematics. It is now in the British Museum where it can be seen, about a foot wide and 15 feet long. It contains calculations written in hieratic script. The Rhind Papyrus was written by a scribe called Ahmes in 1650 BC who stated that many of the results were copied from a papyrus of 1800 BC. It gives tables of fractions and topics on algebra and trigonometry.

The table of fractions expresses **2/n** in the form **1/a + 1/b** for all odd values of **n** from **3** to **101** for example:

$$2/3 = 1/2 + 1/6, \quad 2/5 = 1/3 + 1/15 \text{ etc}$$

The writer of the papyrus had in fact found that

$$\frac{2}{(2n+1)} = \frac{1}{(n+1)} + \frac{1}{(n+1)(2n+1)}$$

The Hieratic Numerals

Thus

Since each numeral has its own symbol, the numbers do not need to be in any particular position.

Exercise 3

Complete

and compare with Qu1 of the previous exercise.

Hieratic fractions

The Hieratic system of numerals also had symbols for fractions but again, they only dealt with unit fractions i.e. fractions that a numerator of one. The "one over" phrase is usually indicated with a dot:

Only unit fractions were used and each had its own symbol with one exception. The fraction **2/3** had its own symbol:

$$2/3 = \text{𓏲}$$

Exercise 4

Perform these additions giving the answers using the Egyptian Hieratic fractions.

[1]

[2]

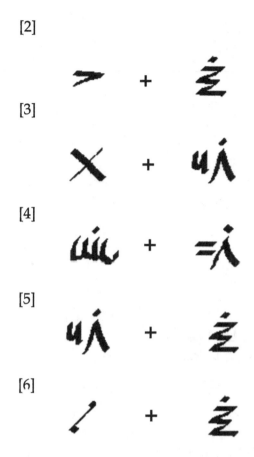

[3]

[4]

[5]

[6]

In 332 BC, Egypt was conquered by the Macedonian king Alexander the Great. He was, in fact, welcomed as a liberator by the Egyptians who had been under Persian domination and in 331 BC he founded Alexandria which replaced Memphis as the capital of Egypt in 320 BC. On the death of Alexander in 323 BC, his generals divided up the Macedonian Empire among themselves and General Ptolemy took over the rule of Egypt bringing Greek learning and culture to the Egyptian capital. Ptolemy 1 was ruler of Egypt from 323 BC to 283 BC. He founded a school of learning at Alexandria that became a magnet for scholars throughout the middle east. Euclid (c330BC-260BC) was one of the first to teach mathematics at the University of Alexandria. The lighthouse at Pharos, an island just off the coast at Alexandria became one of the seven wonders of the ancient world.

===============================

Answers

Exercise 1

1.

2.

3.

Exercise 2

[1]

[2]

[3]

[4]

[5]

Exercise 3

Exercise 4

[1]   [2]   [3]

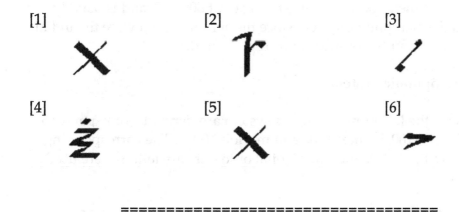

[4]   [5]   [6]

====================================

The Greeks

Although much of the cultural development throughout Europe has evolved from the ancient Greek civilisation the Greek's system of numeration was unsophisticated and cumbersome. The ancient Greeks used two different number systems but neither of these used place values or had a symbol for **zero,** both of which are essential for any sophisticated number system.
The Attic or Herodian system was named after Attica, the name for Ancient Athens or Herodian, Aelius Herodianus , a scholar born in Alexandria in around 180 A.D. Herodian worked in Rome for Emperor Marcus Aurelius as an authority on Greek grammar and writings and died around 250 A.D.
 This system of numeration was in use from about 600 B.C. and is usually referred to as the Acrophonic system since the symbols used were the initial letters of the words for the number to be represented.

The Greek Acrophonic System

The symbols for the key numbers in this system are derived from: pente=5, deka=10, hekaton=100, khilioi=1000 and myrioi=10000. The corresponding initial letters are Γ, Δ, H, X and M. Capital or lower case letters were used, thus we have:

I=i=iota	Γ=π=pi	Δ=δ=delta	H=η=eta	X=χ=chi	M=μ=mu
1 |	5 Γ	10 Δ	100 H	1000 X	10000 M
2 ||	6 Γ|	20 ΔΔ	200 HH	2000 XX	20000 MM
3 |||	7 Γ||	30 ΔΔΔ	300 HHH	3000 XXX	30000 MMM
4 ||||	8 Γ|||	40 ΔΔΔΔ	400 HHHH	4000 XXXX	40000 MMMM
	9 Γ||||	50 ⌐Δ	500 ⌐H	5000 ⌐X	50000 ⌐M

The symbols for 50, 500 , 5000 and 50000 are composite, being combinations of the symbol for 5 with the appropriate power of ten.

Understanding Greek numerals is complicated because different Greek city states used different versions of these symbols and the symbols also varied when used for different measures, e.g. money, weight or distance.

Examples

[1]   5 6 7 8

Γˣ Γᴴ H Γᴬ Δ Δ Γ I I I

[2]   3 8 0 7

XXX Γᴴ H H H Γ I I I

Exercise 1

[1]   What is the Greek Acrophonic for 4444

[2]   What is the Greek Acrophonic for 666

==

The second system that the Greeks used was called the Ionic or Alphabetic system. This system uses the 24 letters of the Greek alphabet together with 3 obsolete Greek letters that were also found in the Phoenician alphabet. The Phoenicians were a Semitic tribe living along the Mediterranean coasts of Syria, Lebanon and Northern Palestine.
Almost all of the Alphabets in use in Europe today are derived from the alphabet invented by the Phoenicians at around 1500 BC.
One of the earliest known uses of the Greek Alphabetic numerals can be seen on coins used in Egypt during the rule of Ptolemy (286-246 BC), after the death of Alexander the Great.

Ionic Greek Numerals

The first 9 letters represent 1 to 9, but with the symbol for 6 represented by the older Greek/Phoenician character digamma (=double gamma).

alpha	beta	gamma	delta	epsilon	digamma	zeta	eta	theta.
A	B	Γ	Δ	E	F	Z	H	Θ
α	β	γ	δ	ε	ƒ	ζ	η	θ
1	2	3	4	5	6	7	8	9

The next 9 letters represent numbers 10 to 90, but again, we have one character from the older Greek/Phoenician alphabet, koppa which represents 90.

Iota	kappa	lambda	mu	nu	xi	omicron	pi	koppa
Ι	Κ	Λ	Μ	Ν	Ξ	Ο	Π	Ϙ
ι	κ	λ	μ	ν	ξ	ο	π	ϙ
10	20	30	40	50	60	70	80	90

The final set of 9 letters represent 100 to 900 with the older Greek/Phoenician character sampi or san representing 900.

Rho	sigma	tau	upsilon	phi	chi	psi	omega	sampi
Ρ	Σ	Τ	Υ	Φ	Χ	Ψ	Ω	Ͳ
ρ	σ	τ	υ	φ	χ	ψ	ω	ͳ
100	200	300	400	500	600	700	800	900

The larger numerals are written on the left.

Examples

[1] 123 = ΡΚΓ or ρκγ

[2] 444 = ΥΜΔ or υμδ

[3] 301 = ΤΑ or τα

Exercise 2

What are these numbers written in Ancient Greek numerals ?

[1] 765 [2] 248 [3] 979

==

Numbers greater than 999 were dealt with using composite characters:

1000	2000	3000	4000	5000	6000	7000	8000	9000
'A	'B	'Γ	'Δ	'E	'F	'Z	'H	'Θ

or 'α 'β 'γ 'δ 'ε 'f 'ζ 'η 'θ

(Note that the dash, or iota, may also be written as a subscript, e.g. 8000 = ₁H)

For numbers greater than 9999 we have:

10000	20000	30000	40000	50000	60000	70000	80000	90000
	β	γ	δ	ε	ς	ζ	η	θ
M	M	M	M	M	M	M	M	M

Examples

[1] $98765 = \overset{\theta}{M}\ \text{'Hψξε}$

[2] $12345 = M\text{'βτμε}$

Exercise 3

Perform the following addition in Greek ionic numerals and then translate into the decimal system:

$$\begin{array}{r} \overset{}{M}\ \text{'β τ μ ε} \\ \overset{\varepsilon}{M}\ \text{'δ τ κ α} \\ \hline \end{array}$$

+

==

Larger Numbers

$\overset{\theta}{M}$ can be regarded as 9×10000 and this simple idea allowed the representation of numbers greater than 90000, for example, 87650000 can be represented by

$$87650000 = \overset{^{i}\eta\psi\xi\varepsilon}{M}$$

For writing *very large* numbers, some Greek mathematicians devised their own systems. We illustrate three of these suggestions using the number 987654321 .

Aristarchus of Samos (c310..c230BC) brought the higher part of the number down in front of the **M**:

9 8765 4321 $= \quad M \overset{\theta}{^{\iota}\eta} \psi \xi \varepsilon M \, {^{\iota}\delta} \tau \kappa \alpha$

Apollonius of Perga (c262..c180BC) suggested using powers of **M** so that

$\overset{\alpha}{M} = M = 10000$

$\overset{\beta}{M} = M^2 \quad \overset{\gamma}{M} = M^3$ **etc** and used the symbol $\chi\alpha L$ as a plus sign thus

9 8765 4321 $= \overset{\beta}{M} \theta \; \chi\alpha L \; \overset{\alpha}{M} \, {^{\iota}\eta\psi\xi\varepsilon} \; \chi\alpha L \; {^{\iota}\delta\tau\kappa\alpha}$

Archimedes (c287..c212 BC)

Along with Newton and Gauss, Archimedes is generally claimed to be one of the most outstanding mathematicians of recorded history. He was born in Syracuse in Sicily but little is known of his life story however the Greek writer Plutarch (AD 46 – 120), describes how Archimedes was killed by a Roman soldier during the capture of Syracuse in the second Punic war between the Greeks and the Romans. A carving of a sphere inscribed in a cylinder was placed on his tomb in Syracuse. It illustrated Archimedes favourite proofs:

[1] the volume of the sphere is two thirds of the volume of the cylinder
[2] the area of the sphere is two thirds of the area of the cylinder

At some time around 250B.C., Archimedes wrote a letter to King Gelon, the king of Sicily, showing how he could calculate the number of grains of sand that would be needed to completely fill the known universe. For this task, Archimedes needed to invent a new way of representing large numbers.

Archimedes and the Sand Reckoner

The **Sand Reckoner** is the name given to the letter that Archimedes sent to King Gelon of Sicily in which he estimates the number of sand grains needed to fill the whole universe.
The largest symbol in the Greek number system was one myriad and so the Greek numbers came to a stop at one myriad of myriads:

one myriad = M = 10000 $\overset{\text{M}}{}$
one myriad of myriads =M.M = 100 000 000, represented by $\overset{\text{M}}{\text{M}}$

The number of grains of sand needed to fill the universe is much larger than this and therefore Archimedes needed to invent a way of describing these huge values.

$\overset{\text{M}}{}$

Archimedes took $\overset{\text{M}}{\text{M}}$ to be the basis of his system for representing large numbers. Unfortunately for Archimedes, he did not have the use of indices for representing powers of numbers. In order to get to grips with his method we will write

$$\Omega = M^2 = 10^8$$

Archimedes defined numbers up to Ω to be numbers of the first order and then, using Ω as a unit, he defined multiples of Ω to be numbers of the second order up to the value Ω^2.
He then uses Ω^2 as a unit for numbers of the third order, thus we have:

First order numbers	1	2	3	4	......	Ω
Second order numbers	Ω	2Ω	3Ω	4Ω	......	Ω^2
Third order numbers	Ω^2	$2\Omega^2$	$3\Omega^2$	$4\Omega^2$	......	Ω^3

·

Numbers of order Ω	$\Omega^{\Omega-1}$	$2\Omega^{\Omega-1}$	$3\Omega^{\Omega-1}$	$4\Omega^{\Omega-1}$	......	Ω^{Ω}

⎱ first period

These are called numbers of the first period.

Let $\Omega^{\Omega} = P$ and we now use P to construct numbers of the **second period**:

Second Period

First order numbers	P	$2P$	$3P$	......	ΩP
Second order numbers	ΩP	$2\Omega P$	$3\Omega P$	......	$\Omega^2 P$
Third order numbers	$\Omega^2 P$	$2\Omega^2 P$	$3\Omega^2 P$	....	$\Omega^3 P$

·
·

Numbers of order Ω	$\Omega^{\Omega-1}P$	$2\Omega^{\Omega-1}P$	$3\Omega^{\Omega-1}P$	....	P^2

⎱ second period

Third Period

First order numbers	P^2	$2P^2$	$3P^2$	......	ΩP^2
Second order numbers	ΩP^2	$2\Omega P^2$	$3\Omega P^2$	......	$\Omega^2 P^2$
Third order numbers	$\Omega^2 P^2$	$2\Omega^2 P^2$	$3\Omega^2 P^2$	......	$\Omega^3 P^2$

·
·

Numbers of order Ω	$\Omega^{\Omega-1}P^2$	$2\Omega^{\Omega-1}P^2$	$3\Omega^{\Omega-1}P^2$	......	P^3

⎱ third period

·
·

and so on for Ω periods:

·
·

Period Ω

First order numbers	$P^{\Omega-1}$	$2P^{\Omega-1}$	$3P^{\Omega-1}$	......	$\Omega P^{\Omega-1}$
Second order numbers	ΩP^2	$2\Omega P^2$	$3\Omega P^2$	......	$\Omega^2 P^{\Omega-1}$
Third order numbers	$\Omega^2 P^2$	$2\Omega^2 P^2$	$3\Omega^2 P^2$	......	$\Omega^3 P^{\Omega-1}$

⎱ period Ω

·
·
·

Numbers of order Ω	$\Omega^{\Omega-1}P^2$	$2\Omega^{\Omega-1}P^2$	$3\Omega^{\Omega-1}P^2$	......	P^{Ω}

The last number in this series is $P^{\Omega} = (\Omega^{\Omega})^{\Omega} = 10^{80\ 000\ 000\ 000\ 000\ 000}$.

Examples

[1]   987 654 321 = 9Ω + 87 654 321

     = 9 second order units + 87 654 321 first order units in period 1

[2]

     $8.10^{25} + 6.10^{20} + 4.10^{15} + 2.10^{10}$

     = $80.\Omega^3$ + 60 000.Ω^2 + 40 000 000.Ω + 200Ω

     =   80 fourth order units
        + 60 000 third order
        + 40 000 200 first order units in period 1

[3] $4.10^{8\ 000\ 000\ 000\ 480}$ = $4.10^{480}.(10^{800\ 000\ 000})^{10\ 000}$ = $4.\Omega^{60}.P^{10\ 000}$

               = 4 units of order 61 in period 9 999

This was the system of enumeration that Archimedes invented in order to describe the number of grains of sand needed to fill his known universe.

Samos

Barely 30 miles long and just 2 miles from the coast of Turkey, lies the island of Samos. Pythagoras was born in Samos in about 580 BC but for now, we are more interested in an astronomer who was born about 250 years later. His name was Aristarchus.

Aristarchus of Samos (c310 — c230 BC) was the first astronomer to state that the Earth revolves around the sun and that the Sun was fixed in the sphere of the stars. He discovered that the Earth was round and was the first person to calculate the relative distances to the Sun and the Moon.
Archimedes used calculations made by Aristarchus in order to calculate "the number of the sand" .. the number of grains of sand needed to fill the Universe.

The Sand Reckoner

Archimedes aims to find an upper bound for the number of grains of sand that will fill the Universe and so any measurement that he has to estimate are always an over-estimates, often by a factor of ten times the recognised value of his time.

Archimedes starts with a poppy seed that he supposes will contain not more than 10 000 = one myriad, grains of sand.

One finger breadth is not more than the width of 40 poppy seeds.

One stadium (a unit of length) is taken as 10 000 fingers.

Notation

In order to explain Archimedes' calculations, we will use **D(x)**, **P(x)** and **Rad(x)** to represent the diameter, perimeter and radius of a sphere or circle **x**.

The size of the Earth

Archimedes' contemporaries estimated the perimeter of the earth to be about 300,000 stadiums. One stadium being about 600 feet, this gives a value of 35 000 miles. The true value is about 24 900 miles. Archimedes takes his estimate for the perimeter of the Earth to be less than ten times this value given by Aristarchus, thus Archimedes takes

$$\textbf{P(Earth)} < \textbf{3000 000 stad} \qquad \textbf{[1]}$$

Since **P(Earth)** = π.**D(Earth)**, Archimedes can then state

$$\textbf{D(Earth)} < \textbf{1000 000 stad} \qquad \textbf{[2]}$$

The size of the Sun and the Moon

Aristarchus had given an estimate of the diameter of the Sun as between 18 and 20 times the diameter of the Moon. The correct value is in fact about 400 times the diameter of the Earth, however, Archimedes assumes that the diameter of the sun is less than 30 times the diameter of the moon and states

$$\textbf{D(Sun)} < \textbf{30.D(Moon)} \qquad \textbf{[3]}$$

Archimedes correctly supposes that the diameter of the Moon is less than the diameter of the Earth, thus

$$D(Moon) < D(Earth) \qquad [4]$$

From [3] and [4] he deduces that

$$D(Sun) < 30.D(Earth) \qquad [5]$$

And using [2], concludes that

$$D(Sun) < 30.1000\ 000\ stad \qquad [6]$$

Archimedes follows Aristarchus in stating that the Sun is fixed in the sphere of the stars and that the Earth moves in a circle around the Sun. He calls the sphere of the orbit of the Earth "**The World**" and now calculates an upper bound for the diameter of the orbit of the Earth.

Aristarchus, by experiment, had estimated that 720 diameters of the Sun would complete the 360 degree circle of the orbit of the Earth and so, Archimedes takes his estimate to be 1000 diameters. He considers a regular **chilleagon** (a 1000 sided polygon) fitted inside the orbit of the Earth and deduces that

$$\text{side of chilleagon} < D(Sun)$$

so that $\qquad$ **P(chilleagon) < 1000.D(Sun)** $\qquad$ [7]

Now consider any circle first with a hexagon and then with an n sided polygon inscribed in it. If n is greater than 6 we have:

$$D(circle) = 2.rad(circle) = \frac{1}{3} . 6.rad(circle) = \frac{1}{3} . P(hexagon) < \frac{1}{3} . P(n\text{-agon})$$

From this with [7], $\qquad$ **3.D(world) < P(chilleagon) < 1000.D(Sun)**

And using [5] we have $\quad$ **3.D(world) < 1000.30.1000 000 stad**

Therefore $\qquad$ **D(world) < 10 000 000 000 stad**

Using his system for representing large numbers we can write as **100Ω stad.**

(Ω = M.M = 1000.1000)
Therefore

$$D(world) < 100\Omega \text{ stad} \qquad\qquad [8]$$

We will now follow his calculation of the number of grains of sand needed to fill the world. He uses the fact that if the diameter of a sphere is multiplied by **k** then the volume of the sphere is multiplied by **k**^{3}.

The Calculation

One poppy seed contains at most 10000 grains of sand
The width of a finger is at most 40 poppy seeds

$\therefore$ a finger sphere　$< 40^3$ poppy seeds

$= 64\ 000$ M grains of sand

$= 6\Omega + 4000$M

$< 10\Omega$ grains

$\therefore$ a 100 finger sphere　$= 100^3$ finger spheres

$< 100.\text{M}.10\Omega = 1000M\Omega$ grains

$\therefore$ a 10000 finger sphere　$= 100^3$ x(100 finger sphere)

$< 100\text{M}.1000\text{M}\Omega = 10\text{M}\Omega^2$ grains of sand

Now a stadium (stad) is 10000 fingers.

$\therefore$ a 100 stad sphere $= 100^3$ x(10000 finger sphere)

< 100 M .$10\text{M}\Omega^2 = 1000\ \Omega^3$ grains of sand

$\therefore$ a 10000 stad sphere　$= 100^3$ x(100 stad sphere)

< 100 M .$1000\ \Omega^3 = 10\ \Omega^4$ grains of sand

$\therefore$ a 100M stad sphere　$= 100^3$ x(10000 stad sphere)

$$< 100 \text{ M} .10 \ \Omega^4 = 1000\text{M}.\Omega^4 \text{ grains of sand}$$

$\therefore$ a 10000M stad sphere $= 100^3 \times (100\text{M stad sphere})$

$$< 100 \text{ M} .1000\text{M}\Omega^4 = 10\text{M}.\Omega^5 \text{ grains of sand}$$

$\therefore$ a 100Ω stad sphere $= 100^3 \times (10000\text{M stad sphere})$

$$< 100 \text{ M} .10\text{M}\Omega^5 = 1000\Omega^6 \text{ grains of sand}$$

Therefore, a sphere with diameter equal to the orbit of the Earth ("the World") will hold not more than $1000\Omega^6$ grains of sand.   [9]

But Archimedes wants to calculate the number for the whole of the Zodiac, which he calls the sphere of the fixed stars. To do this he follows Aristarchus who, according to Archimedes, has brought out a book in which he states that "the fixed stars and the Sun remain unmoved, while the Earth moves around the sun in the circumference of a circle, the Sun lying in the middle of the orbit and that the sphere of the fixed stars, situated about the same centre as the Sun, is so great that the circle in which he supposes the Earth to revolve bears such a proportion to the distance of the fixed stars as the centre of a sphere bears to its surface". Archimedes interprets this to mean that

Earth : World = World : Universe

That is, the ratio of the diameter of Earth to the diameter of the Earth's orbit is equal to the ratio of the diameter of the Earth's orbit to the diameter of the Zodiac.

Now from [8],   **D(world) < 100Ω stad**

and taking **D(earth) = 1000 000 stad** gives

D(world):D(Earth) < 10,000

He therefore deduces that
                D(Zodiac): D(world) < 10,000

Then by [9]   **volume of the Zodiac < $(10,000)^3 \times 1000\Omega^6$**

$$= 10^{12} \times 10^3 \times (10^8)^6$$

$$= 10^{15} \times 10^{48} = 10^{63}$$

Thus Archimedes deduces that 10^{63} grains of sand would fill the universe.

Archimedes' calculation for the diameter of the universe gave a figure that approximates to 2 light years. He could not know that the stars were in fact distant suns but we now know that the nearest star, Proxima Centauri, is 4.2 light years distant with a mass of one tenth of that of the sun.
If we take this distance of 4.2 light years as the radius of the Zodiac then we can repeat Archimedes' calculation and see how near he was to today's estimate:

The sand rekoner 2008

Let the radius of the Zodiac be 4.2 light years = 4.2 x 9.46 x 10^{15} metres

Volume of the Zodiac = $\dfrac{4 \pi}{3} (39.7 \times 10^{15})^3$ = 262095.8 x 10^{45} m^3

Assuming that a poppy seed contains about 10000 grains of sand and that a poppy seed is 1 millimetre diameter, then one cubic metre of poppy seeds would contain 10000 x (1000)^3 = 10^{13} grains of sand.
The universe would therefore contain 262095.8 x 10^{45} x 10^{13} = 2.6 x 10^{63} grains of sand.
Two thousand three hundred years ago, Archimedes arrived at 10^{63} and had to invent a number system in order to perform the calculation…uncanny.

There is a fine description of the Sand Reckoner at

www.calstatela.edu/faculty/hmendel/Ancient%20Mathematics/

==

Greek Fractions

The Greeks had a variety of ways of representing fractions using alphabetic numerals but notations could be contradictory and confusing, for example:

[1] Using a dash:

Remembering that $\mu = 40$ and $\beta = 2$ $\mu\beta'$ could mean $\dfrac{1}{42}$

or it could mean $40\frac{1}{2}$

[2] Using an overline:

Remembering that $\pi = 80$ and $\delta = 4$ $\overline{\mu\beta}\,\pi\delta'$ means $\dfrac{42}{84}$

==

Exercise 4

Perform the following calculations using Greek Ionic numerals:

[1] $\Gamma' + F'$

[2] $\iota\beta' + \delta'$

[3] $\delta' + \overline{\beta}\gamma'$

[4] $\overline{\iota\beta}\,\kappa' + \overline{\iota\gamma}\,\lambda'$

[5] $\overline{\iota\alpha}\,\kappa\delta' + \overline{\theta}\,\kappa'$

==============================

Answers

Exercise 1   [1]   XXXXHHHHΔΔΔΔIIII

[2] $\overline{\Pi}^{H}$ H $\overline{\Gamma}^{Δ}$ Δ Γ I

Exercise 2   [1]   ψξε   or   ΨΞΕ

[2]   σμη   or   ΣΜΗ

[3]   ꙅοθ   or   ϿΟΘ

Exercise 3   ᶠ
M′ϜΧΞϜ = 66666

Exercise 4   [1]   β′   or   Β′

[2]   γ′   or   Γ′

[3] $\overline{\iota\alpha}$ ιβ′ or $\overline{\text{IA}}$ ΙΒ′

[4]   αγ′   or   ΑΓ′

[5] $\overline{\iota\alpha}$ κδ′ + $\overline{\theta}$ κ′ = $\overline{\rho\theta}$ ρκ′

or $\overline{\text{IA}}$ ΚΔ′ + $\overline{\Theta}$ Κ′ = $\overline{\text{PΘ}}$ ΡΚ′

Shang Numerals: 1400BC

Anyang is about 500 miles North West of Shanghai in the Honan province of Eastern China. In 1899 at Xiao Dun, a small village near Anyang, thousands of bones and tortoise shells were discovered with ancient Chinese characters inscribed on them. The site had been the capital of the kings of the Shang Dynasty who ruled from 1400 to 1045 BC. Many of the inscriptions held numeric data such as numbers of prisoners taken in battle or the numbers of animals killed in hunts.

The numbers showed that the scribes were using a decimal system.

This table shows the symbols that the scribes had used:

This number system is not a place value system. The order of the symbols is not important and the following would all be interpreted as three hundred and sixty five:

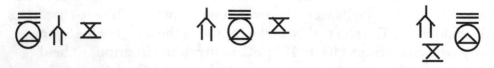

However, we would find it strange if someone wrote that the number of days in a year was " five sixty three hundred ". We will assume that the scribes of the Shang dynasty kept to a fixed order in the way that they wrote down the numeric symbols.

Thus, we will assume that the number 5432 would have been represented as

5 4 3 2

There is no zero in the Shang number system for example:

300 is represented as

3003 is represented as

Exercise on Shang numerals

1. Add

to

2. Add

to

The Counting Board: 400BC

At around 400BC, Chinese traders and started using a counting board for arithmetic calculations. It was a simple board, marked out in rows and columns. Each number was represented in one row using small rods of bamboo or ivory but the columns represented groups of ten. Thus the numbers were being represented in a base ten number system.

Example

Ten thousands	thousands	hundreds	tens	ones
	111	1111	1111	11
		11	111	1111
1	11111	1111	11111	

The diagram illustrates a counting board with the numbers   **3 4 4 2**
                                                       2 3 4
                                                  1 5 4 5

To find the sum total, any column with ten rods in it would have ten rods removed and one rod placed in the next column. Thus the calculation of the total in this case would have proceeded as follows:

Ten thousands	thousands	hundreds	tens	ones
	111	1111	1111	1
		11	111	
	1	11111	1111	
			1	

Ten thousands	thousands	hundreds	tens	ones
	111	1111	11	1
		11		
	1	11111		
		1		

Ten thousands	thousands	hundreds	tens	ones
	111	11	11	1
	1			
	1			

Giving the total as **5 2 2 1.**

The use of the counting board prompted the invention of numerals that seem to match the small rods that were used on the counting board. There were in fact two sets of numerals.

Counting Board Numerals

1	2	3	4	5	6	7	8	9
I	II	III	IIII	IIIII	T	TT	TTT	TTTT
—	=	≡	≣	≣	⊥	⊥	⊥	⊥

The numerals were used with the place value system corresponding to the columns on the counting board. The right hand numeral represented the units, the next column the tens and so on. If only the top set of numerals were used then there could be confusion, for example, | | | might be representing the decimal number 111, or it could be 21, or 12, or 3. To avoid this problem, they alternated which set of numerals for each digit. The units numeral was taken from the top row, the tens from the bottom row and so on so that the number 1234 would be represented as

1234   — || ≡ ||||

Spaces were used where we would use the zero. Thus:

23 =||| **203** || ||| **2003** = |||

Exercises on Counting Board Numerals

1.

        — ‖ ≡ ⦀

   ADD

        = ‖| ≡ ⦀|

2.

        ≡ ⦀| ⊥ 𝕋

   ADD

        ≣ ⊤ ⊥ 𝕋𝕋

Answers

Shang Numerals   [1]

[2]

The Counting Board

    [1]   ≡ ⦀| ⊥ 𝕋𝕋𝕋

    [2] | ‖ ≡ ⦀|

=====================================

Etruscans and Romans

Tuscany, that is Toscana in modern Italy, is a province just to the North of Rome. The main cities of Tuscany are Florence, Pisa and Sienna. Three thousand years ago, this region was called Etruria. The Greek historian Herodotus described how seafarers from Turkey settled the area and by about 650 BC they had developed a rich culture. However, wars with invading Celtic tribes eventually led to Etruria being reduced to a number of city states, one of which was Rome. Rome grew in power and began its attacks on other Etruscan city states in 498 BC and by 264 BC Rome had completed its conquest.

The Etruscans developed a system of numerals from the tally marks cut on pieces of wood or bone, used by the shepherds, thus we have one cut each for 1 to 4, an extra cut at five and cross cuts to mark ten:

I I I I Λ I I I I X would be the tally count for ten sheep.

This method of counting led to the Etruscan numerals

I II III IIII Λ ΛI ΛII ΛIII ΛIIII X

The Romans inherited and developed this system further and used subtraction to make the symbols more concise:

I II III IV V VI VII VIII IX X XI XII XIII XIV XV XVI XVII XVIII

XIX XX XXI XXII XXIII XXIV XXV XXVI XXVII XXVIII XXIX XXX

Numbers were (and still are) made up from the following symbols:

V=5 X=10 L=50 C=100 D=500 M=1000

A bar over the top of one of these indicated multiplication by 1000 thus

$\overline{V}$=5000 $\overline{X}$=10 000 $\overline{L}$=50 000 $\overline{C}$=100 000 $\overline{D}$=500 000 $\overline{M}$=1000 000

The subtraction rule for placing a smaller Roman numeral to the left of a larger one is:

"the symbol **z** cannot come before a symbol larger than **10z**"

Thus, **IV** and **IX** are O.K. but not IL or IC

      I cannot come before **L C D M**

      X cannot come before **D M**

Thus we do not find **99** written **IC** but rather **XCIX**

and **999** is not **IM** but the more complicated **CMXCIX**

Exercise

1. Write the following date using Roman numerals:

1066, 1215, 1492, 1918, 1945, 1990, 1999, 2007

2. Complete the following three columns of the multiplication table

x	X	XI	XII
I			
II			
III			
IV			
V			
VI			
VII			
VIII			
IX			
X			
XI			
XII			

Answers

[1]   1066 = **MLXVI**

   1215 = **MCCXV**

   1492 = **MCDXCII**

   1918 = **MCMXVIII**

   1945 = **MCMLXV**

   1990 = **MCMXC**

   1999 = **MCMXCIX**

   2007 = **MMVII**

[2]

x	X	XI	XII
I	X	XI	XII
II	XX	XXII	XXIV
III	XXX	XXXIII	XXXVI
IV	XL	XLIV	XLVIII
V	L	LV	LX
VI	LX	LXVI	LXXII
VII	LXX	LXXVII	LXXXIV
VIII	LXXX	LXXXVIII	XCVI
IX	XC	XCIX	CVIII
X	C	CX	CXX
XI	CX	CXXI	CXXXII
XII	CXX	CXXXII	CXLIV

===================================

The Maya

The Yucatan peninsular of Southern Mexico juts out into the Caribbean Sea reaching to within 200 miles of Cuba and borders on the steamy jungles of Belize and Guatemala. In parts, the low-lying limestone rock is honeycombed by underground networks of fresh water caves and streams. The terrain is dotted with many circular fresh water lagoons called cenote, where the roofs of the underground caves have fallen in. When viewed from space, the locations of these cenote are seen to follow a circular arc which follows the outline of a huge crater formed about 65 million years ago when a meteorite or comet smashed into the Earth, the event that led to the extinction of the dinosaurs.

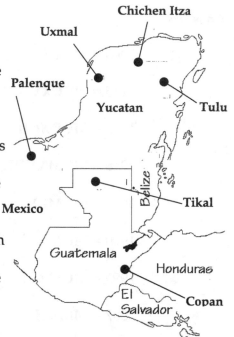

The abundance of clean fresh water filtered through the limestone rocks encouraged the development of early civilizations in the otherwise arid lands of the Yucatan and the forests of Guatemala and Belize. These areas were settled by the Maya from about 2600 BC where they created an outstanding culture of astronomy, sculpture, hieroglyphic writing and an architecture that reached its peak between 200 and 900 AD.

Copan lies in the extreme west of Honduras near the border with Guatemala. An agricultural community settled there and developed the skills needed to build the first known stone architecture of the region in 900 BC. By 160 AD a secure kingdom was established that grew into one of the most important city states that was at its height between 400 and 800AD. Copan is now a UNESCO world heritage site. (see http://whc.unesco.org/en/list/129)

Tikal lies in the north of Guatemala, about fifty miles from the border with Belize. Tikal was a thriving Mayan city from 200 to 900 AD. It covered about six square miles and was home to about 60,000 people. The sophisticated stone architecture was decorated with carvings and wall paintings. The Temple of the Giant Jaguar, completed around 700 AD houses the tomb of the High Priest where he was buried along with hundreds of vases, jade jewels and other treasure.

Some of the finest architecture, carving and painting of the Mayan culture was produced in Palenque in the Tabasco province of southern Mexico. Palenque was the capital of the B'aakal city state which was ruled by Pacal The Great from 615 to 683 AD who left some of the most magnificent sculptures and tomb works of the Mayan culture.   The Temple of Inscriptions was built on the top of a step pyramid at Palenque in 692 AD. A slab of stone in the temple reveals a stone stairway leading down to ground level where we find the shrine of the "divine" Pacal The Great. He had claimed that he could trace his ancestry back the creation of the world on 12th August, 3113 BC.

Ruins of palaces, temples and pyramids can also be found at Tulum on the East coast of the Yucatan, Uxmal in the North West . At Chichen Itza, the Caracol building is thought to be the ruins of a Mayan observatory.
From the tenth century AD, the Mayan empire fell into decline and was over run by the Toltecs of Mexico. The most impressive ruins of the Toltec era are the ruins of the temple to the sun at Chichen Izta. The Toltecs were conquered by the Aztecs of central Mexico who were in turn, defeated by Hernan Cortes in 1521.

The Mayan peoples of Central America were the first to use a place system for numbers together with a symbol for zero, long before its earliest use in India.
While Europe was struggling with clumsy Roman numerals, the Mayans were using a base 20 number place value system. They used the following numerals:

Simple Examples:

2+3   ● ●   +   ● ● ●   =   ————

4+5   ● ● ● ●   +   ————   =   ● ● ● ●
                                    ————

6+7   ● ————   +   ● ●   =   ● ● ●
                     ————   ————
                                   ————

8+9   ● ● ●   +   ● ● ● ●   =   ● ●
          ————   ————   ————
                                      ————
                                       ————

10+11   ════   +   ●   =   ●
           ════   ════
                       ════   ●

The last example here illustrates the fact that the Mayans wrote their numbers from the bottom up! The result of **10+11** is **one twenty** plus **one**

An example of the Mayan base 20 placed value system is given here. The value of the number represented is **178127**:

●   place value 20^4 value = 160000x1 = 160000

● ●   place value 20^3 value = 8000x2 =   16000

————   place value 20^2 value =   400x5 =   2000

● ————   place value 20 value =   20x6 =   120

● ●   units   value =   1x7 = ____7
————   178127

Example 1

The number **456 = 400 + 2x20 + 16** is written in Mayan as

Example 2

A Mayan addition sum

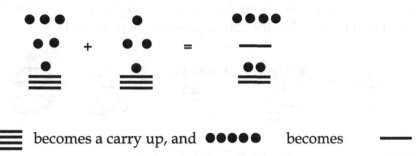

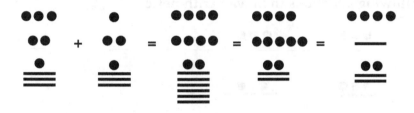 becomes a carry up, and ●●●●● becomes ——

A Mayan child might have carried out the calculation as follows, remembering that twenty carries one dot up to the next level:

Example 3

Perform the addition 888+444 in Maya arithmetic

Solution

$$888 + 444 \qquad \text{(decimal)}$$

$$= 2;4;8 + 1;2;4 \qquad \text{(base 20)}$$

Example 4

Using zero:
(note the importance of the spaces between the "digits")
It is believed that the symbol for zero represents a cocoa pod.

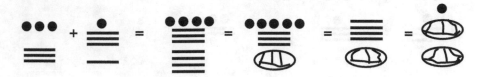

(The decimal version of this sum is 75 + 325 = 400)

Exercises

Perform the following additions in Mayan arithmetic

1

2

3

4

5

6

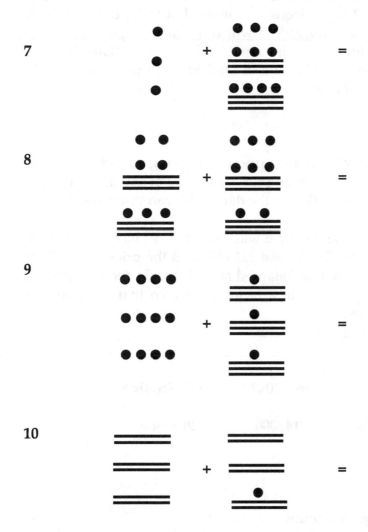

7   +   =

8   +   =

9   +   =

10   +   =

In 1519, Hernan Cortez landed on the Yucatan peninsular at Tabasco, with a fleet of 11 warships. The Mayan culture had been in rapid decline since the invasion by the Toltecs. However, he was welcomed by the local Mayan population and soon married a local girl called Malinche who acted as an interpreter for him. Cortez went on to wage war against the Aztecs at their capital Tenochtitlan which is now called now Mexico City. By 1521 he had conquered the Aztec Empire.

In 1541, a Spanish missionary called Diego de Landa began working with the Mayan peoples to learn of their history and culture. However he did not approve of the Mayan religion and their texts written in hieroglyphics. As a result, he ordered that all Mayan icons and texts should be destroyed. A small number of Mayan texts survived, of which the most important are the Dresden Codex kept in the Sachsiche Landesbibliothek in Dresden, the Madrid Codex, at the American Museum in Madrid and the Paris Codex in the Bibliotheque National in Paris.

The Long Count

The Long Count was used by the Mayan priests for giving dates to buildings or other objects. If we accept the Mayan priests estimate of 360 days in a year then the Long Count gives immediately, the date in Mayan years and days.

The date for the creation of the universe was assumed to have taken place on the date we would give as 12th August 3113 BC and the priests gave their dates as the number of days that had elapsed since then. Instead of keeping to the base 20 system, they slotted in a factor of 360, so that the numbers would record years and days.

The place values for the **Long Count** were:

		18x20			
Units	20	=360	360x20	360x20x20	360x20x20x20
1	20	360	7200	144000	2880000 ………

Example 5

Find this time period using long count:

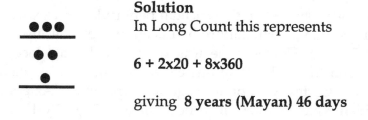

Solution
In Long Count this represents

6 + 2x20 + 8x360

giving **8 years (Mayan) 46 days**

Example 6

Calculate this period in long count:

Solution

In Long Count, this represents

3 + 2x20 + 360(3 + 2x20 + 2x400)

giving **843 years, 43 days**

Example 7

An archaeologist found this date on a
piece of Mayan pottery.
(i) What date is this in years (Mayan)
and (ii) what date is it on our calendar?

Solution

(i) Using the Long Count, this translates to

12 + 1x20 + 360(3 + 14x20 + 8x400)

= **3483 years (M) 32 days from the birth of the universe.**

(ii) The number of days from the birth of the world is

360x3483 + 32 = **1253912 days**

To translate this to our calendar, we divide by the number of days in a year
which is **365.242**

1253912 / 365.242 = **3433 years 36 days**

Considering that the world was created on **August 12th 3113 BC**

3433 – 3113 = 320

so this gives us a date of **August 12th 320 AD plus 36 days**

= 17th September 320 AD

This Mayan date was on a plate that was found in **Tikal.**

Exercise 2

This Mayan date was is on a building in Palenque in Tabasco province. It is the date that the building was completed.
Find the date of completion

(i) in Mayan years and days

(ii) in our calendar

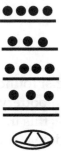

==

Answers

Exercise 1

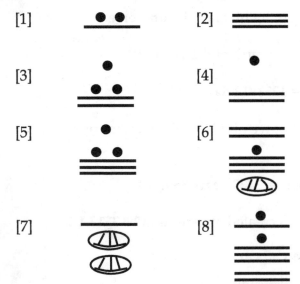

[9]      [10]

Exercise 2

[i]   Using the long count,
        the number of days from the birth of the world is **3769x360 + 260**

        The Mayan date is therefore 3769 years 260 days

[ii]   The number of days from the birth of the world is **1357100 days**

        Divide by **365.242** to get **3715 our years 226 days**

        Subtract the birth date of the world **3715 – 3113 = 602 years**

        Add **August 12th** to get **August (12+226)th**

        Our date is therefore August 238 th AD 602
                                        -30
                            = Sept 208 th AD 602
                                        -31
                            = Oct   177 th AD 602
                                        -30
                            = Nov 147 th AD 602
                                        -31
                            = Dec   116 th AD 602
                                        -31
                            = Jan   85 th AD 603
                                        -28
                            = Feb   57 th AD 603
                                        -31
                            = March 26 th AD 603

        answer (i) **year 3769 day 260**
               (ii) **26th March 603 AD**
        =========================

Hindu Numerals

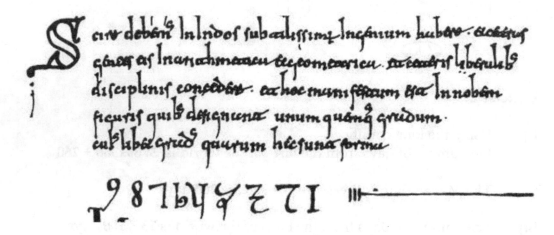

All over the world people use a base ten place system of numerals called Arabic numbers.   Just ten symbols, 0,1,2,3,4,5,6,7,8,9 are sufficient to represent numbers of any magnitude. The symbols are written highest to lowest, left to right: 5432 means five thousands, four hundreds, three tens, two units.

We would not be sure what to make of 2
                                              3
                                              4
                                              5

Fractions are easily dealt with. Thus 0.123 represents one tenth part, 2 one hundredth parts plus 3 one thousandth parts.

Although we call these Arabic numerals, or Hindu-Arabic numerals we know that the Indians were the first to develop a base ten place value system. How the Indian base ten place system was developed is unsure but we do know that it was studied by Arab mathematicians and transmitted from the East Arabian Empire through Persia, across North Africa to southern Spain and thence into Europe.

The most important script to evolve in India was the Brahmi script that first appeared in the 5th century BC. Brahmi numerals were in use from about 300BC. A document discovered in 1881 in Bakhshali in Pakistan 50 miles from Peshawar shows a remarkable ability in algebra and arithmetic. The

Bakhshali document was written sometime between 200AD and 400AD and gives rules for arithmetic calculations including fractions, and a remarkable method for finding square roots.

The Brahmi numerals have been found on inscriptions and coins dating from the first century AD:

—	=	≡	Ұ	Ր	⊄	7	ς	?
1	2	3	4	5	6	7	8	9
⋉	♂	ᴊ	×	ᴊ	⟙	✗	⊖	⊕
10	20	30	40	50	60	70	80	90

100 = 7   1000 = ∮

The units and tens have special symbols whereas the hundreds and thousands were written using place values, thus the number four thousand three hundred and fifty two would appear as:

4352 = Ұ ∮ ≡ 7 ᴊ =

Brahmi numerals were being used in India from about 300 BC to the 4th century AD and our numerals are descended from these Brahmi characters. Of course, the Brahmi numerals evolved and changed with time and by the 4th century AD, they had evolved into another set called the Gupta numerals. These were in general use from the 4th to the 8th century AD. The Gupta dynasty ruled over a large swathe of northern India from the Arabian sea to the Ganges and this "golden age " in India lasted from 300 AD to 500 AD. During this period the Gupta kings were keen patrons of art, mathematics and science.

1	2	3	4	5	6	7	8	9
—	=	≡	૪	Ի	♨	ๆ	ৎ	৲
Gupta numerals around 4th century A.D.								

Brahmagupta

Ujjain is in Madhya Pradesh, central India, about 200 miles East of Ahmadabad. It is one of the seven sacred cities of the Hindus and every 12 years, thousands of worshippers gather there for the Kumb Mela celebrations. Brahmagupta was born in Ujjain in 598 AD and he became the head of the Astronomical Observatory in Ujjain when it was India's foremost centre for mathematical research. In the year 628 Brahmagupta wrote Brahmasphuta siddhanta, "The Opening of the Universe", a textbook on mathematics. In it, he explained the positional base ten number system, defined zero as the result of subtracting a number from itself and with the use of a symbol for zero gave rules for the arithmetic of negative numbers. In this remarkable book Brahmagupta gives rules for the summation of series, gives the area of a cyclic quadrilateral as the product of its diagonals and describes an algorithm for finding square roots.

The "Brahmasphuta Siddhanta" was translated into Arabic in 776AD.

The Gupta script however, also changed with use and eventually evolved into the Nagari script which was in use from the 8th until the 11th century.

1	2	3	4	5	6	7	8	9	0
९	२	३	४	५	६	७	८	९	०

Nagari numerals around 11th century A.D.

Al Khwarizmi

The city of Mecca was conquered by Mohammed and his army in 630 AD and this began the rapid expansion of the Muslim empire throughout the middle East. By 633 AD, the Arabian peninsular was under Islamic rule and by 638 AD the empire had spread to include Palestine and Syria. By 660 AD the Islamic Empire extended from the Mediterranean to India and included Mesopotamia (Iraq), Persia (Iran) and Egypt.

Al Khwarizmi was born in Baghdad in 780 AD at a time when Muslim scholars were able to travel from India in the East to Egypt and further, along the coast of North Africa. He was a mathematician employed in the "house of Wisdom" in Baghdad, and wrote on algebra, geometry and astronomy. His most important work is probably "Hisab al-jabr w'al-muqabala" written in 820 AD, the first text ever written on algebra, which included methods for solving both linear and quadratic equations. "al-jabr" = completion and is the origin of the word Algebra. "al-muqabala"=balancing. These words refer to his method for solving quadratic equations that we would call "completing the square". Al-Khwarismi was one of the greatest mathematicians of his time and among his many works on mathematics and astronomy is "Algorithmi de Numero Indorum", "The Hindu Art of Calculation". The original, in Arabic seems to have been lost but fortunately there is a translation into Latin. In his book he describes methods of calculation using the Indian base ten, place value system using a symbol for zero as a 'place holder'. His book was principally responsible for the spread of calculation with Indian numerals throughout the Islamic empire. At this time the Islamic empire had spread as far as southern Spain. As Arab scholars explained how to use the Indian base ten place system of numerals and how to use decimal fractions the numerals continually changed their form. These numerals come from a book on calculation dated 969AD:

1	2	3	4	5	6	7	8	9	0

Al-Biruni was an Arabic scholar who wrote many books on Hindu mathematics and astronomy. One of his texts written around 1082 gives the following numerals:

1	2	3	4	5	6	7	8	9	0

In a book written around 1300 by al-Banna al-Marrakushi in Morocco, we find the following set of numerals:

1	2	3	4	5	6	7	8	9

The earliest evidence of knowledge of the Indian place system in Europe is contained in a manuscript called the Codex Vigilanus written by a Spanish monk in 978 AD:

Fibonacci

Leonardo of Pisa was born in Pisa, Italy, in 1170. His father Guilielmo Bonacci, held a post in Bejaia in what is now Algeria, representing the merchants of the republic of Pisa. His son Leonardo, better known as Fibonacci, which is a shortened form of the Italian for "son of Bonacci", was educated in Bejaia where his father worked. It was here that he was introduced to the art of calculation using the Indian place system. Leonardo subsequently travelled widely around the Mediterranean with his father and saw how useful the Indian number system would be to the merchants who were struggling to do arithmetic with the Roman number system. Around 1200 AD, he went back to Italy, and in 1202 wrote Liber Abaci, a book of arithmetic and algebra that introduced the Hindu base ten place value system into Europe.

The number system invented by Hindu mathematicians and developed from the Brahmi numerals around the first century AD, spread by the scholars of the Arab empire through the middle East and across North Africa thus entered into Europe and thenceforward to the rest of the world.

Modern Chinese Numerals

The traditional Chinese characters for the numerals 1 to 9 are derived from
the ancient Shang numerals, and are in common use in modern China:

1	2	3	4	5	6	7	8	9	0
一	二	三	四	五	六	七	八	九	零

Numbers are written with a mixture of the base ten place value system and
the use of special characters for ten, hundred, thousand and ten thousand.
One hundred thousand, however, is written as ten, ten thousands.
One million is written as one hundred, ten thousands.
The special characters are:

ten   十

hundred   百

thousand   千

ten thousand   萬

We have, for example

12   十二

34   三十四

The symbols for the powers of ten are:

		index
10 =	十	1
100 =	一百	2
1000 =	一千	3
10000 =	一萬	4
100000 =	一十萬	5
1000000 =	一百萬	6
10000000 =	一千萬	7
100000000 =	一億	8
1000000000 =	一十億	9
10000000000 =	一百億	10
100000000000 =	一千億	11
1000000000000 =	一兆	12
10000000000000 =	一十兆	13
100000000000000 =	一百兆	14
1000000000000000 =	一千兆	15
10000000000000000 =	一京	16
100000000000000000 =	一十京	17
1000000000000000000 =	一百京	18
10000000000000000000 =	一千京	19

Thus, for example, the number 123456789 would be

123456789 =

一億二千萬三百萬四十萬五萬六千七百八十九

If one zero digit or more occurs within a number then, although it is not needed, the zero symbol is inserted:

$$27 = 二十七$$
$$207 = 二百零七$$
$$2007 = 二千零七$$
$$20007 = 二萬零七$$
$$20607 = 二萬零六百零七$$
$$2006007 = 二百萬六千零七$$
$$206007 = 二十萬六千零七$$
$$200600 = 二十萬零六百$$
$$20600 = 二萬零六百$$
$$2060 = 二千零六十$$

Exercises

What are the following in Hindu Arabic numerals?

一

    一百二十三

二

    二百三十

三

    四百四十四

四

    四百零四

五

    九百八十七

六　　二千三百四十五

七　　一千二百三十四

八　　六萬五千四百三十二

九　　七萬六千五百四十三

十　　一百萬

Answers

一　　　123

二　　　230

三　　　444

四　　　404

五　　　987

六　　　2345

七　　　1234

八　　　65432

九　　　76543

十　　　1000000

There is a good number converter at

www.mandarintools.com/numbers.html

CHAPTER 2

Rotations, Triangles and Circles

When the Sumerians settled the fertile lands between the river Tigris and the river Euphrates in what is now Iraq in around 3500 BC they abandoned their nomadic way of life and began to develop a sophisticated culture of writing, art and sculpture that was to become one of the world's first civilizations in the land of Mesopotamia, the place between the rivers. They also developed an understanding of astronomy, science and mathematics that was to attract scholars from all over the Eastern Mediterranean. They invented a method of representing numbers that was to lead to a base 60 number system. Around 2000 BC they were invaded by another tribe called the Babylonians who further developed their culture and founded the city of Babylon in 1900 BC, 50 miles South of the modern city of Baghdad. In 1894 French archaeologists found, at the site of ancient Babylon, the ruins of one of the first schools ever to be built. They found the remains of classrooms for boys and girls and lying in the ruins were pieces of clay tablets that may have been children's class notes. Hammurabi was King of Babylon from 1795 BC to 1750 BC and ruled over an ancient civilization that had laws, a police force, roads and irrigation. Nebuchadnezzar ruled Babylon from 605 to 562 B.C. and constructed the hanging gardens, one of the sevens wonders of the ancient world. From the time of Hammurabi, Babylon was the centre of learning for some of the worlds first mathematicians and astronomers, indeed, Pythagoras who was born in 580 BC on the Greek island of Samos, spent twenty years of his life in Babylon both as a student and a teacher. From their observations of the stars and the changing seasons, they calculated that the Sun moved round the Earth in 360 days, so that the circle of their year was divided into 360 parts. Today, we know otherwise but we still use their figure of 360 divisions or degrees for one revolution so that one hundred and eighty degrees form a "straight angle" and ninety degrees form a corner or right angle:

360^0 180^0 90^0

Greek scholars such as Pythagoras (540B.C.) and Euclid (300 B.C.) studied and developed the mathematics of the ancient Babylonians. They could prove, for example, that the three angles of any triangle always add up to exactly 2 right angles or 180^0. Pythagoras is most famous for the theorem bears his name while Euclid is most well known for his book on geometry, "The Elements", which used to form a major part of the school mathematics syllabus. Euclid's Elements consists of thirteen books that represented a collected and reasoned account of the mathematics of the day. It presents us with what we call a "axiomatic system". An **axiom** is a generally accepted law or maxim, a self evident truth that is assumed without proof.

Axiom : from the Greek **'axios' = a worthy thought**

Examples of Euclid's axioms for plane geometry:

-Two points determined a unique line called the join of the two points.

-Two lines determine a unique point called the intersection of the two lines unless they are parallel.

-Through any point not on a given line, there is exactly one line passing through the point and parallel to the given line (Euclid's fifth postulate).

-The cases of congruence merit more discussion. There are four cases of congruence for triangles. These are supposedly self evident rules that determine if two triangles are exactly the same shape so that one of them could be placed to fit exactly over the other. It is possible that the triangle may have to be turned over.

Suppose that two triangles are labeled ABC and PQR. We will describe the four cases of congruence that make $\angle A=\angle P$, $\angle B=\angle Q$, $\angle C=\angle R$, AB=PQ, BC=QR and CA=RP

The cases of congruence are:

[1] SAS   Two sides and the included angle

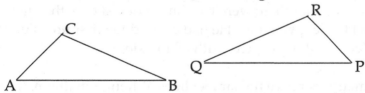

If $\angle A=\angle P$, AB=PQ and AC=PR   then $\angle B=\angle Q$, $\angle C=\angle R$ and BC=QR

[2] ASA   Two angles and a corresponding side

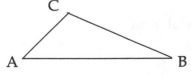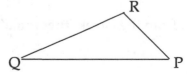

If   ∠A=∠P, ∠B=∠Q and AC=PR   then ∠C=∠R, AB=PQ and BC=QR

[3] SSS   Three sides

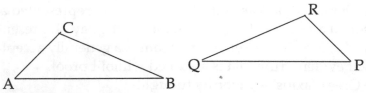

If   AB=PQ, BC=QR and CA=RP   then ∠A=∠P, ∠B=∠Q and ∠C=∠R

[4] RHS   Right angle, hypotenuse and side in a right angled triangle

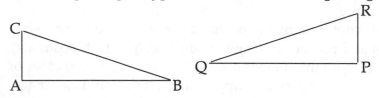

If   ∠A=∠P=90, AB=PQ and BC=QR then , ∠B=∠Q, ∠C=∠R and CA=RP

Euclid's axioms have generated much discussion since he wrote his book around 300 BC and in 1826, Nikolas Lobachevsky (1793-1856) presented a paper to the Physical Mathematical Society of Kazan that demonstrated a consistent geometry in which Euclid's fifth postulate is changed. Lobatchevsky supposed that there were **two** lines not just one, through the given point parallel to the given line. He had created the first **non-Euclidean geometry** by challenging the accepted truth of an axiom.

In **projective geometry** there are no parallel lines although mathematicians often like to say that parallel lines meet at a **point at infinity.**
As long as the axioms are not "self contradictory" propositions that are deduced from those axioms represent a logical mathematical truth. These proved propositions are called **theorems.**

Theorem : from the Greek **'theorema'** = a speculation

The system of theorems that link together and support each other is called a **theory.**

In this case the theory is called **Euclidean Geometry,** a theory that accurately describes the world about us at the normal everyday scale of things. When we consider the scale of the very large, the Galaxies and Stars, we have to appeal to the theory of relativity. When we consider the scale of the very small, electrons, protons and photons, we appeal quantum theory!

Chapter 2 presents a number of theorems of Euclidean Geometry in a logical sequence but sometimes pointing out informal approaches to well known results, as an example of a set of **theorems** building up into a **theory.**

In the following figure, angles α and β are called opposite angles.

Theorem 1. Opposite angles are equal

Proof

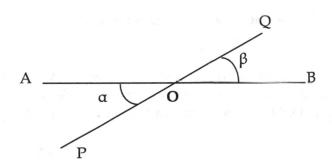

   Suppose that the line AB is rotated about a point O on the line, so that it is rotated to the position PQ. Since the line OA rotates to the position OP and OB is rotated to the position OQ, the measures of rotation (angles) denoted by α and β in figure 1 will be the same. The angles α and β are called opposite angles and this argument shows that opposite angles are equal.

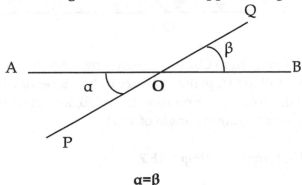

α=β

In the following diagram, AB is a straight line. O is a point on the line and OQ forms two angles θ and φ with the line AB. θ and φ are called adjacent angles standing on the straight line AB.

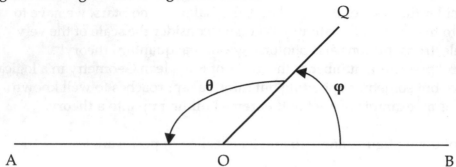

Theorem 2

Adjacent angles add up to 180⁰

Proof

Suppose that O is a point on the line AB, between A and B, and that the portion OB rotates about the point O through an angle φ to the position OQ. Suppose now, that OQ rotates through a further angle θ to the position OA.

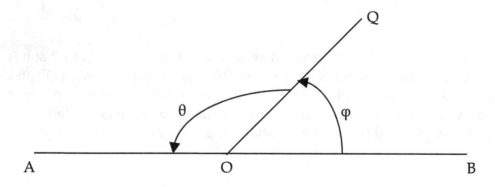

The total rotation from OB to OA is measured by θ+φ, but we know that the total rotation from OB to OA is the measure of half a revolution. A compete revolution is 360 degrees, so the rotation from OB, anti clockwise to OA is half a revolution, or a "straight" angle of 180⁰ .

Therefore   **θ+φ = 180⁰**

Example. The line AB rotates through 60⁰ about the point O between A and B to the position PQ.

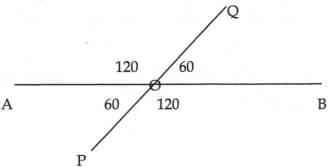

Parallel Lines

Given any line AB, this will point in a unique direction from A to B, which we can indicate with a direction arrow: $\overrightarrow{AB}$.

Let X be some point not on AB then Euclid supposed that there would be just one line through the point X which pointed in the same direction as $\overrightarrow{AB}$. Suppose we call this line XY, then we have the following figure:

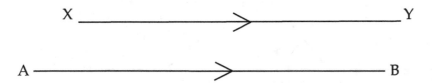

The fact that the two lines AB and XY point in the same direction is indicated by an arrow on each line. We say that the two lines are parallel and we assume that parallel lines never meet and that they are always "the same distance apart".

Suppose now, that we draw a line PQ that crosses each of the parallel lines AB and XY. (such a line is called a transversal)

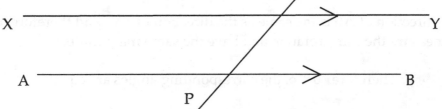

Suppose that PQ makes an angle θ with XY and makes an angle φ with AB:

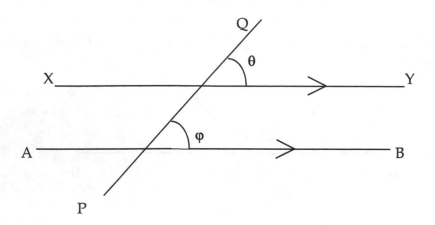

The angles θ and φ are called corresponding angles.

================================

Theorem 3

Corresponding angles are equal

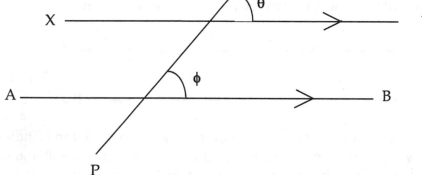

Proof

θ measures the rotation from the direction of $\overrightarrow{XY}$ to the direction of $\overrightarrow{PQ}$

φ measures the rotation from the direction of $\overrightarrow{AB}$ to the direction of $\overrightarrow{PQ}$

But the direction of $\overrightarrow{AB}$ is the same as the direction of $\overrightarrow{XY}$ and therefore θ and φ measure the same rotation and have the same magnitude.

Therefore θ = φ, the corresponding angles are equal

In the following figures, θ and φ are called alternate angles

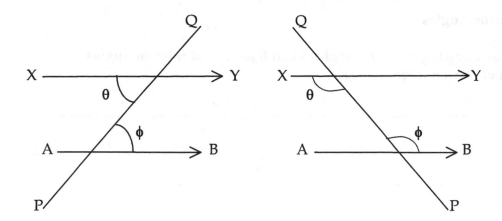

Theorem 4
Alternate angles are equal

Proof

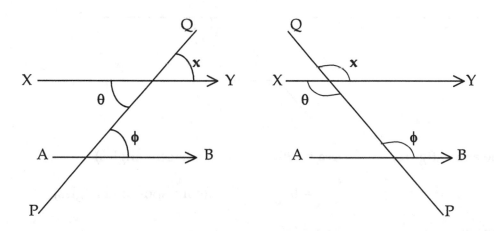

Referring to the figures, $\theta = x$ (opposite angles)

                          $x = \phi$ (corresponding angles)

therefore $\theta = \phi$ (both are equal to x)

The alternate angles are equal
===========================

Interior Angles

In the following diagram, angles **a** and **b** are called **interior angles** (sometimes co-interior angles)

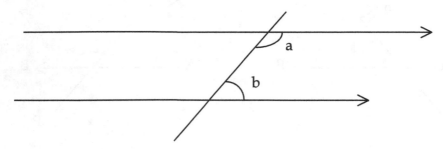

Theorem 5:   Interior angles add up to 180 degrees

Proof

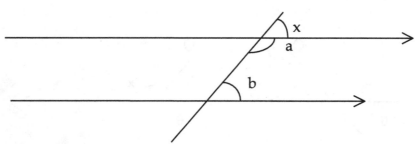

In the above figure,   **x+a = 180**   (adjacent angles)

                              x = b   (corresponding angles)

therefore   **b+a = 180**

Example 1   Calculate angle x

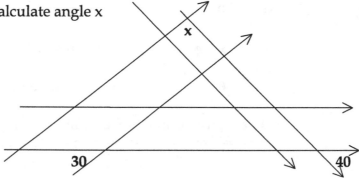

Solution

Draw the line XY parallel to the base line

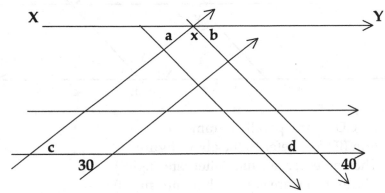

Then **a=c** and **c=30** (alternate angles)

and **b=40** (corresponding angles)

but **a+x+b = 180** (adjacent angles)

therefore   x=180-30-40 = 110 **degrees**

=================================

The rules of the game are that we can only use theorems that we have proved. We should not use the "angle sum of any triangle" because we have not demonstrated the theorem yet.

The Parallelogram

The two pairs of parallel lines in the next figure form a parallelogram. That is, a four sided figure with each pair of opposite side parallel.

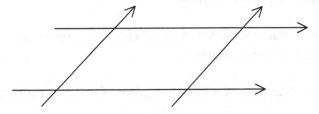

We prove that the opposite angles are equal and that the opposite sides are equal

Theorem 6   Opposite sides and angles of a parallelogram are equal

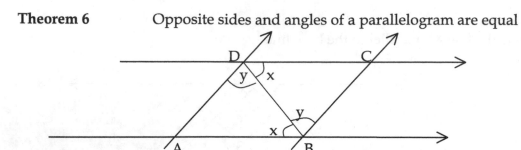

Proof

ABCD is any parallelogram
Join BD and refer to the above figure
The angles x are equal (alternate angles)
The angles y are equal (alternate angles)
BD is common to triangles ABD and CDB
Therefore $\triangle ABD \cong \triangle CDB$ (ASA)
Therefore $\angle A = \angle C$, AB=CD and AD=CB

Further $\angle ABC = \angle CDA$ since both angles are x+y   ■

======================================

(Note: It has been traditional to finish a geometrical proof with the Latin phrase **Q.E.D.** "quod erat demonstrandum" = "which was to be proved" however, we rather use the Halmos symbol ■ . This symbol will appear on occasion but need not be rigorously employed.)

Exercise 1
[1]   Prove that if both pairs of opposite sides of a quadrilateral are equal then the quadrilateral is a parallelogram.

[2]   Prove that if both pairs of opposite angles of a quadrilateral are equal then the quadrilateral is a parallelogram.

[3]   Prove that if one pair of sides of a parallelogram are equal and parallel then the quadrilateral is a parallelogram.

[4]   Prove that the diagonals of a parallelogram bisect each other.

[5]   A parallelogram with equal sides is called a rhombus. Prove that the diagonals of a rhombus are perpendicular.

The Angle Sum of a Triangle.

The Greek Mathematicians of 5th century BC were able to prove that the three angles of any triangle add up to two right angles or 180^0. Here is Euclid's proof that he described in his book on geometry, The Elements, written at about 300 years before the birth of Christ.

Theorem 7   Given any triangle ABC, $A+B+C= 180^0$.

Proof 1 (Euclid)

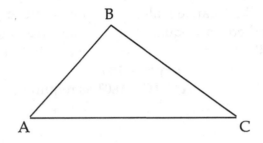

Construction: Draw the line through C parallel to AB

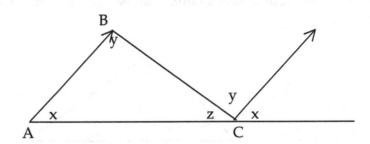

The angles marked x are equal   (corresponding angles)
The angles marked y are equal   (alternate angles)
The angles at C add up to 180^0.   (straight line)

Therefore $x+y+z = 180$
Therefore $A+B+C = 180^0$ as required

■

Proof 2

An alternative construction is to draw a line through B parallel to AC

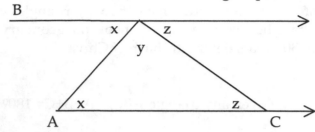

The angles marked x are equal   (alternate angles)
The angles marked z are equal   (alternate angles)
The angles at B add up to 180^0.   (straight line)
Therefore   **x+y+z = 180^0**.
Hence   **A+B+C = 180^0** as required.   ■

Demonstration.

An informal proof of this theorem can be demonstrated using a diagram and a pencil.

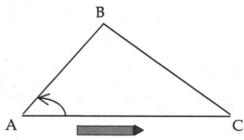

Place a pencil along AC, pointing to C and rotate the pencil about the point A, through angle A, so that it now points to B.

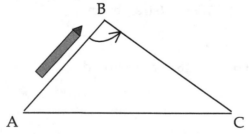

Now rotate through angle B so that the pencil lies along BC:

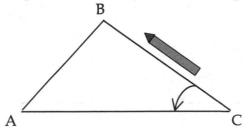

Lastly, rotate through angle C so that the pencil then lies along CA:

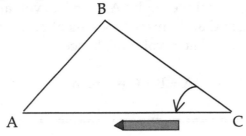

The pencil started on AC pointing at C and finished on CA, pointing at A. Therefore, after turning through the three angles A, B and C, the pencil has turned over, making a half turn.

Therefore,   **A + B + C = half a turn.**

◼

==========================

Alternatively, instead of using a pencil, we can actually walk round the triangle, turning through each of the angles when we get to a vertex:

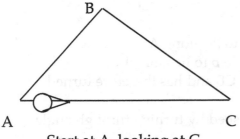

Start at A, looking at C.

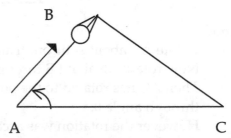

Turn through angle A and walk to B.

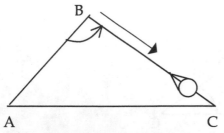

Turn through angle B and
walk backwards to C.

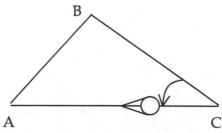

Finally turn through angle C.

Having turned through the three angles A, B and C, you are now looking at A. You started looking at C and finished looking at A and so you have turned through half a complete revolution, hence

$$A + B + C = \tfrac{1}{2}\text{ turn}$$

■

=============================

We can use a similar argument to prove another useful result:

Theorem 8   The exterior angle of a triangle is equal to the sum of the two interior opposite angles.

To Prove, in the figure, that $\theta = a + b$

Proof

        Rotate AC about A through angle **a** to lie along AB.
        Now rotate AB about B through angle **b** to lie along BC.
        Then AC has rotated to the position CB and has therefore turned through angle θ.
        However the rotation was accomplished by turning through angle **a** then through angle **b**,
        therefore $a + b = \theta$ as required.   ■

Alternatively:

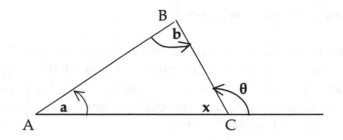

$$a + b + x = 180^\circ \quad \text{(Angle sum of triangle)}$$

$$\theta + x = 180^\circ \quad \text{(straight line)}$$

therefore $\qquad \theta = a + b \quad$ (both are $180^\circ\text{-}x$)

=================================

The Isosceles Triangle.

An isosceles triangle has at least two equal sides.
If all three sides are equal then the triangle is also called equilateral.
We can show that the angles opposite the equal sides in an Isosceles triangle
are equal.

Theorem 9   The base angles of an isosceles triangle are equal.

Given   In triangle ABC, AB=AC

Prove   that $\angle B = \angle C$

Proof
In triangle ABC, AB=AC.
Draw the bisector of angle A,
 meeting BC at M.
 Consider what happens if we fold the
 triangle on the line AM.
 Since the angles marked **x** are equal, when folded, AB will lie over AC.
Also, since the lengths of AB and AC are equal, AB will fold over exactly
onto AC and therefore BM will fold over onto CM.
Therefore, angle ABM folds over exactly onto angle ACM and hence the two
angles must be equal.   ■

Reductio Ad Absurdum

We prove the converse of this theorem using a different method of proof called "reductio ad absurdum". This is a Latin phrase that means reduce to the impossible. In short, we assume the result is not true and deduce from this, something that is impossible, leaving us with the conclusion that the result could not be "not true".

Theorem 10
If two angles of a triangle are equal then the sides opposite these angles must be equal in length.

Given   In triangle ABC, $\angle ABC = \angle ACB = b$

Prove   that AB=AC

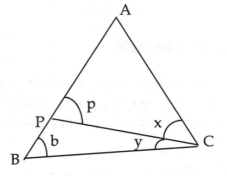

Proof (part 1)
Suppose that AB is longer than AC
so that we can cut off AP=AC along
the line AB.
  Here AP=AC

Then	$x + y = b$	($\angle ABC = \angle ACB$, given)
also	$p = x$	(by theorem 10 , since AP=AC)
hence	$p + y = b$	
but	$p = b + y$	(exterior angle of triangle BPC)
therefore	$b+y+y = b$	
so that	$y + y = 0$	
hence	$y = 0$	

If y=0, then B and P are the same point so AB cannot be longer than AP which was drawn equal to AC. Therefore, AB cannot be longer than AC.

Suppose then that AC is longer than AB so that we can extend AB to AP with AP=AC.

Proof(part 2)
Suppose that AB is shorter than AC
So that we can extend AB to AP with
AP=AC.

Here AP=AC
Let $\angle ABC = b$ and $\angle ACB = b$

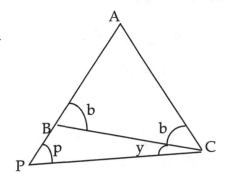

then $b = p + y$ (exterior angle of triangle BPC)
but $p = b + y$ (by theorem 10 , since AP=AC)
therefore $b = b + y + y$
so that $y + y = 0$
hence $y = 0$
If y=0, then B and P are the same point so AB cannot be shorter than AP
which was drawn equal to AC. Therefore, AB cannot be shorter than AC.

Proof(part 3)
By **part 1**, AB cannot be longer than AC and by **part 2** AB cannot be shorter
than AC. We therefore conclude that AB must be equal to AC and this
concludes the proof.   ■

=================================

In about 300 BC, Pappus of Alexandria gave the following proofs of the
isosceles triangle theorems. These proofs were rediscovered by a computer
program written by E. Gelernter in 1960:

The proof given by Pappus uses the fact that a triangle is congruent to itself.
One way of thinking about this fact is to imagine that you are able to go
behind the figure of the triangle and see its mirror image:

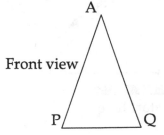 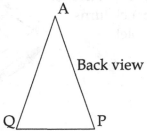

Theorem 9 (Pappus)

Given   In triangle APQ, AP = AQ

Prove   ∠ P = ∠ Q

Proof

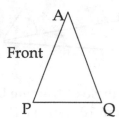

In triangles APQ and AQP

        AP = AQ, AQ = AP and PQ = QP

Therefore triangles   APQ are congruent (SSS)
                and   AQP

Therefore   ∠ P = ∠ Q and ∠ Q = ∠ P   ■

To prove the converse, we observe that given ∠ P = ∠ Q and ∠ Q = ∠ P we
also have **PQ = QP** so that the triangles are congruent (ASA)

Therefore **AP = AQ** and **AQ = AP**   ■

===

Angles of a Quadrilateral

If we rotate our pencil through
the four angles of a quadrilateral
we see that the pencil turns
through one complete
revolution.

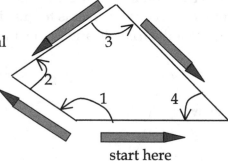

start here

Note also, that if we divide the quadrilateral into two triangles, when we turn through the angles of the quadrilateral, we also turn through the angles of the two triangles that make up the quadrilateral:

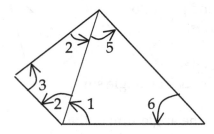

Angles **1+2+3+4+5+6 = 1** revolution

Or, we could write:

1+2+3+4+5+6 = (1+5+6) + (2+3+4) = 2 triangles = 1 revolution

Convex Polygons
A convex polygon has all of its angles less than 180 degrees so that the vertices all point outwards from the inside of the figure. A polygon that is not convex is called a reflex polygon.

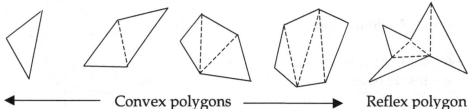

←——————— Convex polygons ———————→ Reflex polygon

The angle sum of a convex polygons can be found by dividing it up into triangles, for example:

Name	Sides	Triangles	Angle sum (revolutions)
Triangle	3	1	½
Quadrilateral	4	2	1
Pentagon	5	3	1½
Hexagon	6	4	2
Heptagon	7	5	2½
Octagon	8	6	3

Theorem 11A The sum of the angles of an **n sided** convex polygon is **(2n-4)** right angles. (Theorem 11 will be generalized later)

Proof 1
Choose any vertex and join it to
The remaining vertices.

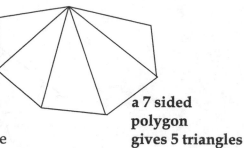

a 7 sided
polygon
gives 5 triangles

This gives **(n-2)** triangles
Each of these triangles has an angle sum

of 2 right angles and the angle sum of the
angle sum of the polygon is equal to the sum
of the angles of the triangles.
Therefore the sum of the angles of the polygon is **2(n-2) = 2n-4 right angles**

Proof 2
Alternatively, we can choose any point inside the polygon and since it is convex, we will be able to join this point up to each of the vertices without the joins crossing any of the sides:

This will form **n** triangles, one on each side.
The total angle sum for the figure is then

angle sum = **2n** right angles

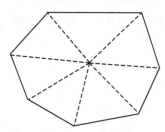

however, four of the right angles
are at the inside point.

Therefore, the angle sum of the
polygon is **(2n – 4)** right angles.

seven triangles form
a seven sided polygon

Exercise 2

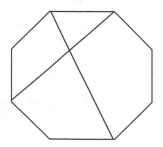

This figure represents a
regular octagon.
What is the angle between
the two diagonal lines
that have been drawn?

Exterior Angles.

If you walk round any closed figure, then you will turn through 1 revolution. For example, if you walk round the circle in the figure below, starting at point P say, looking East, as you walk round, you will soon face North (90^0 turn) then face West (180^0 turn) then South (270^0 turn) then East again (360^0 turn).

Walking round this quadrilateral
Also turns you through 360^0 .

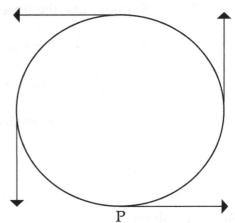

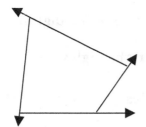

In the next figure, the six sides of a hexagon have been extended to produce what are called the exterior angles of the hexagon. This produces the paddle wheel figure.

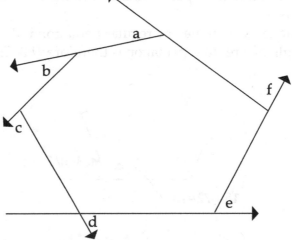

The exterior angles are marked a,b,c,d,e and f.
 If you walk round this figure then you will turn through each of these angles in order and turn yourself through 1 revolution.
Thus a+b+c+d+e+f = 1 revolution.

Hence we have our next theorem:

Theorem 12

The sum of the exterior angles of any polygon is 360⁰ .

Proof 1   Proved by walking round the figure!

Proof 2   Suppose that there are **n** sides   int ∠ / ext ∠

Each exterior angle is 180-(an interior angle)

Add up the **n** exterior angles to get

Sum of exterior angles = 180n –(sum of interior angles)

= 2n –(2n-4) right angles

= 4 right angles   ∎

Regular Polygons

A regular polygon has all its sides and angles equal.

To find the angle of a regular polygon, first find the exterior angle and then subtract from 180°.
For example, the exterior angle of a regular pentagon is 360/5 = 72°.
The interior angle of a regular pentagon is therefore 180-72 = 108°.

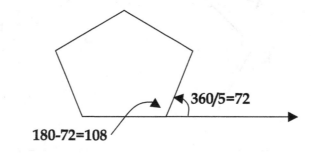

180-72=108   360/5=72

Exercise 3

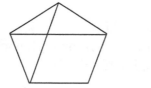

What is the acute angle between the two diagonals of this regular pentagon?

Exercise 4　　　　**Complete the following table:**

Regular figure	Ext. angle	Int. angle
△		
▢		
⬠		
⬡		
(heptagon)		
(octagon)		

Reflex Polygons

It is easy to see how to divide an **n** sided convex polygon into **n-2** triangles, but not so easy for a reflex polygon. These octagons each split into 6 triangles:

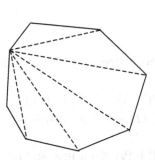

Convex Octagon

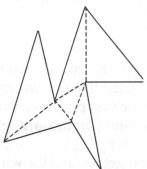

Reflex Octagon

We now return to investigate Theorem 12, for reflex polygons. Since it is not easy to show that any reflex polygon with n sides can be divided into (n-2) triangles, we have to adopt a different approach. Proof by induction is dealt with in chapter 14, however we need to resort to this method here to prove:

Theorem 11 The angle sum of any polygon with sides that do not cross, is always **2n-4 right angles.**

Proof
Firstly, note that we must stipulate that the sides do not cross each other, for:

The angle sum of these two four
sided figures is not the same.

Suppose that the angle sum of any figure of up to K sides is 2K- 4 right angles
Now consider a **K+1** sided figure

For example:

Find a line (here, the thick dotted line) that divides the figure in two but does not cross any of the sides. Let the number of sides of one part be **P+1** and the number of sides of the other part be **Q+1**. Then **P+Q** will be equal to the number of sides of the figure, therefore

$$P+Q = K+1$$

Now each part has at the most **K** sides and therefore according to our supposition, the theorem works for each part.
Therefore the first part has an angle sum of **2(P+1) – 4** right angles
and, the second part has an angle sum of **2(Q+1) – 4** right angles

Add these together, and the whole figure will have an angle sum of
$$2(P+Q+2) – 8 = 2(P+Q) – 4$$

$$= 2(K+1) - 4 \text{ right angles}$$

which we can write as **2N – 4** where **N=K+1**

Therefore, if we assume that Theorem 11 works for polygons of up to **K** sides, then the theorem also works for polygons of **K+1** sides.
Now we know that theorem 11 works for triangles and quadrilaterals, i.e. for **K=3** and **K=4,** and therefore, it will work for **K=5, K=6, K=7 ....**

This proves that **the angle sum of any N sided polygon with sides that do not cross each other,** is **2N – 4** right angles.

==

Exercise 5   Prove that the angle sum of a **P** pointed star is **(P-1)** full turns
For example,
this 4 pointed star
 has an angle sum of 3 full turns.

Negative Rotations

In mathematics, anti-clockwise rotations are taken to be positive and clockwise rotations are taken to be negative.   (Note that this is exactly contrary to navigation where a clockwise rotation round from the North is positive).   We illustrate with an example:

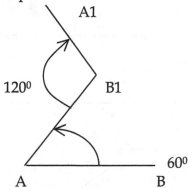

Rotate the line AB about A, through 60^0 to position AB_1

Then rotate AB_1 about B_1 through -120^0 to the position B_1A_1. Combining the two rotations, we have a total angle of $60^0 + (-120^0) = -60^0$. Hence, using the idea of a negative rotation, AB rotates through an angle of -60^0 ending up in the position A_1B_1

The centre of the rotation can be found by drawing the perpendicular bisectors of AA1 and BB1.

The two bisectors meet at the centre of the rotation.

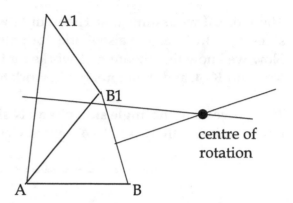

The next figure shows this negative rotation.

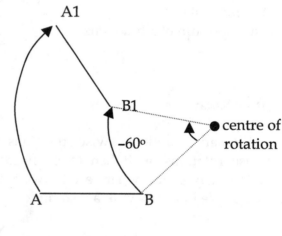

======================================

Exercise 6

[1]

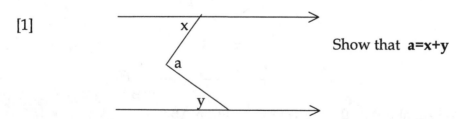

Show that **a=x+y**

[2]

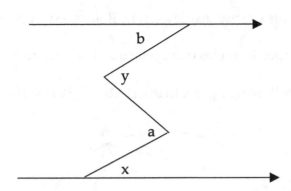

Use the rotating pencil to show that $x+y = a+b$

[3]

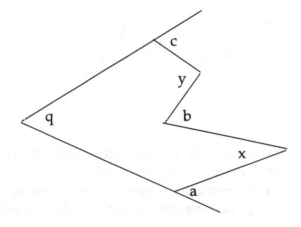

Prove that $q = a - x + b - y + c$

4. If you walk round a figure eight, what angle do you turn through?

5. If you roll a penny round another fixed penny, what angle does it turn through?

==============================

Introducing the Circle

Circle theorems are dealt with in more detail in chapter 20 but here we introduce simple cases of some important circle theorems. The starting point is the isosceles triangle. We only need draw two radii to produce an isosceles triangle.

Definition
If a quadrilateral fits exactly inside a circle, it is called a cyclic quadrilateral.

Theorem 13   Opposite angles of a cyclic quad add up to 180 degrees

To prove, in the following figure, that $\angle DAB + \angle BCD = 180$

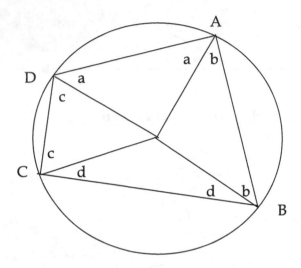

The diagram shows a cyclic quad. with the four equal radii drawn from the centre to the four vertices.   Because the radii are equal, we have four Isosceles triangles so that we can mark the angles a, b, and c equal, as shown. Now from a previous theorem, the angles of the quad. add up to 360^0.
Therefore
$$2a + 2b + 2c + 2d = 360^0.$$
Hence $a + b + c + d = 180^0.$
i.e. $(a+b) + (c+d) = 180^0.$
that is $\angle B + \angle D = 180^0$

and also $(a+d) + (b+c) = 180^0$
so that $\angle A + \angle C = 180^0$

This proves that the opposite angles of a cyclic quad. add up to $= 180^0$.

■

Exercise 7   Consider what happens if two of the vertices of the quadrilateral coincide.

Theorem 14 The angle at the centre = twice the angle at the circumference.

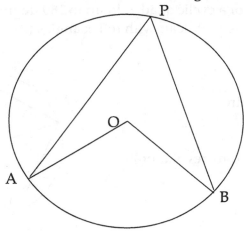

In the above figure, O is the centre of the circle: we prove that ∠AOB = 2∠APB

Join P to the centre so
That we have two
Isosceles triangles

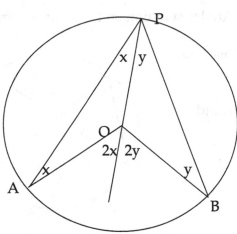

Proof

The angles x are equal   (isosceles triangle: equal radii)
The angles y are equal   (isosceles triangle: equal radii)
The angles at O are 2x and 2y (exterior angle of triangle)

$$\angle AOB = 2x+2y = 2(x+y)$$

$$\angle APB = x+y$$

$$\therefore \quad \angle AOB = 2\angle APB$$

∎

Note how a more general version of Theorem 15 could be used to prove that the opposite angles of a cyclic quad. add up to 180 degrees. We would need a version of Theorem 14 that deals with reflex angles (see chapter 20)

Using theorem 14
the angles at the centre
are 2x and 2y.

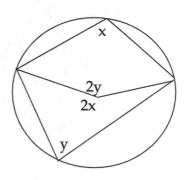

$$2x+2y = 360 \quad \text{(one revolution)}$$

∴ $$x+y = 180$$

==================================

Theorem 15   Angles standing on the same arc are equal

(The same segment theorem)

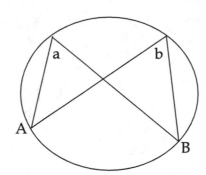

Angles a and b both stand
on arc AB

To prove that a=b

Proof

Join A and B to the centre O

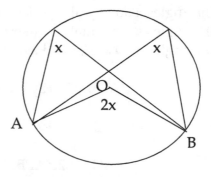

Let $\angle AOB = 2x$
then each of the
angles at the circumference
will be x by theorem 15.

Which proves the theorem.

============================

Theorem 16   The angle in a semi circle is 90 degrees

Given   AOB is a diameter

Prove that ∠APB = 90

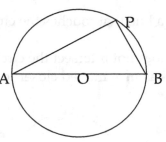

Proof 1
If O is the centre of the circle
then ∠AOB = 2∠APB
by theorem 15, but ∠AOB = 180

$$∠APB = 90$$

=================================

Proof 2

Join P to the centre O

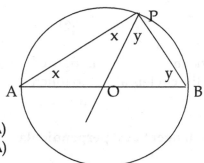

The angles x are equal (isos Δ)
The angles y are equal (isos Δ)

$$2x+2y = 180 \text{ (angle sum of triangle)}$$

∴ $$x+y = 90$$

therefore $$∠APB = 90$$ ∎

====================

Exercise 8   Calculate the angles marked x in these figures

[1]

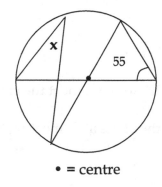

• = centre

[2]

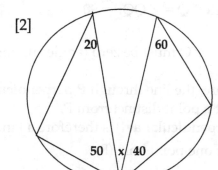

Tangents (Latin: tango = I touch)

A tangent to a circle is a line that touches the circle at one point.

Let **m** be a line that does not intersect the circle and consider what happens as **m** is moved closer to the circle.

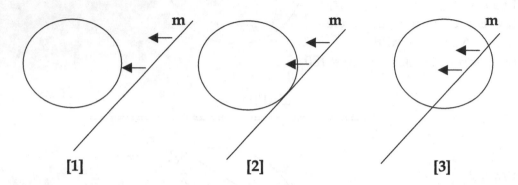

[1]   [2]   [3]

In figure [1] the line **m**   In figure [2] the line   In figure [3] the line
does not intersect the circle is about to touch   intersects the circle
                                     the circle   at two points

Theorem 17   The tangent at P perpendicular to the radius through P

Draw a line through P perpendicular to OP and suppose that this perpendicular meets the circle again at Q.

then   OP = OQ   (radii)

∴   ∠OPQ = ∠ OQP (isosceles triangle)

∴   ∠OPQ = ∠ OQP = 90

∴   ∠O must be zero (angle sum of triangle POQ)

Therefore, the line through P perpendicular to OP does not meet the circle again at a point distinct from P.
The perpendicular at P is therefore a tangent to the circle at P meeting the circle at one point only (P).   ■

Theorem 18   **The alternate segment theorem**

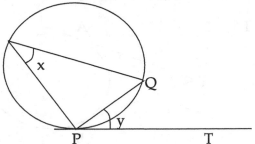

(for a more complete
proof see chapter 22
in volume 2)

PT is the tangent at P and PQ is a chord.
Prove that the angle between the tangent and the chord (y) is equal to the
angle in the alternate segment (x).
Proof

Draw the diameter PA.

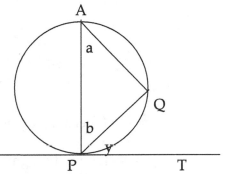

In the given figure: $\angle Q = 90$ (semi circle)

$\therefore$ a+b = 90 ($\angle$ sum of triangle)

but   y+b = 90 (tangent perp radius)

$\therefore$ y = a   (both are 90-b)

but   a = x (same segment: theorem 15)

$\therefore$ y = x   ∎

Exercise 9   Calculate the angles x in the following figures
[1]   [2]

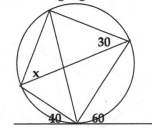

Theorem 19:   Tangents to a circle are equal

To prove that TA = TB

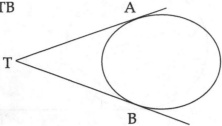

Proof

Draw an angle standing on chord AB

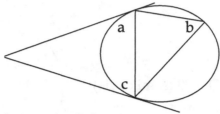

       a = b (alternate segment)

       b = c (alternate segment)

∴   a = c

∴ TA = TB (isosceles triangle)

=================================

Exercise 10   Find the angles of the triangle that is inscribed in the circle [1]

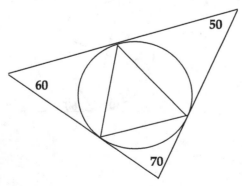

[2]

Find the angles x, y and z

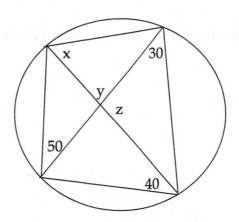

[3]

Given   angles 20
       and 30,

Find   x and y

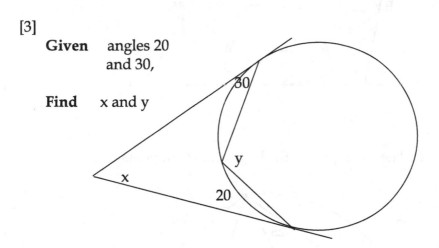

[4]   Find the angles of the inside triangle and the outside triangle

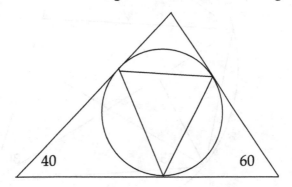

[5]   Given AB=AC, find a, b, c, x, y and z

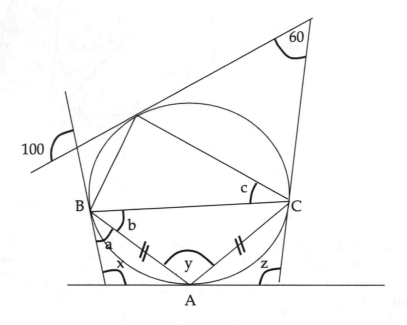

[6]   Calculate the angles of the large and small pentagons

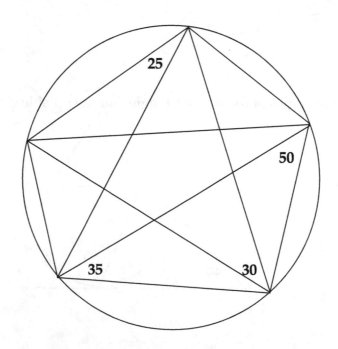

Solutions

Exercise 2   67½

Exercise 3   72

Exercise 4

Regular figure	Ext. angle	Int. angle
triangle	120	60
square	90	90
pentagon	72	108
hexagon	60	120
heptagon	51 3/7	128 4/7
octagon	45	135

Exercise 6   [4]   zero   [5]   2 revolutions

Exercise 7   It gives an alternative proof of the alternate segment theorem

Exercise 8   [1]   40   [2]   10

Exercise 9   [1]   30   [2]   50

Exercise 10
   [1]   55, 60, 65

   [2]   x=60, y=80, z=100

   [3]   x=80, y=130

   [4]   small triangle: 50, 60, 70
          large triangle: 40, 60, 80

   [5]   a=b=c=40
          x=y=z=100

   [6]   Large pentagon   95, 105, 105, 110, 125
          Small pentagon   90, 100, 115, 115, 120

CHAPTER 3

Pythagoras

The Greek island of Samos lies about 1 mile from the Turkish mainland. If you have a map of the area, look about 50 miles south east of Samos and you should find three dots (∴) marking the ruins of the ancient Greek town of Miletus. Miletus was an important centre of trade and learning in the 6th century BC and although it is now some distance inland on the Turkish mainland, due to silt deposits brought down by the river Meander, it was then a thriving Greek port.

Thales (c624..c546 BC) was born in Miletus and quickly became famous for his understanding of mathematics and engineering. He became interested in the mathematics and astronomy of the ancient Babylonians and the Egyptians and visited Egypt where he was able to measure the heights of pyramids and where he studied Egyptian geometry. On his return to Miletus he set up a school to teach the Babylonian and Egyptian discoveries in astronomy and mathematics.

Thales is generally regarded as the founder of Greek philosophy, astronomy and science and he laid the foundations of Greek mathematics. The following theorems in elementary geometry are usually attributed to him.

[1]   Opposite angles are equal

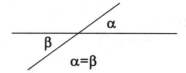

[2]   Base angles of an isosceles triangle are equal

If AB=AC then ∠B=∠C

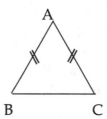

[3]   The angle in a semi circle is 90°

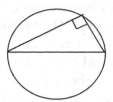

[4]   Thales Theorem   A

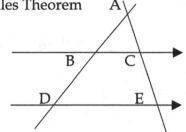

$$\frac{AB}{AD} = \frac{BC}{DE} = \frac{CA}{EA}$$

Pythagoras (c580..c500 BC) was born in Samos about 50 years after the birth of Thales. When in his late teens, he visited Thales in Miletus and was inspired by him to follow a life of study of mathematics and science. From that time on he called himself a philosopher (philo..lover of), (sophos..wisdom) and probably coined the term μαθηματικος = mathematikos or that which is to be learned. In 535 BC, when in his thirties, Pythagoras, like Thales, travelled to Egypt but the outcome of the visit was somewhat different. In 525 BC, king Cambyses II of Persia invaded Egypt and Pythagoras was captured and taken to Babylon as a prisoner of war. While in Babylon, however, Pythagoras was given the freedom to study and teach and was able to learn about Babylonian mathematics and science. In 522 BC, Cambyses died and two years later Pythagoras was able to return to Samos.

Back in Samos Pythagoras founded a school called The Semicircle but in 518 BC he left Samos for Italy. In Italy he established a mystic cult and school called The Society at Crotone, near the base of the big toe on the boot of Italy.

Two hundred years later, the discoveries of the Pythagoreans were collected by Euclid and incorporated into a book called "The Elements" which has been the mainstay of school geometry to the present day.

Pythagoras is most famous for his theorem on right angled triangles but he is also important for introducing the idea of rigorous proof into mathematics and discovering irrational numbers, numbers that cannot be represented as ratios of whole numbers, and for investigations into the nature of the musical scale.

Pythagoras, his String and his Comma

The monochord is a one-string instrument. It consists of a single string, stretched between two pieces of wood called the fixed bridge and the sliding bridge so that it could be plucked to produce a musical note. The quality of the note depends on the material of the string, its thickness and tightness but the pitch of the note is determined by the length of the string. The shorter the string, the higher the pitch of the note.

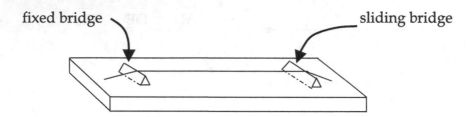

Pythagoras experimented with the monochord and found that certain sounds seemed to go well together (a concord) while others did not (a discord). He found that when two notes were played on similar instruments, there would only be a pleasing effect when the lengths of the two strings were in a simple ratio to each other.

The pitch of the note depends on the number of vibrations the string makes in one second. The note **A** in the middle of the piano keyboard produces 440 vibrations per second (Hertz).

Thus we can write

$$A = 440 \text{ Hz.}$$

When the length of the string is divided in half, the note produced has twice the frequency and we say that the note is an octave higher.

If a length **L** gives a note **A = 440 Hz**

then a length ½ **L** will give a note **A$^+$ = 880 Hz**.

The frequency varies as one over the length $f = \text{constant} \times \dfrac{1}{\textbf{Length}}$

We should remark that there are also a large number of "overtones" produced with any note, most of which are inaudible. Any whole number division of the length of the string produces a note and in theory all of these notes are present. We depict these overtones in the following diagrams.

Note that one wavelength is
always a complete sine wave.
as is shown here:

one wavelength
of a sine wave

In the first figure shows a standing wave that does not travel down the string. The wave depicted is only half a wavelength:

The fundamental note
wavelength W

The first overtone
wavelength W/2

The second overtone
wavelength W/3

The third overtone
wavelength W/4

For our purposes we can ignore the overtones and concentrate on the fundamental note that is produced by the stretched string. Suppose that the open string produces the note **C**, then dividing the length in half will give the note an octave above which we will call **C⁺**.

Length L   **C**

Length ½ L   **C⁺**

Between these two positions for the sliding bridge we can find other fractions between ½ L and L that will produce notes between **C** and **C⁺**

This is how Pythagoras arrived at the notion of a musical scale because there are only a few points of division that give the simple ratios that he wanted. If we examine the positions for the sliding bridge between ½ L and L we find the following fractions with denominators < 7 :

$$\frac{1}{2} \quad \frac{3}{5} \quad \frac{2}{3} \quad \frac{3}{4} \quad \frac{4}{5} \quad \frac{5}{6}$$

Remember, the frequency of the note varies as 1/length. Let the frequency of the note **C** be **f** Hertz then these simple ratios give the following notes:

Length	frequency	Note	Interval
L/2	2f	C+	octave
3L/5	5f/3	A	major 6th
2L/3	3f/2	G	perfect fifth
3L/4	4f/3	F	perfect 4th
4L/5	5f/4	E	major 3rd
L	f	C	unison

Having discovered how to choose notes to make up a musical scale, the number of notes has to be decided. For example the pentatonic scale uses the five notes C, D, E, G and A.

Since we are concerned with using simple ratios between ½ and 1, a good choice for the number of notes in the scale would be twelve since this would provide a good number of simple fractions: twelve evenly spaced notes between ½ and 1 would yield the following positions: ½, 13/24, 7/12, 5/8, 2/3, 17/24, ¾, 19/24, 5/6, 7/8, 11/12, 23/24 and 1 .

For whatever reason, we now use a twelve note scale and name the notes

C   C#   D   D#   E   F   F#   G   G#   A   A#   B   C+

Leaping up in Perfect Fifths

If the open string plays **C** then the simplest ratio, **2/3** will play the note **G**. The leap from **C** to **G** is called a perfect fifth. To jump up a perfect fifth all we need do is take 2/3rds of the length of the string so therefore, in order to find the positions of the notes of the scale by jumping up perfect fifths all we need do is keep on taking 2/3rds of the length of the string but if we go past **C⁺** (at the mid point) we double the length to bring the note down an octave. Taking **C⁺** at the point 0.5 and **C** at the point 1, that is, supposing that the open string length is taken as the unit length, leaping up by perfect fifths (and dropping an octave if we pass **C⁺**) will produce the following positions on the string for the other notes of the scale:

Note	calculation	position on string
C	1	C = 1
G	2/3 = 0.666	G = 0.666
D+	0.666 x 2/3 = 0.444 drop an octave 0.444 x 2	D = 0.888
A	0.888 x 2/3 = 0.592	A = 0.592
E+	0.592 x 2/3 = 0.392 drop an octave 0.392 x 2	E = 0.790
B	0.790 x 2/3 = 0.526	B = 0.526
F+#	0.526 x 2/3 = 0.351 drop an octave 0.351 x 2	F# = 0.702
C+#	0.702 x 2/3 = 0.468 drop an octave 0.468 x 2	C# = 0.936
G#	0.936 x 2/3 = 0.624	G# = 0.624
D+#	0.624 x 2/3 = 0.416 drop an octave 0.416 x 2	D# = 0.832
A#	0.832 x 2/3 = 0.555	A# = 0.555
F+	0.555 x 2/3 = 0.370 drop an octave 0.370 x 2	F = 0.740
C+	0.740 x 2/3 = 0.493 drop an octave 0.493 x 2	C? = 0.986

But the position for **C+** should ideally be **0.5** so that when we drop an octave we get back to **C = 1**. Jumping up by perfect fifths gives a musical scale but the size of the scale is not an octave. The "distance " between two notes is the ratio of the frequencies of the notes and the frequency is proportional to one over the length of the string. The "distance" between **C** and **C?** is therefore given by **1 ÷ 0.986 = 1.014**.

This "distance" between these frequencies is called **Pythagoras's comma.**

The positions of these notes are marked on the next diagram:

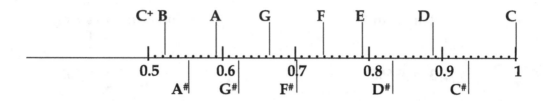

Around the beginning of the 16th century it was decided that the notes of the musical scale should go up in equal jumps or equal temperament.

The Well Tempered Scale
I f the twelve notes are to go up in equal jumps from **C** to **C+** where should the stops then be?
Suppose that the length OC is 1 so that OC+ = ½ then we wish to find the lengths $ON_1, ON_2, ON_3, ON_4, ON_5, ON_6, \ldots\ldots ON_{12}$ for the twelve notes of the scale.
If the frequency of the note OC is represented by C, then the frequency of the note OC+ will be represented by 2C. The "distance" between two notes is the ratio of their frequencies. If we call the notes of the scale $N_1, N_2, N_3, N_4, N_5, N_6 \ldots$ etc with equal temperament, then the frequencies for these notes will be $N_1 = kC$, $N_2 = k^2C$, $N_3 = k^3C \ldots\ldots N_{12} = k^{12}C$ so that the "distances" between N_1 and N_2, N_2 and N_3, N_3 and N_4 and so on, are all equal to k and we will have "equal temperament" .

However we know that $N_{12} = k^{12}C$ should give the note 2C and therefore we know that $k^{12} = 2$, or
$$k = 2^{1/12} \text{ , the twelfth root of 2}$$

Now the length of the string is proportional to one over the frequency and therefore, given that OC = 1, we can find the lengths that will produce the notes N_1, N_2, N_3, N_4, N_5, N_6 ... N_{12} with equal temperament:

The Equal Tempered Scale

	Note	frequency (Hertz)
OC = 1	C	261.6
$ON_1 = 1/k = 0.944$	C#	277.2
$ON_2 = 1/k^2 = 0.891$	D	293.7
$ON_3 = 1/k^3 = 0.841$	D#	311.1
$ON_4 = 1/k^4 = 0.794$	E	329.6
$ON_5 = 1/k^5 = 0.749$	F	349.2
$ON_6 = 1/k^6 = 0.707$	F#	370.0
$ON_7 = 1/k^7 = 0.667$	G	392.0
$ON_8 = 1/k^8 = 0.630$	G#	415.3
$ON_9 = 1/k^9 = 0.595$	A	440
$ON_{10} = 1/k^{10} = 0.561$	A#	466.2
$ON_{11} = 1/k^{11} = 0.530$	B	493.9
$ON_{12} = 1/k^{12} = 0.5$	C+	523.3

So how does the perfect fifth fit into the equal tempered scale?

There are seven jumps (semi tones) from C to G in both the diatonic scale and the equal tempered scale. In the construction of the diatonic scale 12 jumps of a perfect fifth would send us through all of the notes of the scale ending on the high C seven octaves above where we started. In the equal tempered scale seven octaves would be accomplished by doubling the frequency seven times but do we land on the same high C? The frequency of the high C for the diatonic scale is $(3/2)^{12}$ x C. The frequency of the high C for the equal tempered scale is 2^7 x C.

The "distance" between these two notes is the ratio of their frequencies, that is

$$(3/2)^{12} \times C \div 2^7 \times C = 1.013643265 \text{ Hz}$$

This is Pythagoras's comma.

For comparison we show the positions of the notes on both scales:

The Equal Tempered Scale

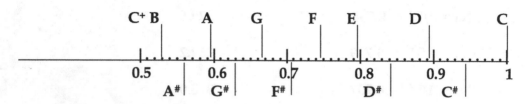

The Scale formed from Perfect Fifths

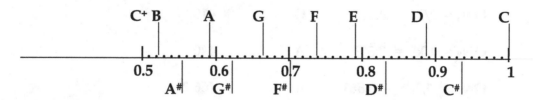

Starting with a frequency of 261.63 for middle C, the frequencies for the notes of the two scales are compared in the following table. The first row is the Diatonic scale and the second row is the equal tempered scale.

C	C#	D	D#	E	F	F#	G	G#	A	A#	B	C+
261.6	279.4	294.3	314.3	331.1	353.6	372.5	392.4	419.1	441.5	471.5	496.7	530.4
261.6	277.2	293.7	311.1	329.6	349.2	370.0	392.0	415.3	440	466.2	493.9	523.2

The "distance" between the two top notes is determined from the ratio of the two frequencies, giving **530.4/523.2 = 1.014, the comma of Pythagoras.**

=================================

The 3,4 5 triangle

If you draw a triangle with sides 3, 4 and 5 units long, then you will find that the angle opposite the longest side (called the hypotenuse), will be exactly 90º (= one corner).
The Egyptians used the 3,4,5 triangle to construct accurate corners for their buildings and pyramids.

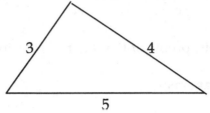

Apart from doubling all the side to get a 6,8,10 triangle, the next simplest Pythagorean triangle is the 5,12 13.
In each case, we see the Theorem of Pythagoras at work, that is, the angle opposite the longest side, the hypotenuse (Greek: hypo..under, teino..stretch), is 90º.

Examples:

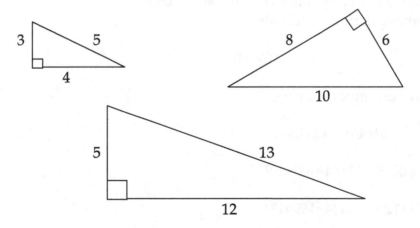

In each of the triangles above, the angle marked with the corner symbol ⌐ is 90º.

Pythagoras's Theorem

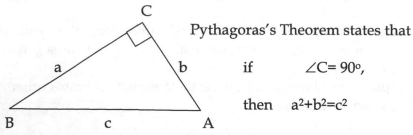

Pythagoras's Theorem states that

if $\angle C = 90°$,

then $a^2 + b^2 = c^2$

We will examine various proofs of Pythagoras's theorem shortly.

The Converse of Pythagoras

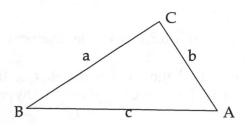

The Converse of Pythagoras's Theorem states that
if, in the above figure $a^2 + b^2 = c^2$

then $\angle C = 90°$.

In our three examples. we have

[1] $3^2 + 4^2 = 9 + 16 = 25 = 5^2$

[2] $6^2 + 8^2 = 36 + 64 = 100 = 10^2$

[3] $5^2 + 12^2 = 25 + 144 = 169 = 13^2$

so that the equation $a^2 + b^2 = c^2$ is satisfied and we deduce from the converse of Pythagoras's Theorem that we have a corner (= 90°) opposite the longest side.

Both Euclid (c.330–c.260 BC) and Pythagoras (c.580–c500 BC) knew how to prove that these results always work, that is,

i) if $\angle C = 90°$ in our triangle ABC, then $a^2 + b^2 = c^2$

and conversely,

ii) if $a^2 + b^2 = c^2$, then we get our 90° corner.

There are many different proofs of Pythagoras now available and we give eight of them.

The proof known to Euclid and Pythagoras is given in Euclid's Elements Book 1 problem 47.

This is an outline of the proof they each knew:

Proof [1]   The Greek's proof

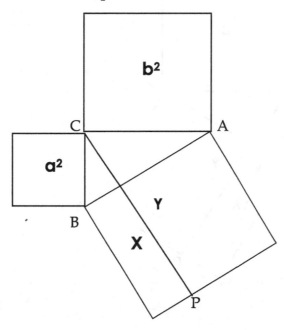

Draw the perpendicular CP, from C, as shown, dividing the square standing on AB into two parts X and Y, then they were able to show that the area of the square standing on BC was equal to the area X, and that the area of the square standing on AC was equal to Y.

Thus $a^2 = X$

and $b^2 = Y$

Adding these two equations gives

$$a^2 + b^2 = X + Y,$$

but $X + Y = c^2$ hence the required result follows.

We now show why $a^2 = X$, in our illustration of the proof of Pythagoras's Theorem.

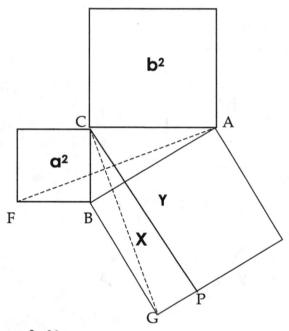

To Prove $a^2 = X$

Proof

$$\triangle ABF = \tfrac{1}{2} a^2 \qquad\qquad (\tfrac{1}{2} \text{ base FB x height a })$$

Now rotate $\triangle ABF$ about B by $90°$.

Now BF rotates to BC and BA rotates to BG

so $\triangle ABF$ rotates to $\triangle GBC$

Therefore, $\triangle$ GBC is also $\frac{1}{2}$ a^2.

But area X is twice $\triangle$ GBC   (base GB and height GP)

Therefore $X = a^2$

Similarly, we can show that $Y = b^2$

Adding these two equations, we have

$$X + Y = a^2 + b^2$$

Hence $c^2 = a^2 + b^2$ ■

Alternatively:
   Euclid used congruent triangles showing that $\triangle ABF \equiv \triangle GBC$

   In triangles $\triangle ABF$ and $\triangle GBC$

                    $AB = GB$ (sides of a square)

                    $BF = BC$ (sides of a square)

                    $\angle ABF = \angle GBC$ (both are 90+B)

   $\therefore$ $\triangle ABF \equiv \triangle GBC$ (the triangles are congruent SAS)

      but $\triangle ABF = \frac{1}{2}$ a^2 and $\triangle GBC = \frac{1}{2}$ X

   $\therefore$ $a^2 = X$

==

Proof [2]   The Chinese Proof

Chinese mathematicians in 300 B.C. also knew of Pythagoras's Theorem, but the Chinese proof was quite different, relying on the algebraic expansion of a bracket.

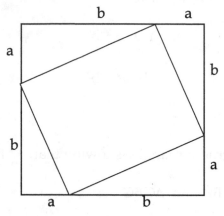

If we stick four triangles onto the c^2 square, we get another square with side (a+b). The area of the large square is then

$$(a+b)^2 = a^2 + 2ab + b^2$$

but this area can also be written down as

$$c^2 + 4 \text{ triangles} = c^2 + 4 \times \tfrac{1}{2} ab$$

$$= c^2 + 2ab$$

Therefore,

$$a^2 + 2ab + b^2 = c^2 + 2ab$$

$$a^2 + b^2 = c^2 \qquad \blacksquare$$

===========================

Proof [3]   **Using Similar Triangles**

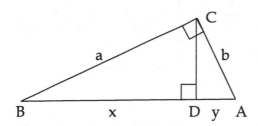

The triangles ABC, CBD and ACD are equiangular and therefore are similar in shape. Therefore, the ratios of corresponding sides are equal:

Thus

$$\frac{x}{a} = \frac{a}{c} \text{ and } \frac{y}{b} = \frac{b}{c}$$

∴ $x = \dfrac{b^2}{c}$ and $y = \dfrac{a^2}{c}$

∴ $x+y = \dfrac{b^2 + a^2}{c}$

but $x+y = c$ and this gives $c^2 = b^2 + a^2$ ■

==

Proof [4]   Using similar Figures

We can see three similar shapes in
this figure:

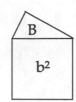
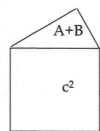
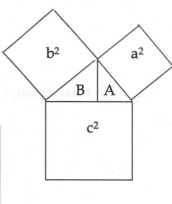

**Since the shapes are similar, the ratio of the square to the triangle is the
same in each figure, thus** $a^2 = kA$

$$b^2 = kB$$

$$c^2 = k(A+B)$$

adding gives $a^2 + b^2 = c^2$ ■

=============================

Proof [5]   Garfield's Proof
This proof is reputed to have been found by James Garfield, 20th president
of the United States of America. He held office for just 4 months being
assassinated in Washington DC railway station on July 2nd 1881.
His figure uses the formula for the area of a trapezium which is
       ½ (sum of parallel sides)x(distance between them)
In this figure,

trapezium = 2triangles + ½ square

$$½ (a+b)(a+b) = 2x ½ab + ½ c^2$$

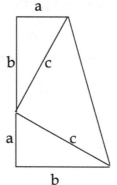

$$(a+b)(a+b) = 2ab + c^2$$

$$a^2 + 2ab + b^2 = 2ab + c^2$$

$$a^2 + b^2 = c^2 \quad ■$$

Proof [6]   By Leonardo Da Vinci (1452 – 1519)

First, ABC, CRQ and DEF are
congruent triangles.

It is easy to see that PCS is
a straight line, because
angle PCS=45+90+45.

Leonardo noticed that
ABPS and DBCE are
congruent shapes.

∴ 2ABPS = 2DBCE

∴   ABPQRS = DBCAFE

∴ a² + b² + 2ABC = c² + 2ABC

∴   a² + b² = c²   ■

=================================

Proof [7]   By Rearrangement

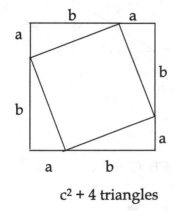

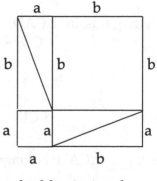

c² + 4 triangles   =   a² + b² + 4 triangles

c²   =   a² + b²   ■

Proof [8]   The Indian proof

Bhaskara gave the following around 1100 AD

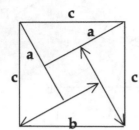

The side of the small square is (b-a)

$\therefore$ c^2 = 4 triangles +$(b-a)^2$

$\therefore$ c^2 = 4 . ½ ab + b^2 –2ab + a^2

$\therefore$ c^2 = b^2 + a^2

■

=================================

The Converse of Pythagoras (Euclid)

Given
$$a^2 + b^2 = c^2$$

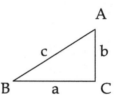

Prove $\angle C = 90°$

Proof   Construct $\triangle CAD$

With $\angle A = 90$ and AD = c

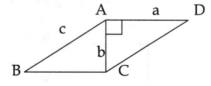

Then by Pythagoras, in $\triangle CAD$,
$$CD^2 = a^2 + b^2$$

$$= c^2 \qquad \text{(given)}$$

Therefore   CD = c

Therefore $\triangle CAD$ is congruent to $\triangle ACB$ (3 sides)

Therefore $\angle BCA = 90$ ■

Pythagorean Triples

(3,4,5), (6,8,10) and (5,12,13) are called Pythagorean triples. They are sets of whole numbers that satisfy Pythagoras's equation. We can generate Pythagorean triples from the equation

$$(m^2 + n^2)^2 = (m^2 - n^2)^2 + (2mn)^2$$

that gives the following right angled triangle:

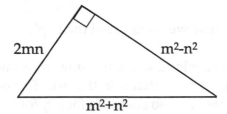

All we need to do is substitute whole number values for **m** and **n** and then work out the sides. The value of **m** runs through the natural numbers, 2, 3, 4, 5,…. and with this **m**, **n** runs through 1,2,3,4… up to one less than **m**.

m	n	m^2	n^2	$m^2+n^2 =c$	$m^2-n^2=b$	$2mn=a$
2	1	4	1	5	3	4
3	1	9	1	10	8	6
3	2	9	4	13	5	12
4	1	16	1	17	15	8
4	2	16	4	20	12	16
4	3	16	9	25	7	24
5	1	25	1	26	24	10
5	2	25	4	29	21	20
5	3	25	9	34	16	30
5	4	25	16	41	9	40
..	..	..	..	..	..	..

We can see these Pythagorean triples appearing in the list:

$$3^2+4^2 = 5^2$$

$$6^2+8^2 = 10^2$$

$$5^2 + 12^2 = 13^2$$
$$8^2 + 15^2 = 17^2$$

$$12^2 + 16^2 = 20^2$$

$$7^2 + 24^2 = 25^2$$
.. .. ..

The question arises: "Can we find all possible Pythagorean triples from this list?"

The answer is certainly **no** because we do not find the triple $9^2 + 12^2 = 15^2$

However it can proved that the triples that are missing from the list are always multiples of a "primitive triple" that is in the list. The primitive triples referred to are triples that have no common factor, for example, (3,4,5). The (9,12,15) triple that is not in the list is simply three times the (3,4,5) triple.
(For a proof of this result, consult a book on the theory of numbers, e.g. "An Introduction to the Theory of Numbers" by Niven, Zuckerman and Montgomery, chapter 5).

Example 1
Every integer solution of $a^2 + b^2 = c^2$ can be put into the form

$$(m^2 + n^2)^2 = (m^2 - n^2)^2 + (2mn)^2$$

Proof
$$\text{Let} \quad a = m^2 - n^2$$

$$\text{and} \quad c = m^2 + n^2$$

then we can solve for m and n:

add $2m^2 = a + c$ so that $m = \sqrt{\frac{1}{2}(a+c)}$

subtract $2n^2 = c - a$ giving $n = \sqrt{\frac{1}{2}(c-a)}$

but clearly, m and n will not always be integer values as in the table. For the

triple $9^2 + 12^2 = 15^2$ we have $\mathbf{m = 2\sqrt{3}}$ **and** $\mathbf{n = \sqrt{3}}$

We cannot find any integer values for m and n that will give this solution. ■

Example 2

Every odd number >1 is a member of a Pythagorean triple

Proof

If p is an odd number then $\mathbf{p^2\text{-}1}$ and $\mathbf{p^2\text{+}1}$ are both even

Therefore $\left(\dfrac{\mathbf{p^2\text{-}1}}{\mathbf{2}}\right)$ and $\left(\dfrac{\mathbf{p^2\text{+}1}}{\mathbf{2}}\right)$ are whole numbers

But $\qquad \mathbf{p^2} \; + \; \left(\dfrac{\mathbf{p^2\text{-}1}}{\mathbf{2}}\right) = \left(\dfrac{\mathbf{p^2\text{+}1}}{\mathbf{2}}\right)$

and this gives a Pythagorean triple

=========================

Example 3

Every even number >2 is a member of a Pythagorean triple

Proof

If p is even then $\mathbf{p = 2k}$ for some integer $\mathbf{k}$

But $(2k)^2 + (k^2\text{-}1)^2 = (k^2\text{+}1)^2$ and this gives a Pythagorean triple.

=========================

Example 4

Find all the Pythagorean triples that form an arithmetic progression

Solution

Let the terms $\mathbf{a - d, a}$ and $\mathbf{a + d}$ form an AP with $\mathbf{a + d}$ being the largest term

Substitute in $x^2 + y^2 = z^2$ to get $(a - d)^2 + a^2 = (a + d)^2$

$$2a^2 - 2ad + d^2 = a^2 + 2ad + d^2$$

giving $\qquad\qquad a^2 = 4ad$ so that $a = 4d.$

The required triples are therefore of the form $\mathbf{3d, \; 4d}$ and $\mathbf{5d}$ ■

Exercise 1

[1]   Find the 15 Pythagorean triples for which all the sides are less than 40.

[2]

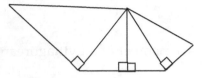

In the above figure,
            (1) all the sides have integer values

            (2) all the sides have different lengths, less than 30

Find the lengths of all the sides in the figure.
What is the total area?

[3]   If (a,b,c) is a Pythagorean triple, prove that at least one of a, b or c is a multiple of 3.

[4]   If (a,b,c) is a Pythagorean triple, prove that at least one of a, b or c is divisible by 5.

[5]   Prove that there cannot be a Pythagorean triple with its terms in geometric progression.

[6]   From example 2 above, every odd number >1 is a member of a Pythagorean triple.
       From example 3 above, every even number > 2 is a member of a Pythagorean triple

       Deduce that every number >2 is a member of some Pythagorean triple.

===================================

Pythagoras in 3D

Pythagoras's Theorem gives us a formula for the length of the diagonal of a rectangle. If the lengths of the sides of the rectangle are x and y, then the diagonal OP in the figure can be found using the result $OP^2 = x^2 + y^2$.

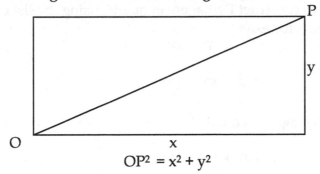

$$OP^2 = x^2 + y^2$$

If O is the origin of coordinates and P is the point P(x,y), then the distance from the origin is given by r where

$$r^2 = x^2 + y^2$$

In three dimensions, we would have a box (more formally called a parallelepiped), with sides x, y and z.

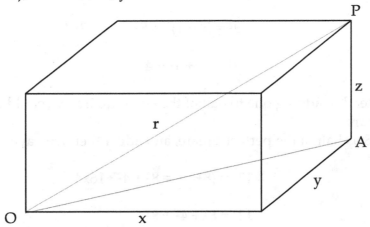

By Pythagoras we have

$$r^2 = OP^2 = OA^2 + z^2$$

Where $\qquad OA^2 = x^2 + y^2$

Hence $\qquad r^2 = x^2 + y^2 + z^2$ ∎

Pythagorean Quads

It is natural now, to ask if we can solve the equation $r^2 = x^2 + y^2 + z^2$ with whole numbers in the same way as we did with the Pythagorean triples.

It is easy, in fact, to construct Pythagorean quads, using the list of Pythagorean triples, for example,

$$5^2 = 3^2 + 4^2$$

and $\qquad\qquad 5^2 + 12^2 = 13^2$

so we have the Pythagorean quad

$$\mathbf{3^2 + 4^2 + 12^2 = 13^2}$$

Another way of generating Pythagorean quads is to take half of an even square, for example

$$\tfrac{1}{2} \times 16 = 8,$$

and then add 1 to the result, giving, in our example, (8 + 1). Now square this bracket:

$$9^2 = (8 + 1)^2 = 8^2 + 1^2 + 2 \times 8$$

thus $\qquad\qquad \mathbf{9^2 = 8^2 + 1^2 + 4^2}$

In fact, instead of adding one to half of the even square, we could add any perfect square.

If we take 8 as half of the perfect square, and add 4 then we have

$$12^2 = (8+4)^2 = 8^2 + 4^2 + 16 \times 4$$

$$\mathbf{12^2 = 8^2 + 4^2 + 8^2}$$

Adding 9 would give

$$17^2 = (8+9)^2 = 8^2 + 9^2 + 16 \times 9$$

$$\mathbf{17^2 = 8^2 + 9^2 + 12^2}$$

Quads and Quins

We have seen that there are many whole number solutions to the equation

$$x^2 + y^2 = r^2 \ ,$$

for example x=3, y=4, r=5,

and also, that we can find an infinite number of solutions to

$$x^2 + y^2 + z^2 = r^2 \ .$$

for example, x=3, y=4, z=12, r=13

There is no difficulty in extending the number of terms on the left hand side of the equation, for example, consider the equation

$$x^2 + y^2 + z^2 + w^2 = r^2$$

we can write

$$51^2 = (50 + 1)^2$$

$$= 50^2 + 1^2 + 2 \times 50$$

$$= 50^2 + 1^2 + 100$$

$$= 50^2 + 1^2 + 4 \times 25$$

$$= 50^2 + 1^2 + 4(3^2 + 4^2)$$

$$= 50^2 + 1^2 + 6^2 + 8^2$$

giving the solution x=50, y=1, z=6, w=8, r=51.

Lagrange

Joseph-Louis Lagrange (1736-1818) was one of the greatest mathematicians of the eighteenth century. His greatest work, on analytical mechanics extended the mechanics of Newton's Principia and in 1793 he presided over the introduction of the metric system in post revolutionary France. In 1770, Lagrange proved this remarkable result:

The Four Squares Theorem (Lagrange 1770)

Every integer can be expressed as the sum of 4 squares (some of which may be zero)

Thus $1 = 1^2 + 0^2 + 0^2 + 0^2$

$2 = 1^2 + 2^2 + 0^2 + 0^2$

$3 = 1^2 + 1^2 + 1^2 + 0^2$

$4 = 1^2 + 1^2 + 1^2 + 1^2 \quad 1 = 2^2 + 0^2 + 0^2 + 0^2$

$5 = 2^2 + 1^2 + 0^2 + 0^2$

$6 = 2^2 + 1^2 + 1^2 + 0^2$

$7 = 2^2 + 1^2 + 1^2 + 1^2$

$8 = 2^2 + 2^2 + 0^2 + 0^2$

. . . .

Exercise 2

In how many ways can 100 be written as

[i] four squares

[ii] four positive squares   (i.e. non zero)

======================================

Lagrange published his four squares theorem in 1770. Also in 1770 Edward Waring gave the following:

Waring's Problem (1770)

Every positive integer is the sum of 9 cubes.

In fact, the only numbers that require 9 non-zero cubes are 23 and 239.

Exercise 3

Write **23** as a sum of 9 cubes.

Exercise 4

Write **239** as a sum of 9 cubes.

Fermat and Diophantus.

Pierre de Fermat was born in France in 1601 and after studying at the University of Toulouse, worked as a civil servant in the French Government. He had always taken a keen interest in mathematics and was fascinated by his copy of a translation of Book 2 of Arithmetica, by Diophantus. Diophantus was a Greek writer who lived in Alexandria at around 250 BC. His most famous work was Arithmetica, a work of 13 volumes on Greek and Babylonian mathematics but unfortunately only 6 of the original 13 volumes remain. Diophantus was fascinated by riddles that had whole number solutions. The following riddle is carved on his tombstone:

"God granted him to be a boy for the sixth part of his life;
 and adding a twelfth part to this, He clothed his cheeks with down;
He married after a seventh part and
five years after his marriage, God granted him a son.
The child died after attaining half his father life
and after four years of grief, Diophantus died."

From this information you are meant to calculate how old Diophantus was when he died.

Fermat owned a copy of book 2 of Arithmetica that contained some problems on Pythagorus's theorem and Pythagorean triples.
Fermat tried to solve equations such as

$$x^3 + y^3 = r^3$$

$$x^4 + y^4 = r^4$$

$$x^5 + y^5 = r^5 \quad \text{and so on,}$$

but he could not find whole number solutions for any of these equations.

In the margin of his copy of book 2, Fermat wrote, "Cubem autem in duos cubos aut quadrato quadratum, et generaliter nullam in infinitum ultra quadratum potestatem in duos eiusdem nominis fas est dividere."

That is, "to write a cube as two cubes or a fourth power as two fourth powers, and in general for any powers greater than two, is impossible".

Fermat added to this brief note :"Cuius re demonstrationem mirabilem sane detex hanc marginis exiguitas non caparet."
i.e. "I have a marvellous demonstration of this, which the margin is too narrow to hold."

This has become known as Fermat's Last theorem:

$$x^n + y^n = r^n$$

cannot be solve in integers if n is greater than 2.

This simple statement had baffled mathematicians for three hundred and sixty years.
No one had found a proof until October 1994, when the British mathematician Andrew Wiles, working at Princeton finally showed that Fermat had been correct.
His proof covers about 200 pages.

On 27 th June, 1997 Andrew Wiles was awarded the Wolfskehl prize worth $50,000 for his efforts. "This was my childhood passion. There's nothing to replace it. I've solved it. I'll try other problems I'm sure but there's no other problem in mathematics that could hold me the way Fermat did.
I had this rare privilege of being able to pursue in my adult life what had been my childhood dream. I know it's a rare privilege, but if you can tackle something in adult life that means that much to you, then it's more rewarding than anything imaginable. I was so obsessed by this problem that for eight years I was thinking about it all the time – when I woke up in the morning to when I went to sleep at night. That's a long time to think about one thing. That particular odyssey is now over. My mind is at rest."
..Andrew Wiles.

Exercise 5

1. If the largest number of a Pythagorean triple is even/odd, the other two numbers cannot differ by an odd/even number.

2. There is no Pythagorean triple with one number equal to the sum or product or difference or quotient of the other two.

3. 3,4,5 is the only Pythagorean triple of consecutive integers.

4. The two smallest numbers in a Pythagorean triple cannot both be odd.

5.   [i] Prove that 1 cannot be a member of a Pythagorean triple.
     [ii] Prove that 2 cannot be a member of a Pythagorean triple.

6. All odd primes are of the form 4n+1 or 4n-1 and those of the form 4n+1 can be written as $x^2 + y^2$ for some integers x and y.

DOWN WITH PYTHAGORAS!

(a cross number in which all of the down clues depend on the Pythagorean triangle ABC with $\angle C = 90$)

ACROSS

1. A power of 2.
3. Square of the fifth prime.
5. A fifth power.
the
6. Half way round the imperial track

7. 26A+29A
8. 1A - 1
9. 7A-20^2
10. Two score and one.

12. 20A+ 30A
13. Legs
15. A perfect cube.
16. A perfect square.

18. Treble seventeen.
20. Times the first three primes.
22. Four dozen for the baker.
24. 26 reversed.
25. Score.
26. sin 7 after a point
27. A multiple of the cube of five.
28. Sum the first five evens.
29. First three non zero digits of sin 34.5.
30. A perfect square.

DOWN

2. If a=7 and b=24 find c^2
3. If c=40 and a=24 find b^2
4. The square of the hypotenuse if other sides are 9 and 40.
5. The area of a 40 by 32 by 24 triangle.
8. c=13, a=5, b = ?
11. c=20, a=16, b = ?
12. c=40, a=25, what is AC^2
14. Find the height if AC=150, and BC=200
17. a=8, c=2a+1, b = ?
19. a=12, c=3a+1, b^2 = ?
20. c=106, a=90, b^2 = ?
21. AC=32, BC=24 find the hypotenuse.
23. AB=290, BC=200, find AC.
25. AB=20, AC=12, find BC^2

Answers

Exercise 1

[1]   (3,4,5), (6,8,10), (9,12,15), (12,16,20), (15,20,25), (18,24,30), (21,28,35), (15,36,39), (5,12,13), (10,24,26), (8,15,17), (16,30,34), (7,24,25), (20,21,29), (12,35,37).

[2]   8,9,12,15,16,17,20,21,29;   area = 420

Exercise 2

$$100^2 = 10^2 + 0^2 + 0^2 + 0^2$$

$$= 9^2 + 3^2 + 3^2 + 1^2$$

$$= 8^2 + 6^2 + 0^2 + 0^2$$

$$= 8^2 + 4^2 + 4^2 + 2^2$$

$$= 7^2 + 7^2 + 1^2 + 1^2$$

$$= 7^2 + 5^2 + 5^2 + 1^2$$

$$= 5^2 + 5^2 + 5^2 + 5^2$$

Giving (i) 7 ways (ii) 5 ways

Exercise 3

$$23 = 2^3 + 2^3 + 1^3 + 1^3 + 1^3 + 1^3 + 1^3 + 1^3 + 1^3$$

Exercise 4

$$239 = 5^3 + 3^3 + 3^3 + 3^3 + 2^3 + 2^3 + 2^3 + 2^3 + 1^3$$

$$= 4^3 + 4^3 + 3^3 + 3^3 + 3^3 + 3^3 + 1^3 + 1^3 + 1^3$$

Exercise 5

[1]   First part:

(a,b,c) is a Pythagorean triple with a<b<c. If c is even prove that b-a is also even.

Proof   Let c=2k and b-a = x

Then $(2k)^2 = a^2 + (a+x)^2$

$4k^2 = 2a^2 + 2ax + x^2$

From this we see that x^2 is even, and therefore x=b-a is even

Second part

(a,b,c) is a Pythagorean triple with a<b<c. If c is odd prove that b-a is also odd.

Proof   Let c=2k+1 and b-a = x

Then $(2k+1)^2 = a^2 + (a+x)^2$

$4k^2 + 4k + 1 = 2a^2 + 2ax + x^2$

From this we see that x^2 is odd and therefore x=b-a is odd

====================================

[2]   If (a,b,c) is a Pythagorean triple with a<b<c we show that we cannot have c=a+b or c=b-a or c=ab or c=b/a

Proof Given $c^2 = a^2 + b^2$, a<b<c

c=a+b gives $(a+b)^2 = a^2 + b^2$ leading to 2ab = 0 which cannot be.

c=b-a gives $(b-a)^2 = a^2 + b^2$ leading to $-2ab = 0$ which cannot be.

c=ab gives $a^2b^2 = a^2 + b^2$ leading to $a^2 = a^2/b^2 + 1$

but $a^2/b^2 < 1$ showing that $1 < a^2 < 2$ and this cannot be

c=b/a gives $b^2/a^2 = a^2 + b^2$ leading to $1/a^2 = a^2/b^2 + 1$ which cannot be because the left hand side is less than 1.

Similarly,
a=b+c gives $c^2 = (b+c)^2 + b^2$ leading to $2b^2+2bc = 0$ which cannot be

a=c-b gives $c^2 = (c-b)^2 + b^2$ leading to $2b^2-2bc = 0$ or $b=c$ which cannot be

a=bc gives $c^2 = (bc)^2 + b^2$ leading to $c^2 = b^2(c^2 + b^2)$ which cannot be because the right hand side will be greater than the left.

a=c/b gives $c^2 = (c/b)^2 + b^2$ leading to $1 = 1/b^2 + (b/c)^2$ or $1 = 1/b^2 + 1/a^2$ which cannot be because both the fractions on the right are less than ½ .

===========================

[3] (3,4,5) is the only triple of consecutive integers

Proof

Let $a^2 + (a+1)^2 = (a+2)^2$

Then $2a^2 + 2a + 1 = a^2 + 4a + 4$

Giving $a^2 - 2a - 3 = 0$

$(a-3)(a+1) = 0$ so that a=3

========================

[4]   The two smallest numbers in a triple cannot be odd

Proof

Suppose that $a = (2p+1)$ and $b = (2q+1)$

Then $4p^2 + 4p + 1 + 4q^2 + 4q + 1 = c^2$

Which shows that c^2 is even so that c must be even.

Let $c = 2k$ then we have $4p^2 + 4p + 1 + 4q^2 + 4q + 1 = 4k^2$

Giving $2p^2 + 2p + 2q^2 + 2q + 1 = 2k^2$ (dividing by 2)

And this cannot be because one side is odd and the other even.

====================

[5]   [i]
Suppose that a=1 then $1 + b^2 = c^2$

which gives $1 = (c-b)(c+b)$ which cannot be because b and c are integers.

[ii]
Suppose that a=4 then $4 + b^2 = c^2$

giving $4 = (c-b)(c+b)$ and the only possible solution in integers is c-b=1 and c+b=4 and this cannot be because c+b=4 would mean that a was the   longest side.

====================

[6]   Suppose that p is an odd prime, then we can mark the positions of
        multiples of 4 (marked as **4k**) in the sequence of numbers as follows:

        p-3 p-2 p-1   p   p+1 p+2 p+3
        4k   .   .   4k   .   .   possibility A
         .   4k   .   .   4k   .   possibility B
         .   .   4k   .   .   .   4k   possibility C

        B is not possible because **p-2** should be odd

        Therefore, we are left with **A:**   **p = 4k-1**

                          or **C:**   **p = 4k+1**

        (The last part is beyond the scope of this book)
        ===============

Down with Pythagoras:

Across:
[1] 16, [3] 12, [5] 32, [6] 220, [7] 688, [8] 15, [9] 288, [10] 41, [12] 94, [13] 11,
[15] 27, [16] 25, [18] 51, [20] 30, [22] 52, [24] 221, [25] 20, [26] 122, [27] 375,
[28] 30, [29] 566, [30] 64

Down:
[2] 625, [3] 1024, [4] 1681, [5] 384, [8] 12, [11] 12, [12] 975, [14] 120, [17] 15,
[19] 1225, [20] 3136, [21] 40, [23] 210,

Diophantus was 84 when he died.

===================================

CHAPTER 4

Squares

The Pythagoras diagram that we met in the last chapter involves drawing squares on the sides of a right-angled triangle. The diagram has some interesting properties which illustrate two important ideas in mathematics (i) generalization and (ii) special cases.

Problem 1 .

We have a right-angled triangle and squares are drawn on each of the three sides.
The centres of the two smaller squares are marked P and Q and the centre of the larger square is marked R.

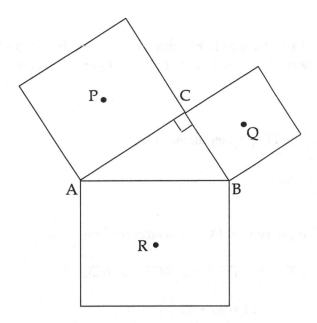

[1] Join PQ and you will find that the line PQ always passes through C.
[2] You will also find that PQ is perpendicular to CR
[3] Measure PQ and CR and you will find that PQ = CR.

We can prove these results using elementary geometry:

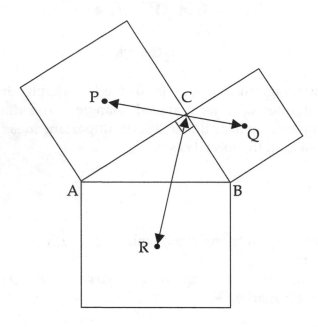

Given $\triangle$ABC is right angled at C and squares are drawn on the sides. The centres of the two smaller squares are P and Q and the centre of the larger square is R.

Prove

> [i]   the line PQ passes through C
>
> [ii] PQ $\perp$ CR
>
> [iii] PQ = CR

Proof

> [i]   To prove that PCQ is a straight line
>
> $\angle$ PCQ = $\angle$ PCA + $\angle$ ACB + $\angle$ BCQ
>
>   = 45 + 90 + 45
>
>   = 180
>
>   Therefore PCQ is a straight line.

[ii]   To prove that PQ ⊥ CR

since ∠ BCA = ∠ BRA = 90, it follows that BCAR is a circle on
diameter BA   (Circle Theorem)
therefore   ∠ BCR = ∠ BAR   (angles on chord BR)

But   ∠ BAR = 45

hence   ∠ BCR = 45

∴   ∠ PCR = 45 + 45 =90 as required

[iii] To prove PQ = CR

Let the sides of the triangle be AB = c, BC = a and CA = b, then, using
Pythagoras in triangle PCA we have

$$PC^2 + PA^2 = b^2$$

Hence $PC^2 = \frac{1}{2} b^2$

Also, in triangle CQB we have

$$CQ^2 + QB^2 = a^2$$

Hence $CQ^2 = \frac{1}{2} a^2$

Thus $PQ = PC + CQ$

$$= (1/\sqrt{2})b + (1/\sqrt{2})a = (1/\sqrt{2})(a + b)$$

Now we use the Chinese construction to find CR.

We draw the square on side AB and complete the large square of side (a+b)
as shown.

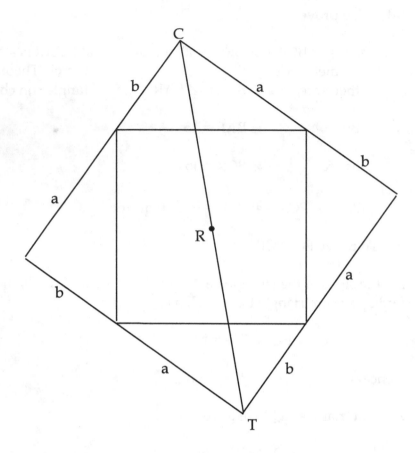

The diagonal CRT is given by

$$CT^2 = 2(a+b)^2 \qquad \text{(Pythagoras)}$$

Thus $\qquad\qquad CT = \sqrt{2}\,(a+b)$

Now CR is half of CT

Hence $\;\; CR = \tfrac{1}{2}\sqrt{2}\,(a+b) = (1/\sqrt{2})\,(a+b) = PQ \quad$ as required

==============================

Generalisation

Although the above proof relies upon angle C being 90^0, results (ii) and (iii) still hold when $\angle$ C is greater or less than 90^0, that is, these results hold for any triangle.

If we make $\angle$ C greater than 90^0 then the line PQ goes above C.
If we take $\angle$ C to be less than 90^0 then we find that the line PQ joining the centres the two squares passes below C, but results (ii) and (iii) still hold.

Problem 2   In the following figures, PQ $\perp$ CR and PQ = CR

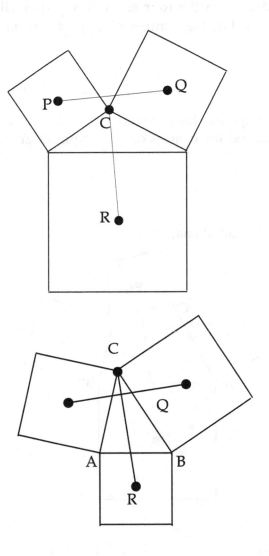

We are clearly dealing here with a more powerful result, but to prove (ii) and (iii) "in general", for any triangles, we need a more powerful method than the proof used in problem 1, which relies upon angle C being 90⁰.
 A proof is given at the end of chapter 25 on complex numbers.
We have seen then that we have two properties of the Pythagoras diagram that turn out to be much more general than at first suspected. Could it be that these two results come from an even more general theorem?
Consider squares drawn on the four sides of any quadrilateral. Again we find that the lines joining the centres of opposite squares are equal and perpendicular.

Problem 3

In the following diagram, four squares are drawn on the sides of a quadrilateral ABCD, and the centres of these squares are marked P,Q R and S as shown.
We find that
        (i) PQ = RS
and   (ii) PQ ⊥ RS
A proof is given at the end of chapter 24

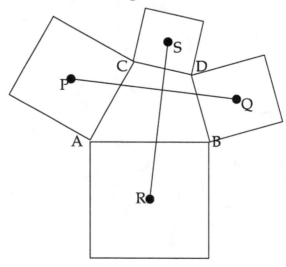

A Special Case

Suppose that we let the side CD get smaller and smaller, with D getting closer and closer to the point C. As C approaches D, the square on CD gets smaller and smaller so that C,D and S eventually end up at the same place.

Problem 4   Let CD → 0 in problem 3

When D reaches the point C, the square with centre S collapses to a point at
C and the quadrilateral ABCD degenerates into the triangle ABC.

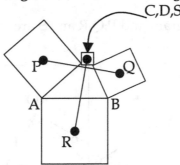

PQ remains equal and perpendicular to SR, but the point S has landed up at
C, therefore, we have the result for the triangle, that PQ is equal and
perpendicular to CR.

Any triangle ABC can be regarded as a quadrilateral ABCD with D infinitely
close to C. The figure for problem 2 can therefore be derived from a
collapsing CD in the figure for problem 3. Problem 2 is a special case of
problem 3.

Another Special case

A further special case of problem 3 arises when the quadrilateral ABCD is a
parallelogram. If ABCD is a parallelogram we get a more specific result.

PQRS is a square

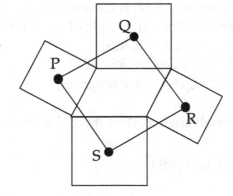

Problem 5

If squares are drawn on the four sides of a parallelogram then the centres of the squares form another square.

Given
ABCD is a parallelogram and P,Q,R and S are the centres of squares drawn on its' sides.

Prove
PQRS is a square

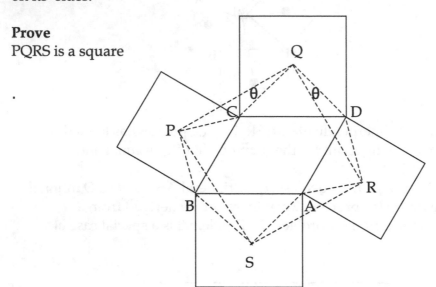

Proof
Consider the dotted triangles PQC, QRD, RSA and SPB.
We show that these triangles are congruent, that is, exactly the same shape.
In each triangle one side is equal to half the diagonal of the one square:

PC=RD=RA=PB

and another side equal to half the diagonal of the other square:

QC=QD=SA=SB

Also, the angles included between these sides are equal because they are made up of 45+a+45, where a is the acute angle of the parallelogram.
The four triangles are thus congruent (**S.A.S**)
Therefore we can conclude that PQ=QR=RS=SP
Further, note that $\angle PQC = \angle RQD = \theta$ (see diagram)
and note that $\angle CQD = 90$.
Rotate $\angle PQR$ about Q through angle θ, then it will lie over $\angle CQD$.
Therefore $\angle PQR = 90$

This is sufficient to prove that PQRS is a square.   ∎

But, there is a **special case** that has to be dealt with:

Problem 6
Suppose that PQ passes through C, then the triangle PCQ collapses into a straight line and the above proof of problem 5 would not hold.

If this happens, then angle C of the parallelogram will be 90° because PCQ is a straight line, and two angles at C are 45° so that $\angle C = 180 - 45 - 45 = 90$. The parallelogram will therefore be a rectangle:

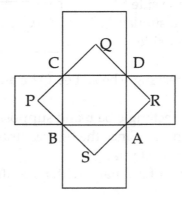

PSQR is then clearly a square because all of its sides are equal and all of its angles are 90°.

If four squares are drawn on the sides of a rectangle then the centres of the squares form another square.

==

Question:

How can you **construct** the square inside this triangle?

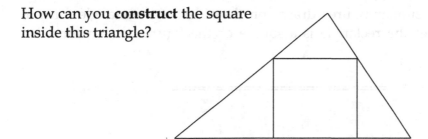

Answer at the end

Can we generalize this result by trying shapes other than squares? Could we draw equilateral triangles or pentagons to find that their centres also form a square?

Suppose that Q is the centre of some
regular figure drawn on one side
and that R is the centre of a similar
shape drawn on the shorter side.

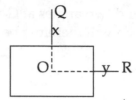

Let the longer side of the rectangle be **a**. Let Q be a distance **x** away from the side of the rectangle.
Let the shorter side of the rectangle be **b** and suppose that R is a distance y away the side of the rectangle. Since the figures drawn are similar, then
$\qquad$ x/a = y/b. Let x=ka and y=kb
If Q and R are to be vertices of a square, with its centre at the centre O of the rectangle then QO = RO.
$$x + \tfrac{1}{2}b = y + \tfrac{1}{2}a$$

$$ka + \tfrac{1}{2}b = kb + \tfrac{1}{2}a$$

$$a(k - \tfrac{1}{2}) = b(k - \tfrac{1}{2})$$

so either a=b in which case the rectangle is a square

or k= ½ in which case the similar figures are squares.

If the centres of similar figures drawn on the sides of a rectangle form a square then either the rectangle is a square or the figures drawn must be squares.

=============================

Problem 7

If a triangle ABC has squares drawn on its three sides as shown then the lines drawn from each vertex to the centre of the opposite square are concurrent.

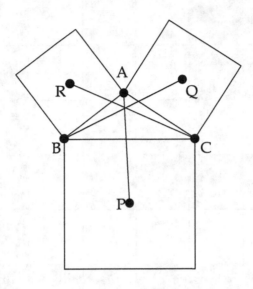

Prove   AP, BQ and CR are concurrent.

Proof
The altitudes of a triangle are the lines drawn from the vertices that are perpendicular to the opposite sides.

Using problem 2, we know that $AP \perp QR$, $BQ \perp RP$ and $CR \perp PQ$. Thus AP, BQ and CR lie along the altitudes of triangle PQR.
The altitudes of any triangle are concurrent (see theorem 5 of chapter 21), therefore AP, BQ and CR are concurrent.

==================================

The sum of the first n odd numbers is a perfect square

=================================

Constructing the square

Start by drawing any small square standing on the base with a corner on the other side:

The square you want will be an enlargement of this one.
So draw the dotted line to find where it hits the other side.

CHAPTER 5

Number Bases, Codes and Factors

The Romans and their Numbers.

Fredius Romanus was a soldier in Britannia who regularly did his shopping in Londinium but he always found it difficult to add up his bills, probably because he had to use the Roman number system.
Fredius used the following symbols:

I	Representing	1
V		5
X		10
L		50
C		100
D		500
M		1000

He did not use a place value system as the Hindus in India did. If a smaller value symbol came before a larger value then Fredius would subtract it but if a smaller value symbol came after a larger value then Fredius would add. Thus for Fredius Romanus,

XL   represents ten from 50 = 40
LX   represents 50 plus ten = 60

The numbers 1 to 20, in Fred's Roman numerals are:

I II III IV V VI VII VIII IX X XI XII XIII XIV XV XVI XVII XVIII XIX XX

The first twenty multiples of ten are

X XX XXX XL L LX LXX LXXX XC C CX CXX CXXX CXL CL CLX CLXX CLXXX CXC CC

Fredius used both silver and copper coins. One 'as' was a copper coin, which was used at the time of the wars with the Phoenecians (~100 BC), who came from the area now occupied by Israel and Lebanon. The **as** weighed about as much as a year two thousand £2 coin.

The silver **denarius** (plural **denarii**), was worth 10 **asses** and one **denarius** was worth **4 sesterces.** Thus Fredius used the following coinage for his shopping:

$$4 s = 1 d$$
$$10 \text{ as} = 1 d$$
$$1 s = 2\tfrac{1}{2} \text{ as}$$

or: **4 sesterces = 10 asses = 1 denarius**

One day Fredius went shopping to stock up his farm and bought:

III chickens @	VI asses each
II sheep @	IX asses each
XV eggs @	II asses each

What was his total bill in denarii, sesterces and asses?

Fredius thought to himself "Why didn't they teach us those numbers they use in that Indian port Muziris when we were in military school. All the cohorts say their way is so much easier to work out and make sure you are not being swizzled"

Fredius was also paymaster for his cohort of soldiers and 123 (CXXIII) of his men were to be paid 45 (XLV) denarii each for work they had put in on the Londinium Wall.

To Fredius, this calculation would have been C X X I I I times X L V and he would have to remember that the X was to be subtracted because it comes before the larger value L.

Fredius could have used a multiplication table such as this:

times	I	V	X	L	C	D	M
I	I	V	X	L	C	D	M
V	V	XXV	L	CCL	D	MMD	V
X	X	L	C	D	M	V	X
L	L	CCL	D	MMD	V	XXV	L
C	C	D	M	V	X	L	C
D	D	MMD	V	XXV	L	CCL	D
M	M	V	X	L	C	D	M

Fred's calculation would have had to be something like this :

```
                          C X X I I I
                  TIMES       X L V
Deal with V                D L L V V V
Deal with L       M M M M M D D L L L
First total       M M M M M M D C C L X V
Deal with X             M   C C X X X
Subtract this from the
first total:    answer   M M M M M D X X X V
```

Thanks to the Hindu's invention of the place system with a symbol for
nothing we use a much more efficient way of calculating with numbers than
the Romans did. We call it the 'place system' since the value of a numeral
depends on its place within the number: 123 means to us, 3 odd ones, 2
groups of ten and 1 group of ten(tens).

We could perform the multiplication 123 x 45 by hand:

$$
\begin{array}{r}
123 \\
\times\ 45 \\
\hline
615 \\
4920 \\
\hline
5535 \\
\end{array}
$$

but our Roman friend did not have such an easy calculation.

Exercise 1

Use Fred's table to find the Roman value of the following and translate each
calculation into English

[1]   VxV   [2] LxV   [3] XxL   [4] VxD   [5] LxD   [6] MxC

Scientists say that if we had evolved with only 4 digits on each hand then we
would probably have ended up grouping our numbers into eights instead of
into groups of ten. The numerals we use in the places value system are
called digits, for the Romans too used to count on their fingers and the word
digit is derived from the Latin word digitus, meaning finger or toe. If we
had only eight digits, then we would be counting:

0, 1, 2, 3, 4, 5, 6, 7, one handful, one handful + 1, one handful + 2 etc.

With one handful being what we call an eight, the place value system would
then give

Octal= base 8

Number	notation in base eight
One	1
Two	2
three	3
four	4
five	5
six	6
seven	7
eight	10
nine	11
ten	12
eleven	13
twelve	14

There is no numeral for eight as this is regarded as one handful that we will call an Oct.

Nine, being one handful plus one, we will call Octyone , twelve Octytwo and so on.

Counting from seven upwards gives:

seven, oct, octyone, octytwo, octythree, octyfour, octyfive, octysix, octyseven, twocty, twoctyone, twoctytwo, twoctythree, twoctyfour, twoctfive, twoctysix, twoctyseven, throcty, throctyone,............,throctyseven, frocty, froctyone, ...,froctyseven, ficty, fictyone,.......fictyseven, sicty, sictyone..........etc.

The base eight number 123_8 represents the number that we would work out as

$1 \times 8^2 + 2 \times 8 + 3 = 64 + 16 + 3 = 83$ DEC   (DEC stands for decimal, base ten),

thus   123 OCT = 83 DEC

As long as we remember to carry handfuls of eight from one column to the next, arithmetic in base eight or octal, presents no problem.

For example, $345_8 + 456_8$ is carried out as

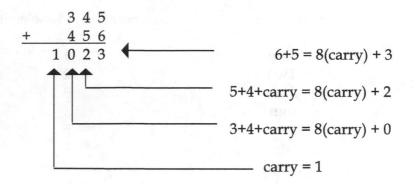

$$3 \quad 4 \quad 5$$
$$+ \quad 4 \quad 5 \quad 6$$
$$1 \quad 0 \quad 2 \quad 3 \quad \longleftarrow \qquad 6+5 = 8(\text{carry}) + 3$$

$$5+4+\text{carry} = 8(\text{carry}) + 2$$

$$3+4+\text{carry} = 8(\text{carry}) + 0$$

$$\text{carry} = 1$$

The equivalent numbers in base ten are

$$3\text{x}64 + 4\text{x}8 + 5 = 192 + 32 + 5 = 229$$

and $4\text{x}64 + 5\text{x}8 + 6 = 256 + 40 + 6 = 302$

so we have two ways of representing the same sum:

in OCTAL	and in DECIMAL
$3\ 4\ 5$	$2\ 2\ 9$
$+4\ 5\ 6$	$+3\ 0\ 2$
$1\ 0\ 2\ 3$	$5\ 3\ 1$

and as a check: $1023_8 = 1\text{x}8^3 + 0\text{x}8^2 + 2\text{x}8 + 3 = 512 + 16 + 3 = 531$ DEC.

Exercise 2

Perform the following Octal addition sums and represent each sum in also in decimal. Then you may check that the two answers represent the same result.

[1] 123
        $+456$

[2] 246
        $+\ 246$

[3] 777
        $+\ 666$

An Algorithm : Base ten to base eight

In order to change a number into base 8, we continually divide by eight and store the remainders:

Example:

Change 2049 DEC into octal

Solution

$$
\begin{array}{rl}
8 \mid 2049 & \qquad . \longrightarrow \text{octal point} \\
8 \mid \underline{256} & \text{r } 1 \\
8 \mid \underline{32} & \text{r } 0 \\
8 \mid \underline{4} & \text{r } 0 \\
8 \mid \underline{0} & \text{r } 4 \\
8 \mid \underline{0} & \text{r } 0
\end{array}
$$

We read off the answer vertically to the octal point

Thus
$$2\,0\,4\,9 \, \text{DEC} = 0\,4\,0\,0\,1 \, \text{OCT}$$

For simple examples, we can usually see the groups of eight without too much tedious calculation thus:

$$1\,0\,0 \, \text{DEC} = 64 + 32 + 4$$

$$= \, 1\,4\,4 \, \text{OCT}$$

Exercise 3

[1]   Use the algorithm to convert 1000 DEC into octal.

[2]   Use the algorithm to convert 8000 DEC into octal.

[3]   Convert 65, 66, 67, 68 into octal.

Divisibility

In base ten, we can check whether a number is divisible by 9, by adding up the digits of the number and checking the digit sum, for example the decimal number 1017 has a digit sum of 9 and so, we conclude that 1017 is divisible by 9.

This is because any power of ten is just one more than a multiple of 9, for example

$$1000 = 999+1.$$
$$100 = 99+1$$
$$10 = 9+1$$

Thus 2000, 200, 20, 2 are each 2 more than a multiple of nine. 3000, 300, 30, 3 are each 3 more than a multiple of nine. The digit sum tells you how much bigger the number is than a multiple of nine.

The same rule applies to 3, for example, 4017 has a digit sum 12 and is divisible by 3 but not by nine.

Divisibility in Octal

How would these divisibility rules translate into octal?
Now each power of eight is just one more than a multiple of 7:

decimal	or, in octal
$8 = 7 + 1$	$10 = 7+1$
$8\times8 = 8\times7 + 7 + 1$	$100 = 77+1$
$8\times8\times8 = 8\times8\times7 + 8\times7 + 7 + 1$	$1000 = 777+1$

and so on, so that if the sum of the digits of an octal number is divisible by 7 then the octal number itself must also be divisible by 7.

Example:

The octal number $1\ 0\ 1\ 5\ 7_8$ has a digit sum of fourteen and so $1\ 0\ 1\ 5\ 7_8$ must be divisible by 7.

Exercise 4

[1]   Which of these decimal numbers is divisible by 9?

     8888, 8765, 8181, 7524

[2]   Which of these octal numbers is divisible by 7?

     6152, 3542, 1562, 2651

Fractions in Base 8

In decimal notation, numerals to the right of the decimal point represent tenths, hundredths, thousandths etc. In a similar way, in base 8 (octal), numerals to the right of the octal or base 8 point represent eighths, sixtyfourths etc. so that the octal number 312.34_8 for example, could be converted into decimal notation by writing

$$312.34_8 = 3 \times 8^2 + 1 \times 8 + 2 + 3 \times 8^{-1} + 4 \times 8^{-2}$$
$$- 192 + 8 + 2 + 3/8 + 4/64$$
$$= 202\,7/16 \text{ or } 202 \cdot 4375$$

In order to convert a decimal whole number into octal, we "divide by eight and store the remainders".

To convert a decimal fraction into base eight, we "multiply by eight and store the overflow".

An Algorithm for fractions: **Decimal to base eight**

Example: Convert the decimal fraction $0 \cdot 3046875$ into octal.

Solution

```
0·  3 0 4 6 8 7 5
 ·            x 8
2   4 3 7 5 0 0 0
             x 8
3   5 0 0 0 0 0 0
             x 8
4   0 0 0 0 0 0 0        0.3046875 DEC = 0.234 OCT
```

The overflows are 2, 3 and 4.

The overflows are stored and do not take part in the next stage of the algorithm multiplication.

The answer can then be read off from the decimal point which has the same position as the octal point for the answer, which in this case, is $0 \cdot 2\,3\,4\,_8$.

Recurring octals

Not every decimal fraction will give a terminating octal fraction. The only decimal fractions that will give a terminating octal fraction are those decimals that happen to be equal to the sum of ordinary fractions that have only powers of 2 as denominators. The next example shows the calculation of the decimal fraction 1/10, resulting in a recurring octal:

Example: Conversion of $0 \cdot 1$ into an octal fraction.

Solution:

$$
\begin{array}{cc}
0\ \cdot & 1 \\
\cdot & \times 8 \\
0 & 8 \\
 & \times 8 \\
6 & 4 \\
 & \times 8 \\
3 & 2 \\
 & \times 8 \\
1 & 6 \\
 & \times 8 \\
4 & 8 \\
 & \times 8 \\
6 & 4 \\
 & \times 8 \\
3 & 2 \\
\multicolumn{2}{c}{\text{etc}}
\end{array}
$$

Reading down from the point, we have the octal answer 0.0631463....and it is clear that there is a recurring pattern:

$0 \cdot 1$ DEC $= 0 \cdot 0\,6\,3\,1\,4\,6\,3\,1\,4\,6\,3\,1\,4\,6\,3\,1\,4\,6\ 3\,1\,4\,6\,3\,1\,4$........octal

Exercise 5

[1]   Convert 0.162109375 OCT into decimal.

[2]   Use the algorithm to convert 0.2, 0.3, 0.4, 0.5, 0.6, 0.7, 0.8 and 0.9
into   octal fractions.

Calculations

As long as we remember to carry bunches of eight from one column to the
next, we can perform additions with octal fractions just as we would in base
ten, as in the following example:

$$
\begin{array}{r}
\text{An octal addition} \\
1\ 2\ 3\ \cdot\ 4\ 5\ 6 \\
+\quad 2\ 4\ \cdot\ 6\ 5\ 4 \\
\hline
1\ 5\ 0\ \cdot\ 3\ 3\ 2
\end{array}
$$

Exercise 6

Perform the following Octal additions:

[1]
$$
\begin{array}{r}
1\ 2\ 3\ \cdot\ 4\ 5\ 6 \\
+\quad 2\ 3\ 4\ \cdot\ 5\ 6\ 7 \\
\hline
\end{array}
$$

[2]
$$
\begin{array}{r}
1 2\ 3\ \cdot\ 4\ 5\ 6 \\
+\quad 6\ 5\ 4\ \cdot\ 3\ 2\ 2 \\
\hline
\end{array}
$$

[3]   Perform this octal subtraction

$$
\begin{array}{r}
2\ 3\ \cdot\ 4\ 5 \\
-\quad 1\ 3\ \cdot\ 5\ 7 \\
\hline
\end{array}
$$

Numbers and Patterns

Choose a three digit number, with the units and hundreds digits different, for example we might choose 529.
Now reverse the digits and subtract.

$$925$$
$$-529$$

Then take the answer, reverse its digits and add. In our example we have

$$
\begin{array}{r}
9\ 2\ 5 \\
-\quad 5\ 2\ 9 \\
\hline
3\ 9\ 6 \\
+\quad 6\ 3\ 9 \\
\hline
1\ 0\ 8\ 9
\end{array}
$$

The final answer will always be 1089.
An interesting number which happens to be 33^2.
Also, if you reverse the digits, you get 9801 which is 99^2.
Can you **prove** that you always get 1089? (Answer at end)

Now would this work in Octal?

Try an example in octal:

$$
\begin{array}{lr}
& 4\ 3\ 2 \\
\text{reverse and subtract} & -2\ 3\ 4 \\
\hline
& 1\ 7\ 6 \\
\text{reverse and add} & +\ 6\ 7\ 1 \\
\hline
& 1\ 0\ 6\ 7
\end{array}
$$

In base 8, we always get the answer 1 0 6 7.

If you reverse these digits, you find that 77x77 = 7 6 0 1 in octal.

If we examine the note that 33^2 = 1089 in base ten,
and rewrite it as 99x11 = 1089
we find again, a result that seems to be "base independent" for in octal,
when we reverse the digits we find that

$$77 \times 11 = 1\ 0\ 6\ 7 \text{ octal}$$

In base 5, you would be correct to guess that we always get 1 0 3 4
and that 44x11 = 1034 (base 5)
and that 44x44 = 4301 (base 5)

More digits

It is natural to ask if there is an extension of this result to 4 digit numbers:
If we have any 4 digit number **abcd** then

(i)   If a>d and c>b the result will be 9999
(ii)   If a>d and c<b the result will be 10890
(iii)   If a>d and b=b the result will be 10989   (proofs at end)

and this result carries over into other number bases, for example, in octal we
find

(i)   If a>d and c>b the result will be   7777 (octal)
(ii)   If a>d and c<b the result will be 10670 (octal)
(iii)   If a>d and b=b the result will be 10767 (octal)

An investigation: What will the results be for a 5 digit number abcde ?

Binary

Decimal means counting in groups of ten.
Binary means that we count in groups of two.
The place values in Binary are: ones, two(ones)=twos, two(twos)=fours,
two(fours)=eights, two(eights)=sixteens and so on so that the place values in
ascending order from right to left are

    65536 32768 16384 8192 4096 2048 1024 512 256 128 64 32 16 8   4   2 1
or
        2^{16} 2^{15} 2^{14} 2^{13} 2^{12} 2^{11} 2^{10} 2^9 2^8 2^7 2^6 2^5 2^4 2^3 2^2 2^1 2^0

In the binary number system we have just two digit 0 and 1 which are called
bits (**binary digits**).

Counting from 1 to 16 in binary will look like this:

1 2 3 4 5 6 7 8 9 10 11 12 13 14 15 16
1 10 11 100 101 110 111 1000 1001 1010 1011 1100 1101 1110 1111 10000

For small numbers if we have to convert from base ten to binary, we usually spot the powers of two, writing the number as a sum thus, to convert 76 DEC to binary (BIN), we can write

$$76 = 64 + 8 + 4 = 1(64) + 0(32) + 0(16) + 1(8) + 1(4) + 0(2) + 0(1)$$

$$= 1\ 0\ 0\ 1\ 1\ 0\ 0\ \text{BIN}$$

An Algorithm : Decimal to binary

The algorithm for conversion from decimal is to divide by 2 and store the remainders.
For example, to convert decimal number 76 into binary using this algorithm gives the following calculation:

```
2 | 7 6   ·
2 | 3 8  r 0
2 | 1 9  r 0
2 |   9  r 1
2 |   4  r 1
2 |   2  r 0
2 |   1  r 0
2 |   0  r 1
2 |   0  r 0
```

Reading off the answer up to the decimal point (which has the same position as the binary point) we have 76 DEC = 01001100 BIN.

Exercise 7   Convert 99 dec to octal using the division algorithm.

Computers and Number bases.

Computer software is of two kinds. Systems software drives the hardware and runs the computer system while applications software operates on data using the system for the benefit of the user. Both programs and data are stored in the memory of the computer using the binary number system. This is because digital computers depend on two valued states, on/off, true/false, 5volts/zero volts; all of which can best be represented by the binary digits (bits), 0 and 1. This is called two state logic. The circuits which make up the computer hardware of a digital computer are designed using two state logic devices.

Everything in the memory of a digital computer, the programs and the data, i.e. numbers, letters, records and files on which the stored programs operate, is stored in bit patterns which we represent as strings of '0' and '1' bits. Raw bit patterns are hard for us humans to remember and to handle without making errors, for example, the string of letters "ABC" would be stored in most computers as the bit string 010000010100001001000011 (using the American Standard Code for Information Interchange -ASCII) and the numbers 1, 2 and 3 would probably be stored as 0000000000000001, 0000000000000010, 0000000000000011 (using 16 bit binary).

We find it difficult to handle these long bit strings and so we have invented various shorthands for the bit patterns. The first shorthand to be commonly used was the Octal code which is a code for patterns three bits long. There are just eight different three bit patterns and each different pattern is given its Octal code:

bit pattern	octal code
0 0 0	0
0 0 1	1
0 1 0	2
0 1 1	3
1 0 0	4
1 0 1	5
1 1 0	6
1 1 1	7

Examples:
The bit pattern
　　　0 0 1 1 0 1 0 1 1 1 1 1 would be coded as 1 5 3 7 octal
The bit pattern
　　　1 0 1 0 1 0 1 0 1 would be coded as 5 2 5 octal.

The bit patterns stored within the memory of a computer could represent characters, machine code instructions, integers, floating point numbers or more complicated structures such as linked lists, queues, stacks or trees.

When you examine a bit pattern in the main storage of a computer, that bit pattern could represent a number a character or other kinds of data or a program instruction.

The same bit pattern can be used or interpreted in a number of different ways.

If the bit patterns are to represent binary numbers, then the octal coding of the bit patterns would of course represent the base 8, or octal representation of the number. The octal codes were intentionally designed that way!

Example: A binary addition and its octal coding

binary addition   octal addition

```
    1 0 1 1 0 0 1 1 1                      5 4 7
  + 0 0 1 1 1 1 0 0 1               +      1 7 1
    1 1 1 1 0 0 0 0 0                      7 4 0
```

The octal code for groups of three bits is rarely used nowadays, it is much more convenient to use codes for groups of four bits. A group of four bits is called a nibble. Two nibbles make a byte (8 bits).

The code for 4 bit patterns is called the Hex code. There are $2^4 = 16$ different four bit patterns and each has its hex code. The first eight codes are exactly the same as the octal codes.

Hex Codes

bit pattern	hex code	bit pattern	hex code
0 0 0 0	0	1 0 0 0	8
0 0 0 1	1	1 0 0 1	9
0 0 1 0	2	1 0 1 0	A
0 0 1 1	3	1 0 1 1	B
0 1 0 0	4	1 1 0 0	C
0 1 0 1	5	1 1 0 1	D
0 1 1 0	6	1 1 1 0	E
0 1 1 1	7	1 1 1 1	F

(the lower case letters can also be used)

Examples:

The bit pattern

    0 0 1 1 0 1 0 1 1 1 1 1 would be coded as 3 5 F hex

The bit pattern

    1 0 1 0 1 0 1 0 1 would be coded as 1 5 5 hex

(leading zeros added to make up the bit pattern to twelve bits.

Base 16 = Hexadecimal

If we had evolved with only four fingers on each hand and four toes on each foot, then the priests and mathematicians of the past may have decided to count in groups of sixteen instead of ten and they may have invented a base sixteen number system.

Base sixteen is a very convenient number base for computer scientists to use since it is easy to convert to and from binary system.

If we count by grouping into piles of sixteen then the ten digits 0..9 are not enough and so we have to invent five more digits to run from eleven to fifteen. Naturally, the early pioneers of computing chose to stick with the numerals 0..9 and chose the beginning of the alphabet for the extra numerals they needed.

The hex codes for groups of four bits are the same as the hexadecimal base 16 digits:

Decimal numbers: 0 1 2 3 4 5 6 7 8 9 10 11 12 13 14 15
Hexadecimal digits: 0 1 2 3 4 5 6 7 8 9 A B C D E F

Examples   Hexadecimal additions:

```
   1 2 3 4        F 1 D 0          B A D
  +5 6 7 8       +D 1 D          +C 0 D
   6 8 A C        F E E D         17B A
```

Exercise 8: Perform the following hexadecimal additions

[1]   A D D   [2]   1 2   [3]   1   [4]   D 0 D 0
    +1 1 1   + B E   + F F F   +D 1 E D

Again, if the bit pattern represents a binary number, then the hex coded version will be the base 16 (hexadecimal) representation of the number.

Example: A binary addition and its hex coding:

```
        binary addition                    hex addition
   0 1 0 1 0 1 1 0 0 1 1 1                     5 6 7
  +0 1 1 1 0 1 1 1 1 0 0 1                    +7 7 9
   1 1 0 0 1 1 1 0 0 0 0 0                     C E 0
```

ASCII codes

In the early days of computing, everything, that is numbers, characters and program instructions had to be entered into the computer as bit patterns by hand. In old 1950s science fiction movies, computers were usually portrayed as a wall of flashing lights which was not too far from the truth for one of the old 1940s mainframe computers. This represents machine code programming at its most basic level. At this level, the bit pattern for every piece of data, number or character, and the bit patterns for each program instruction would be set on a line of on/off switches, then the address of the location in memory, where the bit pattern was to be stored, would be set on another line of switches and finally a load button would be pressed to enter the bit pattern into the memory of the computer. It was not long, before the computer keyboard with its associated software was invented so that data and instructions could be typed in. Naturally, the computer keyboard was based upon the existing typewriter keyboard which meant that the typewriter keyboard had to be encoded into bit patterns, one distinct bit pattern for each key. The need to encode the typewriter keyboard resulted in the invention of the American Standard Code for Information Interchange (ASCII). The first question to be answered was how many bits would be needed for each code so they looked at the typewriter keyboard and counted up how many different characters and symbols there were.

This exercise gives the following kind of calculation:

Letters of the alphabet: upper case   26
                             lower case   26

Numerals   0..9   10

Punctuation: ! " : ; ' , . ?   about 10

Arithmetic symbols - + x = < > ÷ about 10

(oddly enough ÷ got missed out,
and ÷ became / and times became *)

brackets   ( ) { } [ ]   6

Odds and ends: £ $ % & # ~ @   about 10

                                         total 98

Now 6 bits would produce $2^6 = 64$ different bit patterns which is not enough, so they chose 7 bits for each code giving $2^7 = 128$ different codes. Thus the ASCII code started off as a seven bit code for the typewriter keyboard. They were very concerned about keeping memory requirements to the minimum because each bit in the computer's main memory (called in those days core storage) was formed from a small piece of iron wrapped round by a few turns of wire. Main storage in those days was therefore quite expensive. Nowadays memory is cheaper and we don't bother so much! The International Standard code ISO-7 is a seven bit code which is virtually identical to the original ASCII code. A leading 0 bit is usually added to give an eight bit code. This of course then gives 256 different codes, the first 128 being the same as the old ASCII and various processor manufacturers extend the character set to include European symbols (e.g. ü é â ä etc. and various graphic characters such as ■ ▄■ ▌▀┐ └┴ ├─┼ etc.

The hex codes for the alphanumeric characters in the ASCII code are chosen in a fairly logical way as follows:

The numerals 0..9 are coded from 30 hex to 39 hex:

numeral	0	1	2	3	4	5	6	7	8	9
ASCII code	30	31	32	33	34	35	36	37	38	39

The capital letters have codes from 41 to 5A

A	B	C	D	E	F	G	H	I	J	K	L	M	N	O	P	Q	R	S	T	U	V	W	X	Y	Z
41	42	43	44	45	46	47	48	49	4A	4B	4C	4D	4E	4F	50	51	52	53	54	55	56	57	58	59	5A

Changing just one bit converts upper case to lower case for example

'A' has the bit pattern 0 1 0 0 0 0 0 1
'a' has the bit pattern 0 1 1 0 0 0 0 1

just one bit is changed, so that it would be easy to "swop case" if needed:

4 = 0 1 0 0 x x x x   5 = 0 1 0 1 x x x x   upper case

↓   ↓   ↓

6 = 0 1 1 0 x x x x   7 = 0 1 1 0 x x x x   lower case

Lower case letters run from 61 to 7A

a b c d e f g h i j k l m n o p q r s t u v w x y z
61 62 63 64 65 66 67 68 69 6A 6B 6C 6D 6E 6F 70 71 72 73 74 75 7 6 77 78 79 7A

This leaves a few gaps:

From hex 20 to hex 2F we have:
space ! " # $ % & ' ( ) * + , - . /
  20 21 22 23 24 25 26 27 28 29 2A 2B 2C 2D 2E 2F

Which is mainly a jumble of the keys along the top row of the keyboard

The next gap is from hex 3A to hex 40 where we find
 :   ;   <   =   >   ?   @
3A 3B 3C 3D 3E 3F 40

Which are mainly from the bottom right of the keyboard but keeping < = and > together.

There is another gap from hex 5B to hex 60 where we find some 'odds n ends':
 [\   ]   ^   _   `
5B 5C 5D 5E 5F 60

The last gap runs from hex 7B to hex 7F:
 { | }   ~   DEL
7B 7C 7D 7E 7F

Which is a group of keys from all corners of the keyboard, except for 7F = DEL. Now this code has an interesting history. In the early days of computing input into the computer was done using punched cards or paper tape. The letters 'A','B','C' on paper tape would appear as in figure below, with hole punched out for each 1 bit.

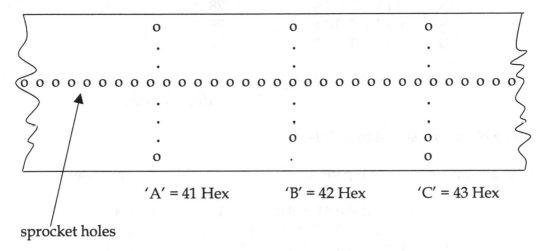

sprocket holes

Punched paper tape with letters 'A','B','C' punched out

If an error was made in the character, then the paper tape would have to be 'backed up', but the bits of paper could not be pasted back into the holes and so the only thing to do was to punch out all the other holes and use 111 1111 = 7F hex, as the deleted character code. This is why the code for deleted character or back delete, is at the end of the ASCII sequence with the code 128 DEC = 7F Hex (= 1111111 BIN).

Data stored in memory :

Letters and Strings

	Bit pattern	hex code
'A'	0100 0001	41
'B'	0100 0010	42
'C'	0100 0011	43
'X'	0101 1000	58
'Y'	0101 1001	59

'Z'	0101 1010	5A
'a'	01100001	61
'b'	0110 0010	62
'c'	0110 0011	63
'x'	0111 1000	78
'y'	0111 1001	79
'z'	0111 1010	7A

"abacus"   61 62 61 63 75 73 00 ◄──────── string terminator

Whole Numbers (stored in 16 bits)

decimal number	bit pattern	hex code
1	0000 0000 0000 0001	0001
2	0000 0000 0000 0010	0002
15	0000 0000 0000 1111	000F
255	0000 0000 1111 1111	00FF
256	0000 0001 0000 0000	0100

Instructions Stored in Memory (from an Intel processor instruction set)

Instruction	Hex code
MOV AX,123	B8 23 01
ADD AX,234	05 34 02
MOV BX,ABC	BB BC 0A
ADD BX,DEF	81 C3 EF 0D
MOV CX,BAD	B9 AD 0B
ADD CX,C0D	81 C1 0D 0C
MOV DX,F1D0	BA D0 F1
ADD DX,D1D	81 C2 1D 0D

There is no real need to study the instruction formats of processor instruction sets: each processor has its own instructions and bit patterns which differ from one brand to another, however, we can see someone's mind at work without too much effort if we look at the move instructions printed above:

MOV AX,123 means move the hex number 123 into 16 bit register AX
MOV BX,ABC means move the hex number ABC into register BX etc.

A register is a fast storage location within the central processing unit (CPU) of the computer. Each is a 3 byte instruction giving 24 bits. The numbers are split into two bytes which are reversed so that the small half of the number comes first. We have

Instruction	opcode	number	bit pattern
MOV AX, 123	B8	23 01	1011 1000 0010 0011 0000 0001
MOV BX, ABC	BB	BC 0A	1011 1011 1011 1100 0000 1010
MOV CX, BAD	B9	AD 0B	1011 1001 1010 1101 0000 1011
MOV DX, F1D0	BA	D0 F1	1011 1010 1101 0000 1111 0001

If we examine the opcodes B8, BB, B9, BA we see they are very similar. In fact the MOV instruction is identified by the first six bits, 1011 10 and the register is then identified by the remaining two bits: AX..00, BX..11, CX..01, DX..10.

These six instructions could be stored in memory as a short program that performs the following hex additions:

CPU register used

AX	**BX**	**CX**	**DX**
123	ABC	BAD	F1D0
+ 234	+ DEF	+ C0D	+ D1D
0357	18AB	17BA	FEED

A memory dump of this short program in main storage of an IBM computer from memory location 1158:0100 would appear as follows:

1158:0100	B82301	MOV AX, 0123
1158:0103	053402	ADD AX, 0234
1158:0106	BBBC0A	MOV BX, 0ABC
1158:0109	81C3EF0D	ADD BX, 0DEF
1158:010D	B9AD0B	MOV CX, 0BAD
1158:0110	81C10D0C	ADD CX, 0C0D
1158:0114	BAD0F1	MOV DX, F1D0
1158:0117	81C21D0D	ADD DX, 0D1D

For more explanation of the Intel CPU registers and the use of the Microsoft debug utility you would need to consult a text on assembly language programming.

Number and factors. (Back to the familiar decimal system)

(Note: Both the word factor and divisor are commonly used in this context and both terms are used here interchangeably without any particular preference.)

Here is a list of the factors of the numbers 1 up to 28:

number	list of factors
1	1
2	1,2
3	1,3
4	1,2,4
5	1,5
6	1,2,3,6
7	1,7
8	1,2,4,8
9	1,3,9
10	1,2,5,10
11	1,11
12	1,2,3,4,6,12
13	1,13
14	1,2,7,14
15	1,3,5,15

16	1,2,4,8,16
17	1,17
18	1,2,3,6,9,18
19	1,19
20	1,2,4,5,10,20
21	1,3,7,21
22	1,2,11,22
23	1,23
24	1,2,3,4,6,8,12,24
25	1,5,25
26	1,2,13,26
27	1,3,9,27
28	1,2,4,7,14,28

The Number of Factors

The factors of 6 are 1,2,3,6. The number of factors of 6 is four since #{1,2,3,6}=4.
(Where the # sign indicates "the number of things in the set".)

The "number of factors" is often called the number of divisors and denoted by d so that we have d(6)=4, d(7)=2 and d(42)=#{1,2,3,6,7,14,21,42}=8.

Example1.

Notice that 6 x 7 = 42 and d(6) x d(7) = d(42).

Does this result always happen?

Example 2.

7 x 4 = 28
The number of factors or divisors of 7 is #{1,7} = 2 so that d(7) = 2.
The number of factors or divisors of 4 is #{1,2,4} = 3 so that d(4) = 3.
The number of factors or divisors of 28 is #{1,2,4,7,14,28} = 6 so that d(28) = 6.

We have   7 x 4 = 28

and also $d(7) \times d(4) = d(28)$.

Example 3.

$$6 \times 5 = 30$$

$d(6) = \#\{1,2,3,6\} = 4 \quad d(5) = \#\{1,5\} = 2 \quad d(30) = \#\{1,2,3,5,6,10,15,30\} = 8$

and we find also that $d(6) \times d(5) = d(30)$.

As long as the two numbers have no **common factors** we find that this result for the number of divisors holds true. If we try the result on $4 \times 6 = 24$ we find

$d(4) = \#\{1,2,4\} = 3, \quad d(6) = \#\{1,2,3,6\} = 4$ and $d(24) = \#\{1,2,3,4,6,8,12,24\} = 8$
$3 \times 4 <> 8$. The result does not work because 4 and 6 have a common factor 2.

Whereas 4 and 5 have no common factor, $d(4) = 3$, $d(5) = 2$ and the factors of 20 are $\{1,2,4,5,10,20\}$ so that $d(20) = 6$ and $3 \times 2 = 6$.

The Number of factors theorem.(without proof)

> Suppose that numbers A and B have no common factor.
> Let A have f factors and B have g factors, then A x B has f x g factors.

Sum of factors:

Example 1

$$6 \times 7 = 42$$

The sum of the factors of 6 is $1+2+3+6 = 12$.
The sum of the factors of 7 is $1+7 = 8$

The sum of the factors of 42 is $1+2+3+6+7+14+21+42 = 96$

and somewhat more surprising, we find that $12 \times 8 = 96$.

Example 2.

$$7 \times 4 = 28$$

The sum of the factors of 7 is $1+7 = 8$
The sum of the factors of 4 is $1+2+4 = 7$
The sum of the factors of 28 is $1+2+4+7+14+28 = 56$ and we find $8 \times 7 = 56$

Example 3.

$$6 \times 5 = 30$$

The sum of the factors of 6 is $1+2+3=6 = 12$
The sum of the factors of 5 is $1+5 = 6$
The sum of the factors of 30 is $1+2+3+5+6+10+15+30 = 72$ and we find $12 \times 6 = 72$.

The sum of factors theorem: (without proof)

> Suppose that numbers A and B have no common factor.
> Let the sum of the factors of A be s and let the sum of the factors of B be t, then the sum of the factors of A x B is s x t.

Sum of squares of factors

For our third example, we add together the squares of the factors of the numbers:

Example 1.

$$6 \times 7 = 42$$

The sum of the squares of the factors of 6 is $1^2+2^2+3^2+6^2 = 1+4+9+36 = 50$
The sum of the squares of the factors of 7 is $1^2+7^2 = 1+49 = 50$
The sum of the squares of the factors of 42 is

$$1^2+2^2+3^2+6^2+7^2+14^2+21^2+42^2 =1+4+9+36+49+196+441+1764 = 2500$$

and here we note that $50 \times 50 = 2500$.

Example 2.

$$7 \times 4 = 28$$

The sum of the squares of the factors of 7 is $1^2+7^2 = 1+49 = 50$
The sum of the squares of the factors of 4 is $1^2+2^2+4^2 = 1+4+16 = 21$
The sum of the squares of the factors of 28 is
  $1^2+2^2+4^2+7^2+14^2+28^2 = 1+4+16+49+196+784 = 1050$
and here we have $50 \times 21 = 1050$

Example 3.

$$6 \times 5 = 30$$

The sum of the squares of the factors of 6 is $1^2+2^2+3^2+6^2 = 1+4+9+36 = 50$
The sum of the squares of the factors of 5 is $1^2+5^2 = 1+25 = 26$
The sum of the squares of the factors of 30 is
$1^2+2^2+3^2+5^2+6^2+10^2+15^2+30^2 = 1+4+9+25+36+100+225+900 = 1300$
and here we note that $50 \times 26 = 1300$

Sum of Squares of Factors Theorem . (without proof)

> Suppose that numbers A and B have no common factor.
> Let the sum of the squares of the factors of A be a2 and let the sum of
> the squares of the factors of B be b2,
> then the sum of the squares of the factors of A x B is a2 x b2.

As you may suspect, these three theorems, the **number of factors theorem,**
the sum of factors theorem and **the sum of squares of factors theorem** are
all special cases of a more powerful result from number theory:

Sum of Powers of Factors Theorem . (without proof)

> Suppose that numbers A and B have no common factor.
> Let the sum of the nth powers of the factors of A be a_n and let the sum
> of the nth powers of the factors of B be b_n,
> then the sum of the nth powers of the factors of A x B is a_n x b_n.

Our theorem on the number of divisors is the special case of this theorem taking n=0.

Our theorem on the sum of divisors is the special case of this theorem taking n=1

The sum of squares of divisors theorem is the special case taking n=2.

The sum of the cubes of the divisors

For illustration, we give an example for the sums of the cubes of the divisors for the multiplication

$$6 \times 5 = 30:$$

The divisors of 6 are 1,2,3,6 so that the

sum of the cubes of the divisors of 6 = $1^3+2^3+3^3+6^3$ = 1+8+27+216 = 252

The divisors of 5 are 1 and 5 so

sum of cubes of divisors of 5 = 1^3+5^3 = 1+125 = 126

now 252 x 126 = 31752.

This should be equal to the sum of the cubes of the divisors of 30.

The divisors of 30 are 1,2,3,5,6,10,15 and 30 so that the sum of the cubes of these divisors is

$1^3+2^3+3^3+5^3+6^3+10^3+15^3+30^3$ = 1+8+27+125+216+1000+3375+27000 = 31752

as the theorem predicts.

Perfect Numbers

Pythagoras classified whole numbers into three groups depending on the sum of the divisors that are less than the number itself. Excessive numbers have the sum of these divisors greater than the number itself while deficient numbers have a sum of the divisors less than the number itself.
Perfect numbers are equal to the sum of their smaller divisors.
Thus, for the number 2 to 28 we have the following Pythagorean classification:

number	list of smaller factors	sum	class
2	1	1	deficient
3	1	1	deficient
4	1,2	3	deficient
5	1	1	deficient
6	1,2,3	6	**perfect**
7	1	1	deficient
8	1,2,4	7	deficient
9	1,3	4	deficient
10	1,2,5	8	deficient
11	1	1	deficient
12	1,2,3,4,6	16	excessive
13	1	1	deficient
14	1,2,7	10	deficient
15	1,3,5	9	deficient
16	1,2,4,8	15	deficient
17	1	1	deficient
18	1,2,3,6,9	21	excessive
19	1	1	deficient
20	1,2,4,5,10	22	excessive
21	1,3,7	11	deficient
22	1,2,11	14	deficient
23	1	1	deficient
24	1,2,3,4,6,8,12	36	excessive
25	1,5	6	deficient
26	1,2,13	16	deficient
27	1,3,9	13	deficient
28	1,2,4,7,14	28	**perfect**

The first two perfect numbers are 6=1+2+3 and 28=1+2+4+7+14.
The first six perfect numbers are 6, 28, 496, 8128, 33550336 and 8589869056.

Perfect numbers have some interesting properties:

(i) they are always equal to the sum of a series of consecutive integers:

$$6 = 1+2+3$$
$$28 = 1+2+3+4+5+6+7$$
$$496 = 1+2+3+4+..........+31$$
$$8128 = 1+2+3+4+5+.......+127$$
$$33550336 = 1+2+3+4+5+......+8191$$

(ii) In general, a perfect number always has the from $2^n(2^{(n+1)}-1)$

(iii) If 2^n-1 is a prime p then $2^{(n-1)}p$ is a perfect number

(iv) Every even perfect number has the form $2^{(n-1)}p$, where p is some prime number.

Exercise 9

[1]   Write the perfect numbers 6, 28, 496, 8128, 33550336 and 8589869056 in the form $2^n(2^{(n+1)}-1)$.

[2]   Find the unique positive 9 digit number such that

(i) all digits 1..9 occur just once
(ii) The number formed by the first n digits is divisible by n

For example: try 126458793
12 is divisible by 2
126 is divisible by 3
1264 is divisible by 4
12645 is divisible by 5
but   126458 is not divisible by 6 so 126458793 is not the number we are looking for.

=====================================

Answers

Fred's bill was 6d 2s 1a

Ex 1

[1] XXV=25   [2] CCL=250 [3] D=500   [4] MMD=2500

[5] $\overline{\text{XXV}}$=25000   [6] $\overline{\text{C}}$=100000

Ex 2

[1]	oct	dec	[2]	oct	dec	[3]	oct	dec
	123	83		246	166		777	511
	+456	+302		+246	+166		+666	+438
	601	385		514	332		1665	949

Ex 3

[1]   1750 OCT   [2]   17500 OCT   [3]   101, 102, 103, 104

Ex 4

[1]   8181 and 7524   [2]   all of them

Ex 5

```
0.2 dec = 0.146314631463....oct
0.3 dec = 0.2314631463146....oct
0.4 dec = 0.314631463146....oct
0.5 dec = 0.4 oct
0.6 dec = 0.463146314631....dec
0.7 dec = 0.5463146314631....oct
0.8 dec = 0.631463146314....oct
0.9 dec = 0.7146314631463....oct
```

Ex 6

[1]   360.245   [2]   1000.000   [3]   7.66

Ex 7

99 dec = 1100011 bin

Ex 8

[1] BEE   [2] D0   [3] 1000   [4] 1A2BD

Ex 9

$$6 = 2^1(2^2 - 1)$$

$$28 = 2^2(2^3 - 1)$$

$$496 = 2^4(2^5 - 1)$$

$$8128 = 2^6(2^7 - 1)$$

$$33550336 = 2^{12}(2^{13} - 1)$$

$$8589869056 = 2^{16}(2^{17} - 1)$$

===================================

Reversing Digits in Base Ten

Choose any three digit number with differing units and hundreds digits:

(i) reverse the digits and subtract
(ii) now reverse the digits and add

Show that the result is always 1089

Solution:

If the number is abc so that its value is $100a + 10b + c$, suppose that $a>c$ then the algorithm gives

$$
\begin{array}{llll}
& 100a & + 10b & + c \\
- & 100c & + 10b & + a \\
\hline
& 100[a-c-1] & + 90 & + [10+c-a] \\
+ & 100[10+c-a] & + 90 & + [a-c-1] \\
\hline
& 100[10] & + 80 & + [9] \quad = 1089 \\
\end{array}
$$

If $c>a$ then the argument is the same as above except that a and c are interchanged.

This algorithm works in any number base, for example, in base 7, the corresponding result would be 1056_7 and in base 5 the result would be 1034_5 .

A proof for any number base n is given here:

Reversing Digits in any Number base

Suppose our number base is base n then we may represent any three digit number in this base as abc_n which will represent c units, b groups of n and a groups of nxn.

Suppose that a>c then our reversing digits exercise will be as follows:

The number abc_n is	$a\,n^2 + bn\ + \qquad c$
reverse the digits to get	$\underline{c\,n^2 + bn\ + \qquad a}$
Subtract	$n^2\,[\,a\text{-}c\text{-}1\,] + [n\text{-}1]n\ +[n\text{-}a\text{+}c]$
reverse	$\underline{n^2\,[\,n\text{-}a\text{+}c] + [n\text{-}1]n + [a\text{-}c\text{-}1]}$
and add	$n^2\,[n\text{-}1]\qquad + 2n^2\,\text{-}2n + [n\text{-}1]$
giving	$n^3 + 0\,n^2 + [n\text{-}2]\,n\ \ + [n\text{-}1]$

which is written in base n as $1\ 0\ [n\text{-}2]\ [n\text{-}1]_n$

============================

These diversions can of course be extended to numbers with more than three digits, but we no longer get unique results. For a five digit number abcde in base ten, we have the following:

If a>e and d>b our result is	99889
If a>c and b>d the result is	109890
If a>c and b=d then the result is	109989

Again, this result can be generalized to any number base.
 If our number in base n is $abcde_n$ then

If a>e and d>b our result is	$[n\text{-}1][n\text{-}1][n\text{-}2][n\text{-}2][n\text{-}1]_n$
If a>c and b>d the result is	$1\ 0\ \ [n\text{-}1][n\text{-}2][n\text{-}1]\ 0\ _n$
If a>c and b=d then the result is	$1\ 0\ \ [n\text{-}1][n\text{-}1][n\text{-}2][n\text{-}1]_n$

============================

CHAPTER 6

Boolean Algebra

George Boole was the son of a cobbler, born in Lincoln, England, in 1815. At the age of sixteen he started work as an assistant teacher and in 1834 he opened his own school in Lincoln. In 1838 he took over the running of Hall's Academy in Waddington, about five miles south of Lincoln. It was about this time that he started studying advanced mathematics and began publishing regularly in the Cambridge Mathematical Journal and in 1864 he published an article on differential equations in the Transactions of the Royal Society, that earned him the Society's Royal medal. In 1849 he was appointed to the post of Professor of Mathematics at Queens College, Cork where he remained for the rest of his life.

He married Mary Everest in 1855 and they had five daughters.

In 1854, George Boole published "An Investigation into the Laws of Thought on which are founded the Mathematical Theories of Logic and Probabilities" and this became the cornerstone of what we now call Boolean Algebra.

He published about 50 papers on advanced mathematics and was awarded honorary degrees from Dublin and Oxford Universities.

In 1857 he was elected Fellow of the Royal Society. He died seven years later at the age of 49.

Predicates and Propositions

Statements such as "He is 20 years old." or "That's huge." cannot be assigned a truth value (true=1) or (false=0) unless the variable "He" in the first statement or "That" in the second statement, is specified or, more formally, assigned a value. Such statements that require the assignment of a value to a variable, before the truth value can be given, are called **predicates**. When a value is assigned to such a variable then the statement is called a **proposition** and we must be able to decide whether the proposition is true or false.

The first statement could be written

> x is 20 years old

and we could then assign the value "Fred" to the variable x to give the proposition

> Fred is 20 years old

In a similar way, the second example statement could be written as the predicate

> y is huge

and, assuming that the adjective huge has been rigidly defined, we could substitute the value "Joanna" for the variable y to give the proposition

> Joanna is huge

which may be either true or false, depending of the definition of the word huge and, presumably, the weight of Joanna.

In such discussions we usually restrict ourselves to a certain class of "things under discussion". This class is called the **universal set**. We may for example, define our universal set to be the set of all living persons or the set of all natural numbers. We will denote the universal set by U unless there is some other commonly used notation. Commonly used symbols for various set are:

> N   the set of natural numbers, 0,1,2,3,4,5,......
> Z   the set of integers, ...-3,-2,-1,0 1,2,3,4.......
> Q   the set of rational numbers or fractions
> R   the set of real numbers

If x is a member of our chosen universal set then a predicate can be represented by p(x). Thus we could write

> p(x) for "x is 20 years old"

Once we have stated our predicate, then there will be a certain set of values for x which will make p(x) a true proposition. This set is called the **truth set** for the predicate p(x) and we write

> $P = \{x \mid p(x)$ is true$\}$

or $P = \{x \mid p(x)\}$

meaning, **P** is the set of values **x**, for which the proposition **p(x)** is true.
We will denote predicates using lower case and the corresponding truth set with upper case.

Example 1
Let the universal set be the set of numerals

> $U = \{0,1,2,3,4,5,6,7,8,9\}$

Let predicates a and b be defined by

> a = "x is odd"   with truth set A
> b = "x is prime"   with truth set B

then

> $A = \{x \mid odd(x)\} = \{1,3,5,7,9\}$
> $B = \{x \mid prime(x)\} = \{2,3,5,7\}$

These sets can be represented using a Venn diagram:

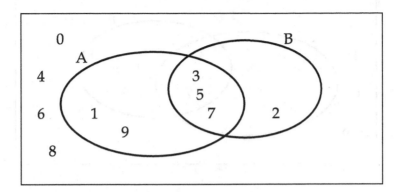

Derived Statements and Derived Sets
(We use example 1 to illustrate the definitions)

AND

Given the two statements or predicates a and b there are some values,
namely 3, 5 and 7 which make both statements true. They the values that are
in the truth sets of both predicates. We define the conjunction (the AND) of
a and b, which is written a∧b to have the truth value 1 when both the
statements a and b are true. The AND of statements a and b is thus defined
by the truth table

a	b	a∧b
0	0	0
0	1	0
1	0	0
1	1	1

The truth set for the statement a∧b is thus

$$\{x \mid odd(x) \wedge prime(x)\}$$
or $$\{x \mid x \in A \wedge x \in B\}$$

This set is called the intersection of the sets A and B and we write

$$A \cap B = \{x \mid x \in A \wedge x \in B\}$$

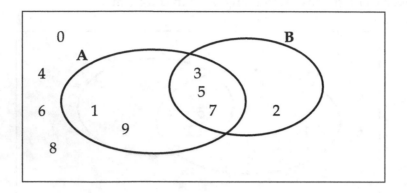

$$A \cap B = \{3,5,7\}$$

The intersection of two sets

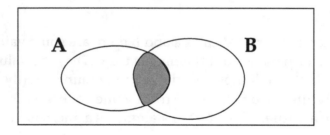

$$A \cap B$$

OR

The OR (disjunction) of the two statements a and b, written **a∨b**, has the truth value 1 when either a or b (or both) has the truth value 1. The truth table for the OR of statements a and b is thus:

a	b	a∨b
0	0	0
0	1	1
1	0	1
1	1	1

In our example the truth set for a∨b is $\{x \,|\, odd(x) \lor prime(x)\}$

which can be written $\{x \mid x \in A \vee x \in B\}$

This set is called the union of the sets A and B and is written

$$A \cup B = \{x \mid x \in A \vee x \in B\}$$

In example 1, $A \cup B = \{1,2,3,5,7,9\}$

The union of two sets

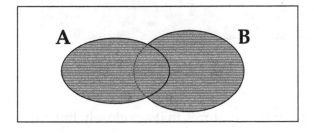

A ∪ B

NOT

The negation of the statement a is

$\sim a = \text{NOT}(x \text{ is odd})$
and similarly $\sim b = \text{NOT}(x \text{ is prime})$

The corresponding truth sets are called the complements of the sets A and B which are denoted by $\overline{A}$ and $\overline{B}$. Thus

$$\overline{A} = \{x \mid x \notin A\}$$

and $\overline{B} = \{x \mid x \notin B\}$

The corresponding truth sets for the statements $\sim a$ and $\sim b$ in example 1 are

$$\overline{A} = \{ 0,2,4,6,8\}$$

and $\overline{B} = \{0,1,4,6,8,9\}$

The Complement of A

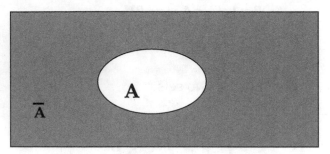

We have here, two algebraic structures that are closely linked:

1. **Set Algebra**
    If A and B are two sets, then the binary operations intersection and union are defined:

    A∩B is the set of elements that are in both A and B
    A∪B is the set of elements that are in either A or B

    The complement of a set is

    $\overline{A}$ the set of elements that are not in A

2. **Propositional Calculus**
    If a and b are two propositions, then the binary operations conjunction (and) and disjunction (or) are defined:

    a∧b is the statement formed from both a and b
    a∨b is the statement formed from either a or b

    The negation of the statement a is

    ¬a which is the statement "not" a

In set algebra, there is an underlying universal set of objects. If x denotes one of these and A is any set, then we must be able to state whether x is a member of the set A or not. Either x is in A (x∈A) or x is not in A (x∉A).

Set algebras may have an infinite number of elements, for example, we could have chosen the underlying universal set to be the set of all natural numbers in which case there would be an infinite number of sets in our algebra. If our underlying universal set was $U = \{1,2,3\}$ then there would only be $2^3 = 8$ sets in the algebra.

In propositional calculus, there is also an underlying set of "things (or persons) we are talking about". If x is one of these things (or persons) and a is any statement, then we must be able to state whether a(x) is true or a(x) is not true. Now if A is the set of values that makes the statement a(x) true, then to say that a(x) is true is the same as stating that x is in the truth set of a (x∈A). If a(x) is not true then x is not in the truth set of a (x∉A).

George Boole realized that both set algebra and propositional calculus were examples of a more general algebraic structure that we now call a Boolean Algebra. Although they were not available to him, electronic switching circuits obey the same rules as George Boole's algebra.

Before building the algebra in general, we briefly look at a rather special application of Boolean algebra in electronic computer switching circuits. This topic is dealt with in more detail in the next chapter.

An Algebra of Noughts and Ones

In a two state switching circuit, we have two voltage levels that represent two "logic levels". In many computers, 5 volts represents "logic 1" and 0 volts represents "logic 0". In switching terminology, "switch is on" is represented by 1 and "switch is off" of represented by 0.

A logic gate is an electronic device that has a number of inputs, for example a and b, or just a and then decides what the out put should be:

inputs { a →
         b → [ ] → output

For our purposes, we only need three kinds of logic gate, the **AND** gate, the **OR** gate and the **NOT** gate.
The rules:

a	b	a+b
0	0	0
0	1	1
1	0	1
1	1	1

a	b	a.b
0	0	0
0	1	0
1	0	0
1	1	1

a	$\bar{a}$
0	1
1	0

These tables define the behaviour of the three basic logic gates:

| OR gate | AND gate | NOT gate |

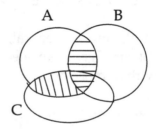

Thus the underlying structure in three very different subject areas, is the same. It is the structure of George Boole's algebra:

In set algebra:

$A \cap (B \cup C) = (A \cap B) \cup (A \cap C)$

In propositional calculus:

Let   p= "I'll have chips" q= "I'll have peas"   r = "I'll have beans"

Then $p \wedge (q \vee r) = (p \wedge q) \vee (p \wedge r)$

I'll have chips with either peas or beans

=   I'll have chips and peas or chips and beans

In switching Circuits:

$a.(b+c) = a.b + a.c$

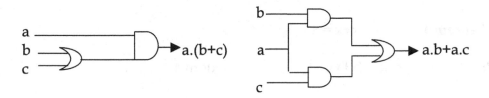

George Boole's Algebra

A Boolean Algebra is a set of elements B say, with two binary operations + and . such that the following rules hold:

1. Closure: if x and y are members of the set B then

    $x+y \in B$ and $x.y \in B$

2. Commutative rules:

    $x+y = y+x$ and $x.y = y.x$

3. Distributive rules:

    (a) $x.(y+z) = x.y + x.z$

    (b) $x + y.z = (x+y).(x+z)$

4. Identity for + $x + 0 = x$

    Identity for . $x.1 = x$

5. Complements: for each $x \in B$ there is $\bar{x} \in B$ such that

    $x + \bar{x} = 1$ and $x.\bar{x} = 0$

========================

From these rules alone, we can deduce a number of theorems. In particular, we can deduce the associative rules and De Morgan's rules, but before we can prove these important results we need to prove a number of preliminary results.

Theorem 1. $x+x = x$

Proof $x+x = (x+x).1$ axiom 4

         $= (x+x).(x+\bar{x})$ axiom 5

         $= x+(x.\bar{x})$ axiom 3b

         $= x + 0$ axiom 5

         $= x$ axiom 4

===================================

Theorem 2. $x.x = x$

Proof $x.x = x.x + 0$ axiom 4

         $= x.x + x.\bar{x}$ axiom 5

         $= x.(\bar{x}+x)$ axiom 3a

         $= x.1$ axiom 5

         $= x$ axiom 4

===================================

Note on duality: the five axioms are completely symmetrical in + and . and in 0 and 1. This means that if we prove any theorem, we can automatically deduce another theorem by swopping + with . and 0 with 1.

This is often called the principle of **Duality.**

Theorem 3   x.0 = 0

Proof   x.0 = 0 + x.0   axiom 4

                       = x.$\overline{x}$ + x.0   axiom 5

                       = x.($\overline{x+0}$)   axiom 3a

                       = x.$\overline{x}$ axiom 4

                       = 0   axiom 5

====================================

Theorem 4   x+1 = 1

Proof   dual of theorem 3

====================================

Theorem 5   x(x+y) = x

Proof   x(x+y) = (x+0).(x+y)   axiom 4

                     = x+0.y   axiom 3b

                     = x+0   theorem 3

                     = x   axiom 4

====================================

Theorem 6   x+xy = x

Proof   dual of theorem 5

====================================

Theorem 7   the associative rule for +

$$x+(y+z) = (x+y)+z$$

Proof   let $A=x+(y+z)$ and $B=(x+y)+z$

then $AB = [x+(y+z)][(x+y)+z]$

$\quad\quad\quad = [x+(y+z)](x+y) + [x+(y+z)].z$ axiom 3a

$\quad\quad\quad = [x+(y+z).y] + [xz+(y+z)z]$ axioms 3b, 3a

$\quad\quad\quad = [x+y] + [xz+z]$ theorem 3

$\quad\quad\quad = [x+y] + z$ theorem 4

thus $AB = B$

also $BA = [(x+y)+z][x+(y+z)]$

$\quad\quad\quad = [(x+y)+z].x + [(x+y)+z].(y+z)$ axiom 3a

$\quad\quad\quad = [(x+y).x + z.x] + [z + y(x+y)]$ axioms 3a, 3b

$\quad\quad\quad = [x+xz] + [z+y]$ theorem 3

$\quad\quad\quad = x+[y+z]$ theorem 4

thus $BA = A$

    but $AB=BA$ by axiom 2, therefore $A=B$ ∎

================================

Theorem 8:   the associative rule for .

$$x(yz) = (xy)z$$

Proof   this is the dual of theorem 7, but we present a similar proof

    let $A=x.(yz)$ and $B=(xy).z$

then $A+B = [x(yz)]+[(xy)z]$

$$= [x(yz)+(xy)][x(yz)+z]$$ by axiom 3b
$$x(yz)+ab=[x(yz)+a][x(yz)+b]$$

$$= [x.(yz+y)][(x+z).(yz+z)]$$ axioms 3a, 3b with commutative rule

$$= [x.y][(x+z).z]$$ theorem 4

$$= [x.y] + z$$ theorem 3

thus $A+B = B$

also $B+A = [(xy)z]+[x(yz)]$

$$= [(xy)z+x][(xy)z+(yz)]$$ axiom 3b

$$= [(xy+x).(z+x)][z (xy+y)]$$ axioms 3b, 3a

$$= [x.(z+x)][z.y]$$ theorem 4

$$= x[yz]$$ theorem 3

thus $B+A = A$

but $A+B = B+A$ by axiom 2, therefore $A=B$ ■

============================

Theorem 9

The complement of x is unique.

Proof

Let $\bar{x}$ be a complement of x

Suppose that b is also a complement of x so that

$$x+b = 1$$

and $x.b = 0$ using axiom 5.

We will show that b and $\bar{x}$ are the same element.

$$b = b.1 \qquad \text{axiom 4}$$

$$= b.(x+\bar{x}) \qquad \text{axiom 5}$$

$$= b.x + b.\bar{x} \qquad \text{axiom 3}$$

$$= 0+b\bar{x} \qquad b.x=0, \text{ given above}$$

$$= x.\bar{x}+b.\bar{x} \qquad \text{axiom 5}$$

$$= \bar{x}.(x+b) \qquad \text{axioms 2 \& 3a}$$

$$= \bar{x}.1 \qquad x+b=1, \text{ given above}$$

$$= \bar{x} \qquad \text{axiom 3}$$

■

====================================

Having proved that the complement of an element is unique, we now prove two results made famous by George Boole's friend, the French mathematician Augustus De Morgan (1806-1871).

Asked at the breakfast table what you would like to drink with your scrambled eggs, you reply
" I don't like coffee or tea "

By this I would take it to mean that you do not like coffee and also, you do not like tea, so your reply could have been
" I don't like coffee and I don't like tea "
with the same meaning.

Let the statement   c = " I like coffee "
and let   t = " I like tea "
then your reply can be written
NOT(c OR t)

This has the same meaning as the statement
NOT(c) AND NOT(t)

Although we have described your preference in terms of statements, De Morgan's Laws belong to the general theory of Boolean Algebras.

Theorem 10   De Morgan's First Law

$$\overline{x+y} = \overline{x}.\overline{y}$$

Proof

We show that $\overline{x}.\overline{y}$ is the unique complement of $(x+y)$, using the definition of the complement from axiom 5.

$(x+y) + \overline{x}.\overline{y} = (x+y)(x+\overline{x}) + \overline{x}.\overline{y}$ axiom 5, $x+\overline{x} = 1$

$\qquad\qquad = x.x + x.\overline{x} + y.x + y.\overline{x} + \overline{x}.\overline{y}$ axiom 3a, extended

$\qquad\qquad = x(x+\overline{x}) + yx + \overline{x}(y+\overline{y})$ axiom 3a

$\qquad\qquad = x + \overline{x} + yx$ axiom 5, $x+\overline{x} = 1$

$\qquad\qquad = 1+yx$ axiom 5

$\qquad\qquad = 1$ theorem 4

also $(x+y).(\overline{x}.\overline{y}) = x.\overline{x}.\overline{y} + y.\overline{x}.\overline{y}$

$\qquad\qquad = 0.\overline{y} + 0.\overline{x}$ axiom 5, $a.\overline{a} = 0$

$\qquad\qquad = 0 + 0$ theorem 3

$\qquad\qquad = 0$ axiom 4, $x+0 = x$

We have proved that $(x+y) + [\,\overline{x}.\overline{y}\,] = 1$

and $(x+y).[\,\overline{x}.\overline{y}\,] = 0$

therefore $[\,\overline{x}.\overline{y}\,]$ is the unique complement of $(x+y)$,   axiom 5 + theorem 9

$$\overline{(x+y)} = \overline{x}.\overline{y}$$

==

Theorem 11   De Morgan's Second Law

$$\overline{x.y} = \overline{x} + \overline{y}$$

Proof

(this, of course, is the dual of theorem 10, however, we give here a proof written out in full)

We show that $\overline{x}+\overline{y}$ is the unique complement of $(x.y)$, using the definition of the complement from axiom 5.

$(x.y) +(\overline{x}+\overline{y}) = (x.y) + (x+\overline{x})(\overline{x}+\overline{y})$	axiom 5, $x+\overline{x} = 1$
$= x.y + \overline{x}.x + x.\overline{y} + \overline{x}.\overline{x} + \overline{x}.\overline{y}$	axiom 3a, extended
$= x(y+\overline{y}) + \overline{x}(x+\overline{x})+\overline{x}.\overline{y}$	axiom 3a
$= x.1 + \overline{x}.1 + \overline{x}.\overline{y}$	axiom 5
$= x+\overline{x} + \overline{x}.\overline{y}$	axiom 4
$= 1 + \overline{x}.\overline{y}$	axiom 5
$= 1$	theorem 4

also, we have

$(x.y).(\overline{x}+\overline{y}) = x.y.\overline{x} + x.y.\overline{y}$	axiom 3a
$= 0.y + x.0$	axiom 5,
$= 0+0$	theorem 3
$= 0$	axiom 4

we have proved that $(x.y) + [\,\overline{x}+\overline{y}\,] = 1$

and $(x.y).[\,\overline{x}+\overline{y}\,] = 0$

therefore $[\,\overline{x}+\overline{y}\,]$ is the unique complement of $(x.y)$,   axiom 5 + theorem 9

$$\overline{x.y} = \overline{x}+\overline{y}$$

============================

Exercise 1

1. Simplify $x.y + x.z + \overline{x}.\overline{z}$

2. Simplify $\overline{x}.\overline{y}.(x+y) + x.y + \overline{x}.\overline{y}$

3. Prove $x.(y+z) + y.(z+x) + z.(x+y) = x.y + y.z + z.x$

4. Prove $x.\overline{y} + y = x + y$

Truth Tables

A statement in propositional calculus has a truth value of either 1 or 0. One way of proving identities in propositional calculus is to use a truth table and show that the two expressions have the same truth value for all values of the variables that are used. Note that the negation of a statement x is usually denoted by ¬x.

Example
   If a and b are any two statements then

$$\neg(a \vee b) = (\neg a \wedge \neg b)$$ (this is equivalent of De Morgan's first law)

Proof (using a truth table)

a	b	a∨b	¬ (a∨b)	¬a	¬b	¬a ∧ ¬b
0	0	0	1	1	1	1
0	1	1	0	1	0	0
1	0	1	0	0	1	0
1	1	1	0	0	0	0

We see that ¬(a∨b) and ¬a ∧ ¬b have the same truth values for all four possibilities given in the table so that we can write ¬(a∨b) = ¬a ∧ ¬b.

The same method can be used to prove identities between expressions involving sets:

Theorem

$$\overline{A \cup B} = \overline{A} \cap \overline{B}$$

(This is De Morgan's first law for two sets)

Proof using a truth table:

We have two sets A and B, and for any element x, belonging to the universal set, there are just four possibilities. In this truth table, an entry of 1 indicates that x is a member of the given set (x ∈ set) while an entry 0 in the truth table indicates that x is not a member of the set (x ∉ set).

	A	B
x is in neither set	0	0
x is in B but not in A	0	1
x is in A but not in B	1	0
x is in both sets	1	1

To prove the result we set out a truth table similar to the one we used for showing that ~(a∨b) = ~a ∧ ~b except that a and b are replaced by A and B and the logic symbols are replaced by the corresponding set symbols:

A	B	A∪B	$\overline{A \cup B}$	$\overline{A}$	$\overline{B}$	$\overline{A} \cap \overline{B}$
0	0	0	1	1	1	1
0	1	1	0	1	0	0
1	0	1	0	0	1	0
1	1	1	0	0	0	0

We see from this truth table that whenever x is a member of the set $\overline{A \cup B}$ then x is also a member of the set $\overline{A} \cap \overline{B}$.

Also, whenever x is a member of the set $\overline{A} \cap \overline{B}$ then x is also a member of the set $\overline{A \cup B}$. More formally, we would write

$$x \in \overline{A \cup B} \quad \Rightarrow \quad x \in \overline{A} \cap \overline{B}$$

and

$$x \in \overline{A} \cap \overline{B} \quad \Rightarrow \quad x \in \overline{A \cup B}$$

hence $\overline{A \cup B}$ and $\overline{A} \cap \overline{B}$ have exactly the same members and so are equal sets.

Exercise 2

Use truth tables, as above to show that that
 (i) $\neg(a \wedge b) = \neg a \vee \neg b$.

 (ii) $\overline{A \cap B} = \overline{A} \cup \overline{B}$

Implication

Given two statements a and b, which may be either true or false, the derived
statements a∧b, a∨b, ¬a (that is a AND b, a OR b and NOT(a)) have the
natural language truth values defined by the following truth table:

a	b	a∧b	a∨b	¬a
0	0	0	0	1
0	1	0	1	1
1	0	0	1	0
1	1	1	1	0

The truth table for the statement
          "if a then b"
is not so obvious.
This statement is the same as the statement "a implies b" and is written
formally $a \Rightarrow b$.
If the implication "$a \Rightarrow b$" is to be true, then clearly, whenever statement a is
true then statement b must be true, otherwise, the implication would be
false. We can thus confidently fill in two rows of the truth table:

a	b	a⇒b
1	1	1
1	0	0

but what the truth values should we assign to $a \Rightarrow b$ when a is false?
Consider the following:

     a = "you win the race"
     b = "I'll give you the prize"

then **a⟹b** can be written

"If you win the race then I'll give you the prize"

The question now is, if I say the statement **a⟹b** under what conditions can you declare that I have uttered a falsehood?

Consider now, the four cases:
1. You win and I gave you the prize — **a⟹b was true**
2. You won but I did not give you the prize — **a⟹b was false**
3. You did not win, I did not give you the prize
   In this case, I have not told a lie, hence — **a⟹b was true**
4. You did not win but I gave you the prize
   In this case also, I have not lied as I did not say what exactly I would
do if you did not win. Since I have not lied — **a⟹b was true**

This argument leads us to define the truth table for the implication statement as follows:

a	b	a⟹b
0	0	1
0	1	1
1	0	0
1	1	1

If the truth set for the statement a is contained in the truth set for statement b (written A⊆B) then the statement **a⟹b** will be "universally true".
In a Venn diagram this would show as:

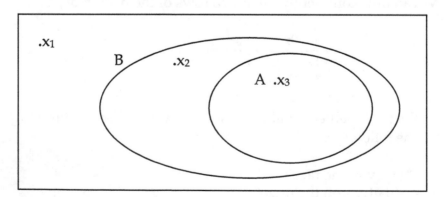

A is the truth set for the statement a and **B** is the truth set for statement b.
If the value x makes the statement a(x) true then x ∈ A. In this Venn diagram,
whenever x ∈ A we have x ∈ B also, so that the statement b(x) will be true.
This is what we mean in natural language when we say "a implies b".
The value x_1 gives $x_1 \notin A \wedge x_1 \notin B$ which is an example of row 1 in the truth
table.
The value x_2 gives $x_2 \notin A \wedge x_1 \in B$ which is an example of row 2 in the truth
table.
The value x_3 gives $x_3 \in A \wedge x_3 \in B$ which is an example of row 4 in the truth
table.

The only missing case is $x \in A \wedge x \notin B$ which is row 3 in the truth table.

The truth set for **a⇒b** is shaded in this figure:

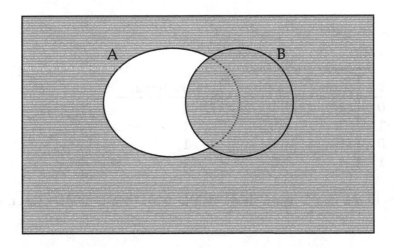

This set (shaded) can be described as $B \cup \overline{A}$

so that the statement **a⇒b** is the same as **b OR (NOT a)**.

Example

Prove that $(a \Rightarrow b) \vee (b \Rightarrow a)$ is universally true.

Proof 1

The truth set for the statement $\mathbf{a \Rightarrow b}$ is $B \cup \overline{A}$

so that the truth set for the statement $(a \Rightarrow b) \vee (b \Rightarrow a)$ is

$$(B \cup \overline{A}) \cup (A \cup \overline{B})$$

$$= (B \cup \overline{B}) \cup (A \cup \overline{A})$$

$$= 1$$

∎

Proof 2 (using a truth table)

We have

a	b	a⇒b	a⇒b	(a⇒b)∨(b⇒a)
0	0	1	1	1
0	1	1	0	1
1	0	0	1	1
1	1	1	1	1

and the entries in the last column show that $(a \Rightarrow b) \vee (b \Rightarrow a)$ always has the truth value 1.

Equivalence

If **a** and **b** are two statements such that both of the statements **a**⇒**b** and
b⇒**a** have truth values 1, then we the truth sets A and B are such that A⊆B
and also B⊆A. This means that the two sets A and B are identical.
The truth set for the statement (**a**⇒**b** ∧ **b**⇒**a**) is the following:

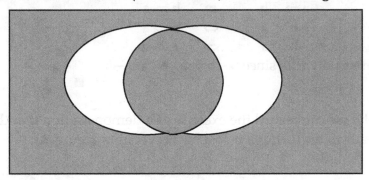

We write the statement   (**a**⇒**b** ∧ **b**⇒**a**) as   **a** ⇔ **b** which is called the
equivalence statement. It can be interpreted as meaning that either **a** and **b**
are both true or **a** and **b** are both false. Thus **a** ⇔ **b** has the same meaning as
(**a**∧**b**)∨ (¬ **a** ∧¬ **b**)

Propositional Calculus
In the classical two valued propositional calculus every formula is either true
or false (and not both!).
The statements **p** and **q** are semantically equivalent if and only if **p** ⇔ **q** is a
tautology. This means that **p** and **q** will have identical columns in a truth
table. For example, the exercise
$$¬a \Rightarrow ¬b$$
$$⇔ \qquad b \Rightarrow a$$
can be regarded either as a single formula
$$¬a \Rightarrow ¬b \qquad ⇔ \qquad b \Rightarrow a$$
for which you are required to demonstrate its universal truth:

Proof 1
$$\qquad\qquad ¬a \Rightarrow ¬b \quad ⇔ \quad b \Rightarrow a$$
if and only if   ¬b ∨ **a**   ⇔   **b** ⇒ **a**
if and only if   ¬b ∨ **a**   ⇔   **a** ∨ ¬b
which is universally true since   ¬b ∨ **a** = **a** ∨ ¬b   ■

We often see a symbol called the "turnstile" ⊧ in such proofs. This symbol comes from the "meta-language" to enable us to talk about objects in the language of the propositional calculus. It has a role similar to the "therefore" ∴ symbol used in elementary algebra. Thus, the first proof would be set out as

$$
\begin{array}{llll}
& \neg a \Rightarrow \neg b & \Leftrightarrow & b \Rightarrow a \\
\models & \neg b \vee a & \Leftrightarrow & b \Rightarrow a \\
\models & \neg b \vee a & \Leftrightarrow & a \vee \neg b
\end{array}
$$

which is universally true since　$\neg b \vee a = a \vee \neg b$ ∎

Alternatively, we can regard the exercise as a demonstration that the formula $\neg a \Rightarrow \neg b$ leads to the formula $b \Rightarrow a$ by a reversible argument:

Proof 2

$$
\begin{array}{ll}
& \neg a \Rightarrow \neg b \\
\Leftrightarrow & \neg b \vee \neg(\neg a) \\
\Leftrightarrow & \neg b \vee a \\
\Leftrightarrow & b \Rightarrow a
\end{array}
$$
∎

Of course, we can always resort to a sledge-hammer (& less satisfying) proof by using a truth table:

a b	$\neg a$	$\neg b$	$\neg a \Rightarrow \neg b$	$b \Rightarrow a$
0 0	1	1	1	1
0 1	1	0	0	0
1 0	0	1	1	1
1 1	0	0	1	1

The identical columns show that　$\neg a \Rightarrow \neg b \Leftrightarrow b \Rightarrow a$

∎

Exercise 3
1. Prove that　$a \Rightarrow (b \Rightarrow a)$　is universally true.
2. Prove that　$(p \vee q) \wedge \neg(p \wedge q) \Leftrightarrow (p \wedge \neg q) \vee (\neg p \wedge q) = 1$
3. Prove that　$((r \wedge q \Rightarrow q) \wedge \neg(\neg r \wedge p)) \Rightarrow (p \Rightarrow q) = 1$
4. Prove that　$(a \Rightarrow (b \Rightarrow c)) \Rightarrow ((a \Rightarrow b) \Rightarrow (a \Rightarrow c)) = 1$
5. Show that　$(\neg a \Rightarrow \neg b) \Leftrightarrow (b \Rightarrow a)$

Logic and Paradoxes

The teacher writes on the blackboard "This statement is not true".
Assuming that all statements are either **true** or **false**, we can argue

1. If the statement is true then "This statement is not true"

2. If the statement is not true then it is not true that the statement is not true, i.e. the statement is true.

The teacher tells the students to remember the rule:
                    "All rules have exceptions"
If this is true then there must be an exception to this rule, i.e. there must be a rule that has no exception.

The teacher then says "I am not telling the truth". Is that true?

Statements about statements or rules about rules are often termed Russell paradoxes after the English mathematician and philosopher Bertrand Russell. A German mathematician, Gottlob Frege had just completed a massive tome on the foundations of arithmetic in which he defined the integers using the idea of a set of all sets. The number one was defined as the set of all sets containing just one element. Thus

$$\text{One} = \{ \{1\}, \{ \}, \{\text{table}\}, \{ \{\text{table, chair}\} \}, \{ \{\text{all sets}\} \}, \ldots \}$$

So One was a set and so we could think of a set containing this one element One and this set would be a member of the set One.

Russell's Paradox

Consider the set **S = the set of all sets that do not belong to themselves**

If $S \in S$ then S does not belong to S

If $S \notin S$ then S is a member of the set S

The paradox seems to arise because we are talking about a set that could be a member of itself.
Russell discovered these paradoxes just as Frege's book was completed and unfortunately it meant that Frege's book was mathematically worthless.

Some adjectives are "self referential" i.e. they can refer to themselves, so that, for example 'short' is a short word or 'simple' is a simple word. Such adjectives will be called "autological", meaning that they can refer to themselves. On the other hand red is not a red word, hard is not a hard word and these adjectives will be called heterological word, meaning that they do not refer to themselves.

Is "heterological" a heterological word or an autological word?

If heterological is heterological then it refers to itself and is therefore autological.
If heterological is autological then it refers to itself and is therefore heterological.

Some sets are members of themselves.
For example, many sets have an infinite number of members, e.g. the integers {1,2,3,4,5, ...}, the set of all polygons, the set of all circles,..etc. There are, of course, an infinite number of these sets.
Let S be the set of all those sets with an infinite numbers of members. Since S has an infinite number of members, S is a member of S. S is a member of itself.

Multiple choice Test

1. The answer to this question is
    (a)   b
    (b)   a
    (c)   Impossible to answer

2. Including this question how many questions have you got correct so far?

    (a)   1½
    (b)   0
    (c)   don't know

3. True or False:
    (a)   (b) is true
    (b)   (a) is false

Answers to Exercise 1: [1] x [2] 1

CHAPTER 7

Logic Circuits

Both the central processing unit and the memory of a modern digital computer are composed of circuits made up of two state logic devices called logic gates. These circuits move bit patterns from one location to another, store bit patterns in the computers memory and perform calculations. In short, they perform all the functions needed in a modern PC. The two states are often represented by voltages, 5 volts for logic 1 and 0 volts for logic 0. In this chapter, we always refer to logic 1 and logic 0 regardless of how they are actually represented in the electronic circuits that make up the computer system. The basic logic gates are the AND gate, the OR gate and the NOT gate but it is in fact possible to make these basic gates out of a single kind of gate (either the NAND gate or the NOR gate) so that a computer could simply be regarded as a bunch of NAND gates. In a similar way, your brain could be regarded as a bunch of interconnected neurons, each neuron having the same basic structure with a number of inputs called dendrites, a cell body that processes the inputs and a single output called the axon. The brain has about one hundred thousand million neurons with about one hundred thousand billion connections and is the most complex computing device you are ever likely to use.

The AND, OR and NOT gates are two state devices having a number of inputs and one single output. Both the inputs and the output can only have two values, logic 1 and logic 0. The value of the output is determined by the inputs and the easiest way to define the different kinds of logic gates is to use a truth table.

The Two Input AND Gate

inputs		output
A	B	A.B
0	0	0
0	1	0
1	0	0
1	1	1

The output of an AND gate is always the product of its inputs.

The 2 input OR Gate

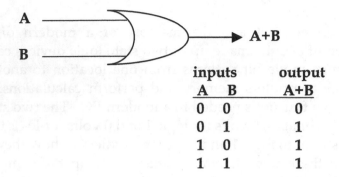

inputs		output
A	B	A+B
0	0	0
0	1	1
1	0	1
1	1	1

The output of an OR gate is always the sum of the inputs provided that we remember that in George Boole's algebra, **1+1=1**.

The NOT Gate or Inverter

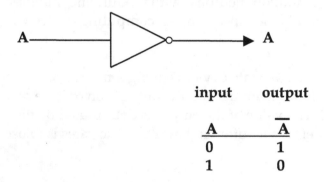

input	output
A	$\overline{A}$
0	1
1	0

Making Logic Gates

A simple switch is the basic element of any logic circuit or switching circuit, forming the hardware for a digital computer. We measure our logic values 0 or 1 as voltages, often with 0 Volts for logic 0 and 5 Volts for logic 1.

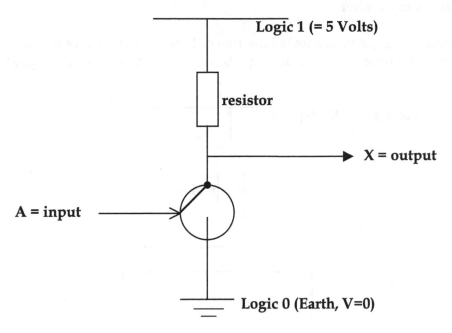

Suppose that we have a switch that is operated by an input A so that a value A= logic 1 will close the switch and a value A=logic 0 opens the switch. The output of the device is the voltage measured at **X** in the figure above.

If the switch is open, with A=0, then no current flows through the device from the 5 volts line to Earth and so the voltage at any point of the device above the switch is at 5 volts, so that the output X = 1.

If the switch is closed, with A=1, then a current flows through the device from the 5 volts line to Earth and there is a voltage drop across the resistor and if we can ignore the resistance of the switch itself, then the output X will be "pulled down" to ground or logic 0. Thus we have the following table:

input A	output X
0	1
1	0

The switch behaves like an inverter or NOT gate.

Switches in Parallel

The following diagram shows how two of these inverters may be connected in parallel. There are two input signals, **A** and **B**, and one output signal **X**

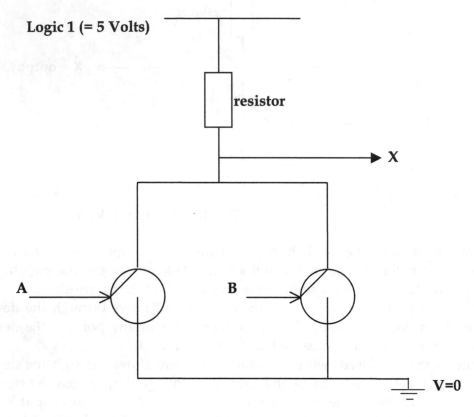

With two such switches connected in parallel, as shown in the above diagram, if either of **A** or **B** is a logic 1, then the output **X** will be pulled down to ground (logic 0). The output **X** will only be at logic 1 if both switches **A** and **B** are open (A=0 and B=0). Thus we have the following table for the operation of the device:

inputs		output
A	B	X
0	0	1
0	1	0
1	0	0
1	1	0

This is the truth table for the NOR (not or) gate, which is represented by this symbol:

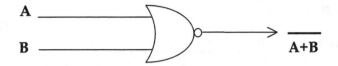

The OR Gate

In order to build an OR gate, the output of the NOR gate must be inverted

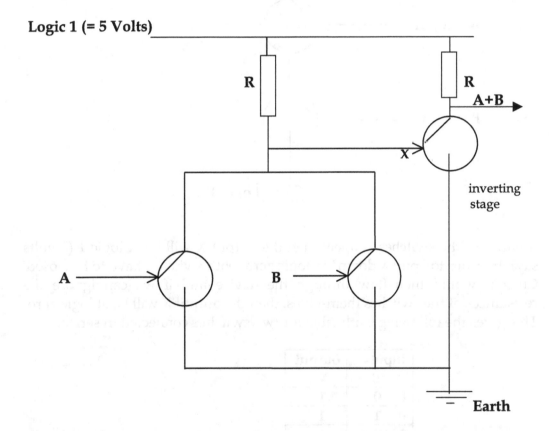

We can see that there are two "gate delays" for this logic diagram for the OR gate rather than just one for the NOR gate and therefore **the OR gate is slower than the NOR gate**.

Two switches in series

The following diagram shows how two switches, used as inverters, can be connected in series.

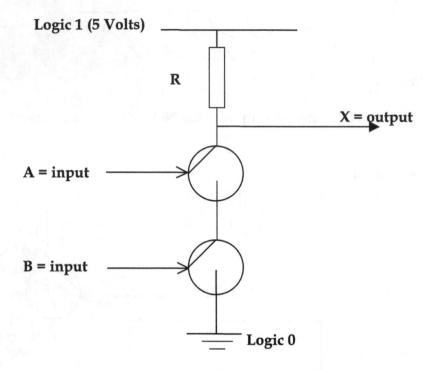

If either of the switches is open, then the output X will be at logic 1 (5 volts say). In order to "pull X down" to logic zero, both switches have to be closed. Current would then flow through the device and if we can ignore the resistances of the switches themselves, then the output X will be at logic zero. This gives the following truth table for two switches connected in series.

inputs A B	output X
0 0	1
0 1	1
1 0	1
1 1	0

The truth table for a **NOT AND = NAND** gate.

The symbol for the **NAND** gate is shown here:

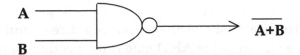

If we wanted to construct an AND gate in this way, we would need to invert the output of the NAND gate and we would need an extra inverting stage to invert $\overline{A+B}$ into **A+B**.

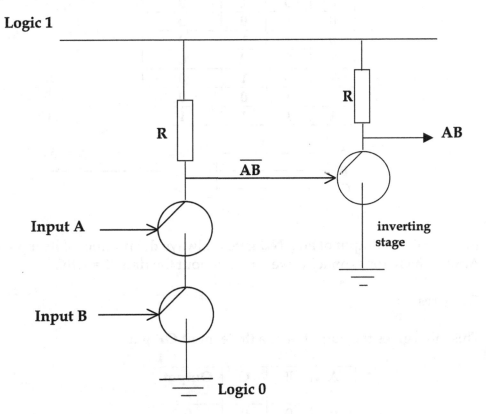

This logic circuit has twice the "gate delay" of the NAND circuit so that the AND gate is slower than the NAND gate.

==================================

Products and Sums

AND gates

The following truth table shows the output of a three input **AND** gate.
We note that the output of the **AND** gate is the product of the inputs.

A	B	C	Output X
0	0	0	0
0	0	1	0
0	1	0	0
0	1	1	0
1	0	0	0
1	0	1	0
1	1	0	0
1	1	1	1

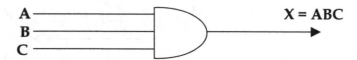

$X = ABC$

In general, the output of an **AND** gate is always the product of its inputs
A.B.C which, for simplicity, we write without the dots **X = ABC** .

OR gates

This table gives the output **X** of a three input **OR** gate.

A	B	C	Output X
0	0	0	0
0	0	1	1
0	1	0	1
0	1	1	1
1	0	0	1
1	0	1	1
1	1	0	1
1	1	1	1

In Boolean Algebra there are only two values, 0 and 1. The arithmetic is just
the same as for ordinary numbers provided that we remember that in George
Boole's algebra, **1 + 1 =1.**

It is not difficult to show that all the usual rules for arithmetic hold for
Boolean expressions if we use **1 + 1 = 1** where needed.

The output of the **OR** gate, shown in the above table can then be written in
Boolean algebra as the sum of the inputs A, B and C.

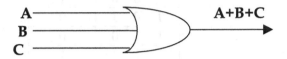

The output of an **OR** gate is the Boolean sum of its inputs.

===========================

Making Logic Circuits

Using AND, OR and NOT gates, plus a little Boolean Algebra, it is a simple
matter to produce logic circuits for any logic function we wish to design.

Example 1

Given three inputs A B and C, design a logic circuit for an even parity check
bit .

Solution

We have three logic inputs **A,B** and **C** which have values 0 or 1. We must
design a circuit with one output **P** that ensures that the total number of one
bits among **A, B, C** and **P** is even.

There are $2^3 = 8$ different combinations for the values of **A, B** and **C** so, a
truth table showing all possible values of **A, B** and **C** that can be input would
have $2^3 = 8$ rows. We need to construct a circuit to produce an output **P** that
will be 1 if the number of 1 inputs is odd and 0 if the number of 1 inputs is
even. The output **P** will then be an even parity check bit for the inputs **A, B**
and **C**. It ensures that the total number of 1 bits in **A, B, C** and **P** is even.

The truth table for the parity check bit.

Inputs A B C	Output P
0 0 0	0
0 0 1	1
0 1 0	1
0 1 1	0
1 0 0	1
1 0 1	0
1 1 0	0
1 1 1	1

The "**Boolean Sum of Products**" expression for the output **P** is

$$P = \overline{A}\,\overline{B}\,C + \overline{A}\,B\,\overline{C} + A\,\overline{B}\,\overline{C} + A\,B\,C$$

Each term in this expression generates a 1 bit for a particular row in the truth table.

P=1 in the second row and this is generated by the term $\overline{A}\,\overline{B}\,C$ when the inputs are **0 0 1.** For any other inputs this term is zero.

P=1 in the 3rd row and this is generated by the term $\overline{A}\,B\,\overline{C}$ when the inputs are **0 1 0** and for any other inputs this term is zero.

P=1 in the 5th row and this is generated by the term $A\,\overline{B}\,\overline{C}$ when the inputs are **1 0 0** and for any other inputs this term is zero.

P=1 in the last row and is generated by the product term **A B C** when the inputs are **1 1 1** and otherwise, this term is zero.

For other rows in the table the values of all the product terms are zero.

The Boolean Sum of Products expression thus gives the correct value for P in all the rows of the truth table.

Whatever the values of the inputs A, B and C, this Boolean expression gives the correct value for P.

Remember, the output of an AND gate is the product of its inputs and the output of an OR gate is the Boolean sum of the inputs.

The logic circuit we need can be constructed from four AND gates, each with three inputs

$$P = \overline{A}\,\overline{B}\,C + \overline{A}\,B\,\overline{C} + A\,\overline{B}\,\overline{C} + A\,B\,C$$

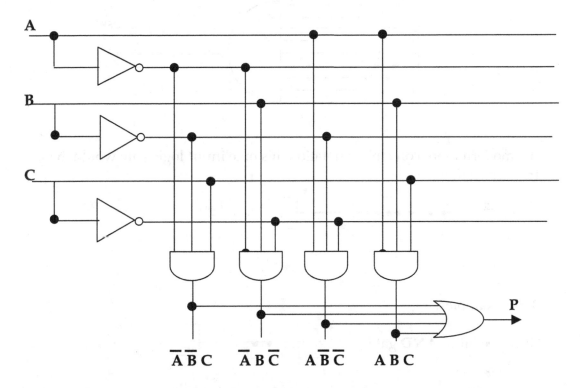

$$\overline{A}\,\overline{B}\,C \qquad \overline{A}\,B\,\overline{C} \qquad A\,\overline{B}\,\overline{C} \qquad A\,B\,C$$

The outputs of the **AND** gates are the product terms. The output of the final **OR** gate is the sum of these product terms giving **P**

A Simplified Notation

Logic diagrams can become very clumsy when the number of inputs to logic gates increases for example, a nine input and gate would have to be drawn drawn like this:

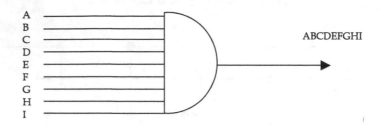

In a modern more convenient notation this nine input logic gate would be drawn as follows:

Other examples:

The three input **AND** gate

The four input **OR** gate

The number of blobs on the input wire indicates how many separate inputs we really have. This convention leads to much simpler diagrams.

Using these conventions the logic diagram for our three input parity generator is now :

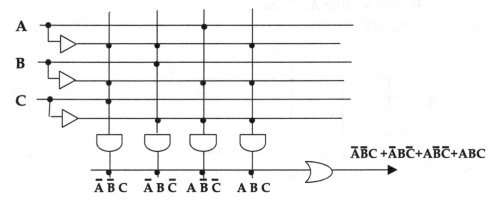

$$\overline{A}\overline{B}C + \overline{A}B\overline{C} + A\overline{B}\overline{C} + ABC$$

$\overline{A}\,\overline{B}\,C \qquad \overline{A}\,B\,\overline{C} \qquad A\,\overline{B}\,\overline{C} \qquad A\,B\,C$

Example 2
Design a logic circuit with three inputs that tells us if the binary input number is prime or not.

Solution
In this example, the three bit input represents a number written in binary form. The three bit binary input therefore represents a number that could have any value from 0 to 7.
The truth table for this example:

ABC	Decimal value	Output P	Product terms
000	0	0	
001	1	0	
010	2	1	$\overline{A}B\overline{C}$
011	3	1	$\overline{A}BC$
100	4	0	
101	5	1	$A\overline{B}C$
110	6	0	
111	7	1	ABC

The Boolean sum of products needed to generate prime number indicator is:

$$P=\overline{A}\overline{B}\overline{C}+\overline{A}B\overline{C}+A\overline{B}C+ABC$$

Giving the following logic circuit:

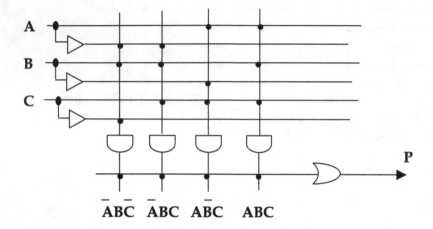

Exercises

Draw logic circuits for 3 inputs A, B and C, with one output P such that

1. P=1 if the binary input has more one bits than 0 bits

2. P=1 if the binary input represents a number divisible by three

3. P=1 if the binary input is between 2 and 6

(solutions at end of chapter)

========================

There is nothing to stop us having more than one output from one logic circuit.

Example 3

Design a logic circuit that has four inputs A, B, C and D with two outputs P and Q. The output P=1 if the binary input is divisible by 4. The output Q=1 if the binary input is divisible by 6.

Solution

The truth table for the circuit is as follows:

ABCD	P	Product for P	Q	Product for Q
0000	1	$\overline{A}\,\overline{B}\,\overline{C}\,\overline{D}$	1	$\overline{A}\,\overline{B}\,\overline{C}\,\overline{D}$
0001	0		0	
0010	0		0	
0011	0		0	
0100	1	$\overline{A}B\overline{C}\,\overline{D}$	0	
0101	0		0	
0110	0		1	$\overline{A}BC\overline{D}$
0111	0		0	
1000	1	$A\overline{B}\,\overline{C}\,\overline{D}$	0	
1001	0		0	
1010	0		0	
1011	0		0	
1100	1	$AB\overline{C}\,\overline{D}$	1	$AB\overline{C}\,\overline{D}$
1101	0		0	
1110	0		0	
1111	0		0	

The Boolean sum of products expression for **P** is

$$P=\overline{A}\,\overline{B}\,\overline{C}\,\overline{D}+\overline{A}B\overline{C}\,\overline{D}+A\overline{B}\,\overline{C}\,\overline{D}+AB\overline{C}\,\overline{D}$$

The Boolean sum of products expression for **Q** is

$$Q=\overline{A}\,\overline{B}\,\overline{C}\,\overline{D}+\overline{A}BC\overline{D}+AB\overline{C}\,\overline{D}$$

giving the following logic circuit:

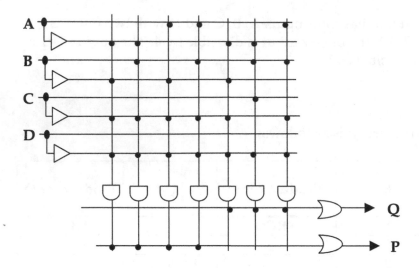

If P=1 then the input ABCD is divisible by four.
If Q=1 then the input ABCD is divisible by six.

==

Programmable Logic Devices (PLDs)

The circuit we have just designed has four inputs, **A,B,C and D** with two outputs **P and Q**. The part of the circuit that feeds into the row of **AND** gates is called **the AND array.** The part of the circuit that feeds into the **OR** gates is called **the OR array.** The function of the circuit is determined by the connections in the **AND array** and the connections in the **OR** array.

Example 4 . Using the above design, draw a logic circuit that has **3** inputs **A, B** and **C** inputs and **2** outputs P and **Q,** so that **P=1** if the binary input is prime and **Q=1** if the binary input is divisible by three.

Solution

Truth tables for P and Q. P=1 when **ABC** is prime. **Q=1** when **ABC** is divisible by **3.**

ABC	P (prime)		ABC	Q	
000	0		000	1	
001	0		001	0	
010	1		010	0	
011	1		011	1	
100	0		100	0	
101	1		101	0	
110	0		110	1	
111	1		111	0	

The Boolean sum of products terms are therefore:

$$P=\bar{A}B\bar{C}+\bar{A}BC+A\bar{B}C+ABC \qquad Q=\bar{A}\bar{B}\bar{C}+\bar{A}BC+AB\bar{C}$$

Giving the following logic circuit

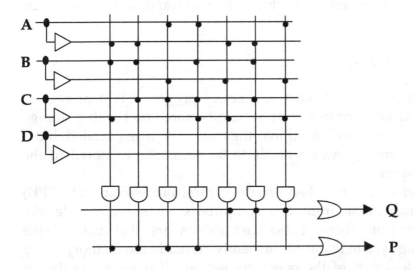

We note that the **D** input line is not use and is therefore redundant. However, these arrays can be mass manufactured and this makes it cheaper to use standard sizes of **AND array and OR array.** Inputs that are not needed do not connect through to the output lines. These chips are produced with all of the connections present formed by fusible links. Programming the chip then consists of blowing the connections that are not needed.

Since there are just two arrays, the **AND array** and the **OR array,** in these programmable devices, there are three possibilities that may be chosen in their manufacture:

[i]   The **AND array** programmable with the **OR array** fixed.
        These chips are called **programmable array logic chips (PAL)**

[ii]   The **AND array** fixed, (holding all possible combinations the inputs,
        so the that **AND array** acts as an address selector)
        the **OR array** is programmable.
        This kind of chip is a programmable read only memory (**PROM**)

[iii]   Both the **AND array** and the **OR array** are programmable.
        This kind of chip is called a programmable logic array (**PLA**) chip.

Modern programmable chips are electrically erasable so that they can be re-programmed by applying a voltage to a certain programming pin. These Programmable Electrically Erasable Logic devices (**PEELs**) avoid the wastage incurred if one-time programmable chips are used and later need to be updated.

Read Only Memory (ROM)

The operating system of a computer is a set of programs that drives the computer system. Most of the operating system is stored on backing storage (e.g. external disk drives) and loaded into the computer when needed, but a small part of the operating system needs to be permanently stored in the memory of the computer.

When a computer is first switched on, the central processing unit (**CPU**) starts to fetch instructions from the computer's memory, decode the instructions and execute them. These instructions are that part of the computers operating system that is permanently stored in the memory. They make up the **resident** part of the operating system. If it were possible to change these instructions, then the computer would not operate in the same way, the next time it was switched on. These instructions can be read by the **CPU,** but the user cannot wipe them out or change them. The resident part of the operating system is stored in **Read Only Memory.**

To illustrate, read only memory, we will see how bit patterns are permentantly stored in a **ROM** chip that has data stored in just eight locations and each location will store one byte of data (one byte=8 bits). In order to read a location in memory, we must supply the address of the location we want to read. Addresses of memory locations are just bit patterns. In order to "address" any one of eight different locations we need just three bits. In our 8 byte ROM chip, the **AND array,** acts as an address decoder that selects the location that we want to read.

The Address Decoder

In this illustration we suppose that address **ABC = 100** is the address that needs to be read so that inut **A** carries a logic **1** with **B and C** carrying logic 0.

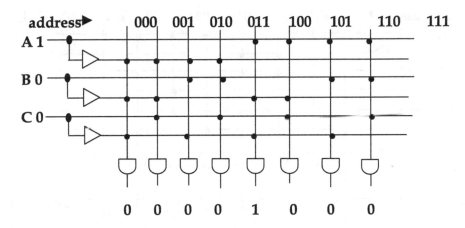

If we look at the line headed **100,** we see that each of the three inputs to the **AND** gate at the bottom will be **1.** Therefore, the output of this **AND** gate will be **1.** All of the other **AND** gates will output zero.

Thus the address **100** is the address that is selected.

The contents of the memory

address	Data stored	Letter
000	0100 0001	A
001	0100 0010	B
010	0100 0011	C
011	0100 0100	D
100	0100 0101	E
101	0100 0110	F
110	0100 0111	G
111	0100 1000	H

The data stored are the ASCII codes for the first eight capital letters.

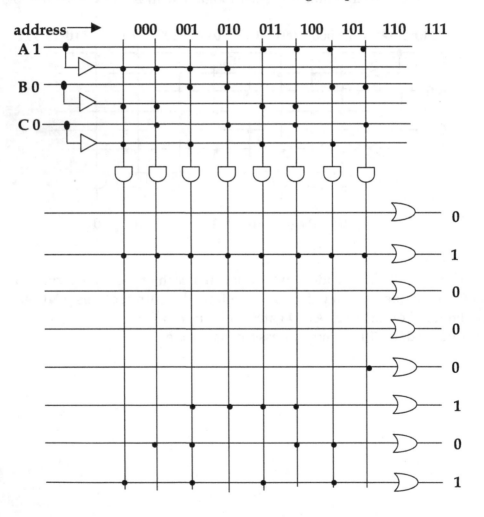

The **100** address line is selected (set = **1**). This logic **1** is passed through the **OR** gates to give the data byte **0100 0101** which represents the letter **E**.

Circuits that do Arithmetic

Computers are rarely used to compute nowadays. They are more frequently used as a library, a post office, a music player, television or chat room. However, within the central processing unit, there are circuits that do perform the original purpose of a computer. That is to do arithmetic.

Example of a binary addition:

```
    1   1   1  ←——————— carry bits
  0   1   0   1
+ 0   1   1   1
  1   1   0   0
```

This binary addition has four stages. Three of these stages produce a **carry** bit to the next stage. The next circuit performs one stage of the binary addition. It is called a **full adder,** that adds two binary digits and a "carry in" bit to produce a **sum bit** and a **carry out.**
(If there was no carry in bit then it would be called a half adder)

The truth table for the **full adder** :

Bits to be added A B	Carry In C	Carry Out Cout	sum
0 0	0	0	0
0 1	0	0	1
1 0	0	0	1
1 1	0	1	0
0 0	1	0	1
0 1	1	1	0
1 0	1	1	0
1 1	1	1	1

Let the bits to be added **A and B** and let the **carry in** bit be denoted by **C**.
Then with then the Boolean sum of products for the **sum** output is

$$\text{Sum} = \overline{A}\,\overline{B}C + \overline{A}B\overline{C} + A\overline{B}\,\overline{C} + ABC$$

The **carry out** is determined by

$$\text{Cout} = \overline{A}BC + A\overline{B}C + AB\overline{C} + ABC$$

Thus the logic circuit for the **full adder** is:

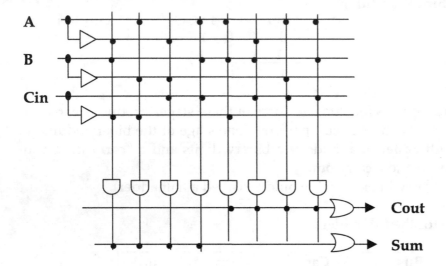

If we represent this full adder circuit by a single block with three inputs **A,B**
and **Carry in,** and two outputs, **Carry out** and **Sum:**

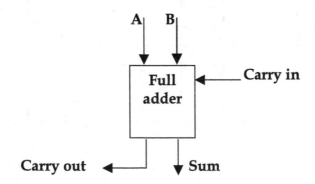

It is easy now, to see how eight **full adders** can be linked together to build a parallel adder.

An **8 bit parallel adder** is illustrated below:

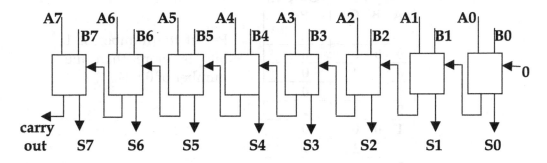

The carry in at the first stage is set to zero.

The following 8 bit binary addition demonstrates the operation of the 8 bit parallel adder:

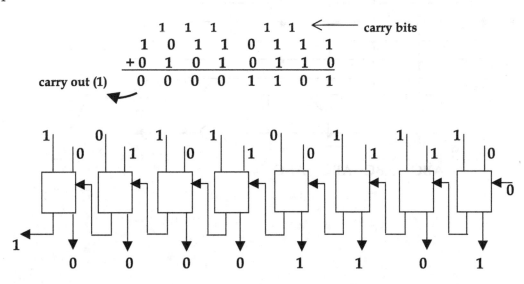

In the next chapter, we will see how the parallel adder can be improved so that it can also perform subtractions but before that can be done, it is necessary to see how negative numbers are represented in the computer system.

====================================

Solutions to Exercises

1.

Inputs A B C	Output P
0 0 0	0
0 0 1	0
0 1 0	0
0 1 1	1
1 0 0	0
1 0 1	1
1 1 0	1
1 1 1	1

P=1 if the number of 1 bits is more than the number of 0 bits.

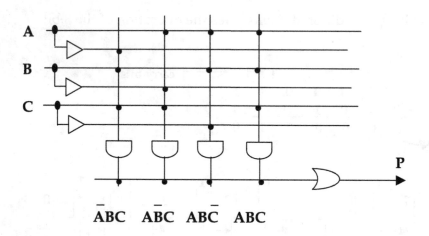

$\overline{A}BC$ $A\overline{B}C$ $AB\overline{C}$ ABC

2.

Inputs A B C	Output P
0 0 0	1
0 0 1	0
0 1 0	0
0 1 1	1
1 0 0	0
1 0 1	0
1 1 0	1
1 1 1	0

P=1 if the binary input is divisible by 3

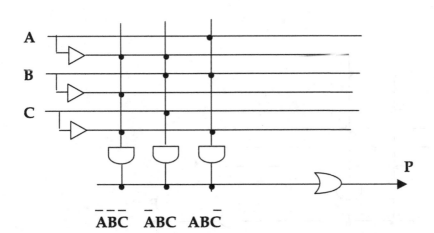

$\overline{A}\overline{B}\overline{C}$ $\overline{A}BC$ $AB\overline{C}$

3.

Inputs A B C	Output P
0 0 0	0
0 0 1	0
0 1 0	0
0 1 1	1
1 0 0	1
1 0 1	1
1 1 0	0
1 1 1	0

P=1 if the binary input is between 2 and 6

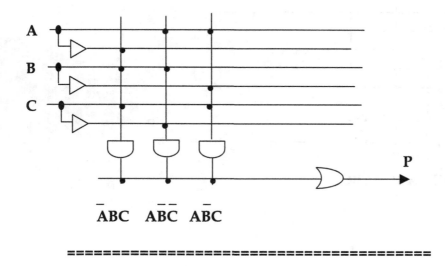

$\overline{A}BC$ $A\overline{B}\overline{C}$ $A\overline{B}C$

===

CHAPTER 8

Complements and Clocks

Konrad Zuse (1910-1995) graduated from the Technical University in Berlin-Charlottenburg in 1935 and when he was finishing his degree in civil engineering, computing was limited to mechanical calculators. Punched card input devices were cutting edge technology. In his engineering studies Zuse had to perform many routine calculations using these hand operated devices and he dreamed of the day when calculations could be done by machine. In 1936 he started work on designing a series of binary stored program computing machines. His first machine, the Z1 was an electronic binary computer that used relays and read instructions from punched paper tape and by 1941 he had upgraded the Z1 to the Z2 and the Z3, the world's first electronic stored program machine using floating point binary arithmetic. Sadly, Zuse's workshops where he built his machines, were destroyed by bombs at the end of the second world war but he survived and fled to Switzerland in 1945 with parts of his latest Z3 machine where he was able to build the Z4. In 1950 a Z4 was installed at the Federal Polytechnical Institute in Zurich. Zuse can be credited, in 1948, with the development of the first high level computing language that he called "plankalkul" but there were no compilers or interpreters available to implement the language. He founded a computing company called Zuse KG which built 250 computers and which was taken over by Siemens in 1967. Konrad Zuse, the greatest of the early computer pioneers, died in Germany in 1995, aged 85.

The memory of a modern home computer system stores the executable programs that drive the various devices that make up the system as well as the programs that the user of the system wishes to run. Computer programs can be divided into three broad categories: systems software, applications software and user written software. The programs that drive the hardware devices, disk drivers, screen drivers, printer drivers etc. form part of the operating system that is necessary in order to be able to use the computer. Applications software is written by software houses for sale in the market place and includes word-processing software, spreadsheets, database applications, accounts packages, computer graphics and design packages.

Ten years ago 90% of computer use involved data processing and little time was spent doing actual calculations. If a survey of computer usage were to be carried out today, a sizable chunk of computer time would be found to involve surfing the inter-net and checking e-mails. Why then do we call these machines computers if so little of the time is taken up by computing? The answer to this question, like the destruction of the Z1, can also be pinned on the second world war. If a new tank or gun was designed for the war effort, then details of the muzzle velocity and angles of elevation would be given to the ladies in the computing rooms (the men were all fighting the war), who, using Newton's formulae for projectiles, would be able to calculate the range of the shell. The first computers were those good ladies with their hand-operated calculating machines.

We have seen in chapter 7 how logic circuits can be built that perform binary additions. In this chapter we investigate how these circuits can be adapted to perform subtractions. Other tasks, such as multiplication and division and floating point operations for fractions are beyond the scope of this book.

But first, we need to see how negative numbers can be stored in the memory of an electronic computer.

Storing Negative Numbers

In ordinary arithmetic, we distinguish between positive and negative numbers using plus and minus signs however, using the plus and minus signs requires different algorithms for different kinds of calculation but it is possible to simplify some of the rules:

$$
\begin{array}{r}
+345 \\
+ \quad +456 \\
\hline
801
\end{array}
$$
is a straight forward addition.

But
$$
\begin{array}{r}
+345 \\
- \quad +456 \\
\hline
\end{array}
$$

would have to be done using rule such as "smaller from the larger with the

sign of the larger" as we might use in school.

It is inconvenient to have to use explicit signs for the storage of numbers in the memory of a computer but fortunately there are ways of avoiding having to use plus and minus symbols.

We can also avoid having different methods for additions and subtractions. Instead of subtracting we add the negative.

Nines Complements in Base ten.

We have to remember to use a given fixed length storage for numbers. To illustrate we will use four decimal numerals for both positive and negative numbers.

The number **345** would then be represented as **0 3 4 5**.

The number **456** would be represented as **0 4 5 6**.

The negative of a number is found by subtracting each digit from **9** and this gives the nines complement of the number. Thus the negative of **0 4 5 6** would be represented by **9 5 4 3**.

In the nines complements representation, negative numbers always start with a 9 digit and positive numbers start with a 0 digit.

The subtraction **345 – 456** is then carried out as **345 + (-456)** which appears in nines complements as

$$
\begin{array}{r}
0\,3\,4\,5 \\
+\quad 9\,5\,4\,3 \\
\hline
9\,8\,8\,8
\end{array}
$$

The result **9 8 8 8** represents a negative number and we find the ordinary decimal representation by subtracting each digit from 9 and writing in the minus sign giving **– 0 1 1 1**.

Example 2:
Perform the subtraction **45 – 76** using four decimal digits and nines complements for representation of negative numbers.

Solution:

45 is represented by $\qquad$ 0 0 4 5
76 is represented by $\qquad$ 0 0 7 6
-76 is represented by $\qquad$ 9 9 2 3
The subtraction **45 – 76** then appears as

$$
\begin{array}{r}
0\,0\,4\,5 \\
+\quad 9\,9\,2\,3 \\
\hline
9\,9\,6\,8
\end{array}
$$

9 9 6 8 represents the negative number **- 0 0 3 1**

Example 3 (using wrap around carry):
Perform **56 – 23** using nines complements.

Calculation:

56 is represented by	0 0 5 6
23 is represented by	0 0 2 3
-23 is represented by	9 9 7 6

$$\begin{array}{r} 0\,0\,5\,6 \\ +\quad 9\,9\,7\,6 \\ \hline 1\,0\,0\,3\,2 \end{array}$$

In this case we have a carry out of the last digit position and as we are only using four digits in our representation this "carry one" will be lost. But hold on! The answer we have is 32 and it should be 33! If there is a carry from the last (leftmost) digit position in nines complements then we have to add the carry to our result (this is called "wrap around carry").

$$\begin{array}{r} 0\,0\,3\,2 \\ +\qquad 1 \\ \hline 0\,0\,3\,3 \text{ (correct answer)} \end{array}$$

Example 4 (using wrap around carry)
Perform the subtraction 369 – 135 using four digit nines complements.

Solution:

Decimal		Nines complements

$$\begin{array}{cccc} & 3\,6\,9 & & 0\,3\,6\,9 \\ - & 1\,3\,5 & + & 9\,8\,6\,4 \\ \hline & 2\,3\,4 & \text{wrap around} & 0\,2\,3\,3 \\ & & \text{carry} & +1 \\ \hline & & & 0\,2\,3\,4 \end{array}$$

Nines complements with wrap around carry always gives the correct answer.

Example 5

Perform this subtraction (-2) – (4) using 6 digit nines complements

Solution

Decimal	Nines complements (6 digits)
- 2	9 9 9 9 9 7
<u>- 4</u>	<u>9 9 9 9 9 5</u>
<u>- 6</u>	(1) 9 9 9 9 9 2
	+ 1
	9 9 9 9 9 3

The nines complements answer is   9 9 9 9 9 3
representing the decimal number   – 0 0 0 0 0 6.

Example 6.

Perform the subtraction +6 – 6 in nines complements using a six-digit nines complements representation.

Solution

Decimal	Nines complements (6 digits)
6	0 0 0 0 0 6
<u>- 6</u>	<u>9 9 9 9 9 3</u>
<u>0</u>	<u>9 9 9 9 9 9</u>

The nines complements answer 999999 represents the decimal number - 000000

Exercise 1   Translate this ten digit, nines complements sum into ordinary arithmetic.

$$9 9 9 9 9 9 9 9 7 9$$
$$9 9 9 9 9 9 9 9 9 8$$
$$9 9 9 9 9 9 9 9 7 7$$
$$\underline{\hspace{3cm} + 1}$$
$$9 9 9 9 9 9 9 9 7 8$$

Clock Arithmetic and Nines Complements

Nines complements representation of positive and negative numbers can be illustrated using a "nines complements clock". The following diagram shows the nines complements clock for a two-digit representation.

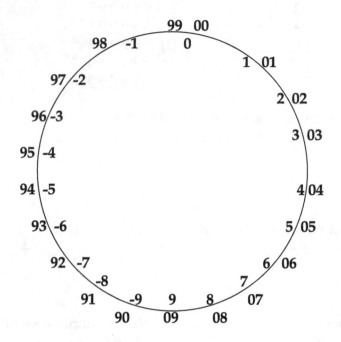

The signed integers are on the inside of the clock and the 2 digit nines complements numbers are on the outside.
The numbers 0 has two representations: 00 and 99 representing +0 and –0.

Example 7:

In two digit nines complements, the subtraction   5   is represented as   05
                                                  - 3   + 96

In nines complements, the - 3 is represented as **(99 - 3)**
and this gives the subtraction **5 – 3** in the form **05 + (99 – 3) = 99 + 2** . Both **99** and **00** represent zero so we step two places round the clock from the zero position to **02**.

Example 8:

The subtraction 3   in 2 digit nines complements is   0 3
             - 5   +9 4

or, 03 + (99 – 5) = 99 – 2 so we step back two places from the zero position to –2.

A problem arises when there is a carry out from the last digit position:

Example 9: Representation of (-3) – (-5)

      Decimal representation   Nines Complements

             -3   9 6
          +   -5   9 4
             -8   **(1)** 9 0

We hope for **99 – 8 = 91** which is the nines complements representation of **-8** but the carry out from the last position means that the nines complements sum is **98– 8**. The remedy is to add this carry to the nines complements answer. The "wrap around carry" converts the **98 – 8** to **99 – 8** as required.

      Decimal representation   Nines Complements

             -3   9 6
             -5   9 4
             -8   ┌ **(1)**   9 0
                       └──→   +1
                                 9 1

Radix – 1 subtraction

This method of dealing with negative numbers and subtraction works in any number base if we remember to add the wrap around carry. We will examine **7's complements** (for base 8), **15's complements** (for base 16), **ones complements** (for base 2) and **nines complements** (for base 10).

7's complements in Octal

The following example is a subtraction using sevens complements in base 8 (octal).

Example 10:

In base 8 "radix-1" complements results in sevens complements arithmetic. In this arithmetic, negative numbers are represented by subtracting each octal digit from seven. In this "sevens complements", negative numbers always start with a leading 7 digit. This example uses four octal digits.

Octal subtraction	sevens complements for base 8

$$
\begin{array}{r}
0056 \\
-0234 \\
\hline
\end{array}
$$

$$
\begin{array}{r}
0056 \\
7543 \text{ (add the negative)} \\
\hline
7621 \\
\end{array}
$$

$$\text{convert to octal: answer} = \quad -0156$$

The leading 7 tells us that the answer is negative. Converting to octal using sevens complements again gives the octal answer - 0 1 5 6.

Example 11:

The subtraction - 2 – 3 in base 8 sevens complements using four digits with wrap around carry;

Octal sum	sevens complements for base 8

$$
\begin{array}{r}
-0002 \\
-0003 \\
\hline
-0005 \\
\end{array}
$$

$$
\begin{array}{r}
7775 \\
7774 \\
\hline
(1)\ 7771 \\
\rightarrow +1 \\
\hline
7772 \\
\end{array}
$$

The leading 7 tells us that the answer is negative so we convert to a negative octal number by subtracting each digit from 7.   **Ans   - 0 0 0 5**

Fifteens complements in Hexadecimal

In base 16 (Hex), "radix – 1" results in fifteens complements arithmetic. Using four Hex digits, the negative numbers are;

-1	F F F E
-2	F F F D
-3	F F F C
-4	F F F B
-5	F F F A
-6	F F F 9
-7	F F F 8
-8	F F F 7
-9	F F F 6
-10	F F F 5
-11	F F F 4
-12	F F F 3
-13	F F F 2
-14	F F F 1
-15	F F F 0

Zero has the two representations   -0   F F F F
              and   +0   0 0 0 0

Example 12

The sum –2 – 3 in Hex fifteens complements using four hex digits and wrap around carry, would appear as

Decimal	fifteens complements
-2	FFFD
-3	FFFC
-5	(1) FFF9
	+1
	FFFA

The leading **F** indicates that the answer is negative so we convert from fifteens complements to Hex by subtracting each digit from **F(=15) to** get the Hex answer **–0005**

Subtracting using Ones complements in base 2

 To find the negative of a binary number, each binary digit, or bit, is subtracted from 1. This gives the same result as if the bits were flipped, changing 1 to 0 and 0 to 1.
Using 8 bits, the binary representation of –2 is – 0000 0010
In ones complements, -2 is represented by   1111 1101

 Similarly, the binary representation of –3 is   – 0000 0011
In ones complements, -3 is represented by   1111 1100

Again, as in any radix-1 representation, a wrap around carry must be implemented if necessary, thus the calculation **–2 –3** would appear as:

Decimal	ones complements	
-2	1111 1101	
-3	1111 1100	(add bits)
-5	(1) 1111 1001	
	+1	
	1111 1010	

the leading **1** bit indicates that the answer is negative. To find the signed binary answer we perform a ones complement and add the negative sign:

Binary answer **– 0000 0101**

========================

Notice that there is a direct translation between fifteens complements and ones complements (here, using four hex digits and 16 bits):

Decimal	ones complements	fifteens complements
-2	1111 1111 1111 1101	FFFD
-3	1111 1111 1111 1100	FFFC
-5	1111 1111 1111 1010	FFFA

In the early days of computing, ones complements was used to represent negative numbers and so perform subtractions without the need of hardware for both a parallel adder and a separate parallel subtractor.

Performing a ones complement on a bit pattern is equivalent to flipping all the bits. It is easy to flip the bits of a bit pattern using **XOR** gates:

Control	b	XOR	
bit	0 0	0	if the control bit is zero then
	0 1	1	bit **b** is unchanged.
	1 0	1	if the control bit is **1** then
	1 1	0	bit **b** is flipped (inverted)

Thus, a 4 bit number could be 'ones complemented', i.e. flipped, when required using a control line with an **XOR** gate for each bit:

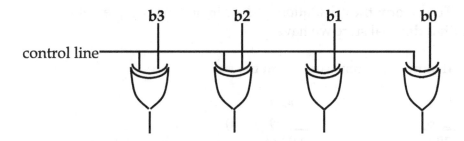

If the control bit is **1** then the four bit pattern **b3 b2 b1 b0** is inverted.
If the control bit is **0** then the bit pattern is unchanged.

============================

Exercise 2   Translate this eight digit sevens complements calculation for octal numbers into (i) binary ones complements (ii) Hex fifteens complements and (iii) decimal arithmetic

$$
\begin{array}{r}
7\,7\,7\,7\,7\,7\,5\,3 \\
7\,7\,7\,7\,7\,7\,7\,6 \\
7\,7\,7\,7\,7\,7\,5\,1 \\
\hline
+1 \\
\hline
7\,7\,7\,7\,7\,7\,5\,2
\end{array}
$$

The Radix Complement

Tens Complements

The tens complements of a decimal number is simply, one more than the nines complements. Using a four digit decimal store, the number **+246** would be stored as **0246**. The negative of this number in tens complements representation is then found as follows:

number	0246
nines complement	9753
add one	+1
tens complement	9754

Example 13: To perform the calculation **543 – 246,** in tens complements using a four digit decimal store we have;

decimal sum	tens complements sum
543	0543
-246	9754 (add)
297	(1) 0297 (ignore the carry out)

Example 14: Perform the calculation **–2 – 3** using four digit tens complements

Solution (we also show here the nines complements for comparison):

decimal	four digits	nines comp.	tens comp.
– 2	– 0002	9997	9998
– 3	– 0003	9996	9997
– 5	– 0005	(1) 9993	(1) 9995
		add carry	ignore carry
		9994	9995
to find the signed		nines comp	tens comp
decimal equivalent		– 0005	-0005

The tens complement result is therefore **9995**.

The advantage of tens complements over nines complements is; if there is a carry out of the most significant digit position, in nines complements we have to add the carry out to the result, whereas, in tens complements we can ignore the carry out.

Also, in nines complements there are two representations for zero, here, for example +0 is represented by **0000** and –0 is represented by **9999.**

In four digit tens complements, zero is represented by **0000,** which is clearly more satisfactory.

The Tens Complements Clock

The two digit tens complements clock shows the relationship between tens complements and clock arithmetic.

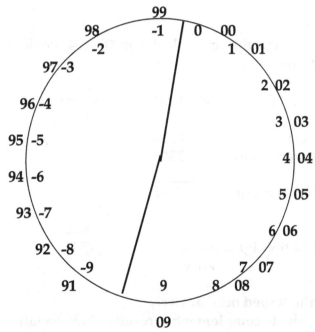

In this clock arithmetic, positive numbers start with a **0** digit and negative numbers start with a **9** digit.

It is easy to see how this works in theory, as the negative numbers are represented by **(100-1), (100-2), (100-3), . . .** etc. Thus, when we perform say –2 –4 on this clock

$$
\begin{array}{ll}
-2 & \text{represented by } 100-2 \\
\underline{-4} & \text{represented by } 100-4 \\
-6 & \quad\quad (100) + \overline{100-6} \text{ (and ignore the extra hundred)}
\end{array}
$$

Eights Complements in Octal Arithmetic

Eights complements is the 'radix complements' method of representing negative numbers in Octal (base eight).
The eights complement of an octal number is therefore the sevens complement plus one.

Example 15: Using a four digit octal store, show how the octal number –234 would be represented.
Solution

Octal number	0234
sevens complement	7543
	+1
eights complement	7544

Example 16: show how the octal calculation **56 –234** would be performed in eights complements.
Solution

First we find the negative of **234** in eights complements;

Octal	0234
Sevens comp.	7543
	+1
eights comp.	7544

	0056
add the eights comp.	7544
answer	7622

to find the signed octal answer,
take the eights complement: result   –0156 (octal)

Example 17: Perform the octal calculation –2 –3 in four digit eights complements

Solution

First we convert the negative numbers into 8's comlements:

	–2		–0002		–3		–0003
7's comp			7775				7774
			+1				+1
8's comp			7776				7775

calculation		7776
+		7775
	(1)	7773

Ignoring the carry out, the eights complements answer is **7773**
The result is negative, so to find the signed octal result, we take the eights complement of the answer

	7773
7's comp	0004
	+1
Answer in signed octal is	–0005

===================================

Sixteens Complements in Hexadecimal

In base sixteen, (hex), the sixteens complement of a hexadecimal number is the fifteens complement plus one. To find the fifteens complement we subtract each hex digit from fifteen.

Using four hex digits, the negative numbers are represented as:

-0001	FFFF
-0002	FFFE
-0003	FFFD
-0004	FFFC
-0005	FFFB
-0006	FFFA
-0007	FFF9 etc.

This is the representation you will usually find on a screen dump of the memory of a computer system.

Example 18: Perform the calculation –2 –3 in sixteens complements:

Solution

$$
\begin{array}{r}
\text{FFFE} \\
+\quad \underline{\text{FFFD}} \\
\end{array}
$$

(1)   FFFB (the carry is ignored)

=============================

Twos Complements in Binary Arithmetic

The radix complement in base 2 (binary) is the twos complements representation which is used to represent negative integers in modern computes. The twos complement of a bit pattern can be found by first forming the ones complement (by flipping all the bits) and then adding 1.

Example 19: Find the twos complement of the bit pattern **1011 0000**
Solution

Bit pattern	1011 0000
Ones comp	0100 1111
	+1
	0101 0000

We can write down the twos complement of a bit pattern "at sight" if we notice the "ripple carry" that occurs when we add a **1** bit to a row of **1** bits:

$$
\begin{array}{ccccccccc}
0 & 1 & 0 & 0 & 1 & 1 & 1 & 1 \\
 & & & & & & & +1 \\
\hline
0 & 1 & 0 & 1 & 0 & 0 & 0 & 0 \\
\end{array}
$$

ripple carry to here

This gives us the rule:

  To twos complement a bit pattern, copy to the first 1 bit then flip the rest.

Example 20. The 2's complement of a bit pattern

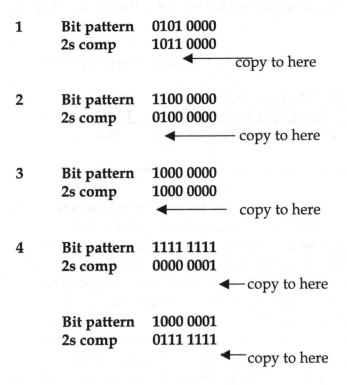

1   **Bit pattern**   **0101 0000**
    2s comp   **1011 0000**
                                ← copy to here

2   **Bit pattern**   **1100 0000**
    2s comp   **0100 0000**
                              ← copy to here

3   **Bit pattern**   **1000 0000**
    2s comp   **1000 0000**
                              ← copy to here

4   **Bit pattern**   **1111 1111**
    2s comp   **0000 0001**
                            ← copy to here

    Bit pattern   **1000 0001**
    2s comp   **0111 1111**
                            ← copy to here

As with the other radix complements we have examined, if there is a carry out of the most significant bit position (i.e. the left hand) then it is ignored.

Example 21: The 2s complements version of the calculation **–2 –3** using 8 bits:

Decimal	binary	2s complement
-2	-0000 0010	1111 1110
-3	-0000 0011	1111 1101
-5	-0000 0101	(1) 1111 1011

Exercise 3   Translate this sixteen bit twos complements calculation into (i) sixteens complements   (ii) decimal arithmetic

```
      1111 1111 1110 1100
      1111 1111 1111 1111
(1)   1111 1111 1110 1011
```

Note that there is a direct translation between sixteens complements and 2s complements. Using sixteen bit storage for numbers the above calculation would appear as:

Decimal		2s complement	sixteens complement
-2		1111 1111 1111 1110	FFFE
-3		1111 1111 1111 1101	FFFD
-5	(1)	1111 1111 1111 1011	(1) FFFB

==============================

Thus, in 2s complements representation for negative numbers we have a convenient theoretical method for representing negative numbers and hence for performing subtractions that avoids the need to explicitly store plus or minus signs. We now examine how this can be implemented in the hardware of a computer, specifically in the arithmetic Logic Unit (ALU) of the central processing unit (CPU) that drives the computer. One circuit can be used for both addition and subtraction. All we need is a control line that tells the ALU whether to perform an addition or a subtraction. If this control line (the add/subtract line) carries a 1 then a subtraction is performed using XOR gates to flip the bit pattern that represents the number to be subtracted. If the add/subtract line is 0 then the adder/subtractor circuit operates as a parallel adder.

Exercise 4

The decimal calculation 65 – 68 when translated into Hexadecimal is

0041	in sixteens	0 0 4 1
-0044	complements this reads	+F F B C
-0003		F F F D

if we complement the sixteens
complements answer an add the
minus sign we get - 0 0 0 3

Perform the following decimal calculations as above, using sixteens complements

(i)   23 - 25   (ii) – 48 + 49 (iii)   (iv) – 65 - 49

We illustrate with the logic circuit for a 4 bit parallel adder/subtractor:

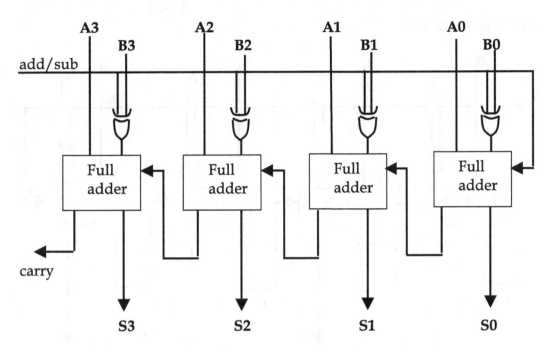

If the add/subtract line is **1** then the **XOR** gates on each **B** input inverts **B** input bit giving the **ones complement,** however, the add/subtract bit is also added to the first full adder stage via its **carry in (cin)** input and this converts the **ones complement** into a **2s complement.** Adding the **2s complement** performs the subtraction.

If the add/subtract line is **0** then the **XOR** gates pass the **B** inputs unchanged, to the full adders and a parallel addition is performed.

The following diagram shows how the subtraction   0 1 1 1
                                                 –0 1 0 1
would be dealt with:

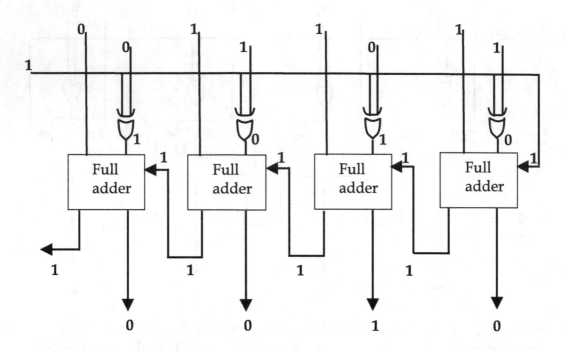

The hardware has performed the subtraction 7–5 = 2

==

Answers to Exercises

Exercise 1   - 20
                         <u> -1</u>
                          -22

Exercise 2 77777753   111 111 111 111 111 111 101 011   FFFFEB
               <u>77777776</u>   <u>111 111 111 111 111 111 111 110</u>   <u>FFFFFE</u>
                77777751   111 111 111 111 111 111 101 001   FFFFE9
               <u>   +1</u>   <u>   +1</u>   <u>   +1</u>
                77777752   111 111 111 111 111 111 101 010   FFFFEA

Each Calculation represents (-20) + (-1) = -21

Exercise 3

               FFEC   -20
              <u> FFFF</u>   <u>-1</u>
       (1)   FFEB   -21

Exercise 4

                                        sixteens
            dec   hex   comp
(i)   23   0 0 1 7   0 0 1 7
           <u>-25</u>   <u>-0 0 1 9</u>   <u>F F E 7</u>
            -2   -0 0 0 2   F F F E

(ii)   -48   -0 0 3 0   F F D 0
           <u>+49</u>   <u>+0 0 3 1</u>   <u>0 0 3 1</u>
            + 1   0 0 0 0   (1) 0 0 0 1

(iii)   - 70   - 0 0 4 6   F F B A
           <u>- 10</u>   <u>- 0 0 0 A</u>   <u>F F F 6</u>
            - 80   - 0 0 5 0   (1) F F B 0

(iv)   - 65   - 0 0 4 1   F F B F
           <u>- 49</u>   <u>- 0 0 3 1</u>   <u>F F C F</u>
            -114   - 0 0 7 2   (1) F F 8 E

CHAPTER 9

Mathematical Thinking

The composer had invited a few friends to his flat to hear a new piano piece that he had just finished writing.

When he had finished playing one of his friends said, "That was beautiful but what does it mean?" whereupon the composer sat back at the piano and played the piece again.

A friend who had not studied mathematics since 'O' level, once asked me, "How do mathematicians think?". The best way learn how to think mathematically is not to talk about maths but be like the composer who did not try to talk about his music but simply performed it. To improve your mathematical thinking you need to perform mathematics but performing mathematics is not solving tedious equations or working out boring calculations. By mathematical thinking, we do not mean recalling a formula and plugging in numbers to get an answer. On The Weakest Link Ann Robinson is often asking questions like "In mathematics, what is forty seven plus seventy four?", but this is not mathematics either. It's like asking "In English Literature, spell Mississippi".

Mathematical thinking sometimes involves unusual ways of looking at the real world and dealing with its puzzles but often has nothing to do with the reality but attempts to use pure logic solve purely abstract puzzles. Sometimes the solution to a mathematical problem might come in useful, but the mathematician would not be bothered if it did not. The concern of the mathematician would only be that the steps in the solution are logically correct. To be a mathematician you must enjoy the challenge of solving new puzzles and problems – crosswords, crossnumbers, sudoku, the Enigma in the New Scientist, the Griddler in the Sunday Telegraph. The danger is that one might become so absorbed in a puzzle that one forgets that one has to pay the mortgage and eat in order to survive!

Some people may have the view that mathematical thinking involves formulae and equations that were forgotten many years ago when they decided not to do mathematics at sixth form but suppose you are doing a crossword and you think "If I put beans for 'horsy dinner' then that will be an s but I think it should be a y, so I'll try curry". That's mathematical thinking, it has the logical structure of "If **X** then **Y** but **Y** is false therefore **not(X)**.

The examples we discuss in this chapter are meant to illustrate methods of mathematical deduction, proof, generalization and special cases to show that mathematical thinking is not necessarily dependent on algebraic formulae and complicated equations but is so often driven by simplicity, symmetry and the search for patterns.

Here, we discuss a number of puzzles, games and problems that introduce an element of mathematical thinking in their solution.

They may not seem very "mathematical" in the traditional sense but they illustrate what we mean by mathematical thinking.

The first half of the chapter poses and discusses a number of problems. The second half is on puzzles and games.

The problems that we use to illustrate mathematical thinking come from the following list and each problem is discussed afterwards in its own separate section.

Mathematical puzzles, problems and games -- a list.

[1] If you add the same number to the top and to the bottom of a fraction, does the fraction get bigger or smaller?

[2] A roll of paper of length 5 metres is wound round a central core of radius 2 cm to form a roll of radius 10 cm.
How many small rolls, on a core of radius 1 cm. can be made with outside radius 5 cm and how long would each of these rolls be?

[3] If I roll one penny round another, what angle will it turn through?

[4] How can you stop a 4 legged table wobbling on a bumpy floor?

[5] Prove that you can always find a pair of perpendicular lines to quarter any given area.

[6] How can you hit a snooker ball so that it bounces off of three sides and rolls through its starting position?

[7] What shape is a football?

[8] If you have a pair of scales, how many weights would you need to weigh all the whole number weights from 1 to 100?

[9] You have 9 balls and a pair of scales. The balls are all the same except for one which is heavier than the others.
How can you find the odd ball in the fewest weighings?
Would it matter if the odd ball was lighter?
What if you did not know if it was heavier or lighter?

[10] Can you win at Nim?

[11] Put the numbers 1..6 round a triangle so that the numbers along each side add up to the same total. In how many ways can this be done?

[12] Put the numbers 1..8 round a square so that the numbers along each side add up to the same total.

[13] How many 3x3 magic squares are there?

[14]

A college student had ran into debt and needed more money in order to survive to the end of term. He wanted to send a note home to his father for some extra fund to pay his bills and fees, but he knew that there would be a family upheaval if his mother found out by how much he had overspent his allowance. His Father loved mathematical puzzles and he had to use the post as there was no telephone and so he sent his Father the following note:

"Dear Father,
Not much to report from this end except that I came across this palpably apposite cryptogram while browsing in the college library:

$$\begin{array}{r} \text{S E N D} \\ + \quad \text{M O R E} \\ \hline \text{M O N E Y} \end{array}$$

I hope you enjoy solving the puzzle and I look forward to receiving your solution,

Your Devoted Son,
Fred,"

These puzzles, generally called cryptograms, have been very popular in especially in Victorian times. Our impecunious friend was confident that his father would "get the message".
The rules for a good cryptogram are
     (i) each letter stands for a different numeral (0..9)
     (ii) each numeral 0..9 occurs exactly once
     (iii) no number begins with 0.

[15] The Pineapple Theorem.

The skin of a pineapple is made up of hexagonal segments. Some of the segments are in rows, going up the pineapple. Some segments go in rows round to the right and some go in rows round to the left. If V is the number of rows going up the pineapple, R is the number of rows going right and L is the number going left, then **V=L+R**

On Question 1

Circular Arguments and Reversible Steps

We are told that we should make sure that we never assume what we are trying to prove in a mathematical proof otherwise we get into a circular argument that goes round in circles getting no where.

However, if the steps in an argument are all reversible, then it is quite O.K. to start at the answer and work backwards.
Mathematicians have a special symbol for a reversible step, it is "**iff**" which means if and only if. Alternatively, we can use the symbol $\Leftrightarrow$.

Problem 1. If you add the same number to the top and to the bottom of a fraction, does the fraction get bigger or smaller?

Discussion:
First make the question more precise:
If a, b and x are all positive numbers is the $\underline{a+x}$ greater than $\underline{a}$?
                                                                            b+x   b

Solution $\dfrac{a}{b} < \dfrac{a+x}{b+x}$

$\Leftrightarrow$ $ab+ax < ab+bx$

$\Leftrightarrow$ $ax < bx$

$\Leftrightarrow$ $\dfrac{a}{b} < 1$ (dividing by **bx**)

reversing the argument, we conclude that if $\underline{a}$ is less than 1, then adding the
                                                                      b
positive number **x** to top and bottom increases the value of the fraction.

For example, adding 1 to the top and bottom of 2/3 increases its value to ¾ .

Exercise 1: If a, b and x are positive numbers, and a/b > 1, show that adding
the positive number x to top and bottom decreases the value of the fraction.

Combining these two results, we can therefore say that, if all the numbers
involved are positive, then adding the same number to the top and bottom of
a fraction, increases its value when it is less than 1 but decreases its value
when the fraction is greater than 1.
A different approach to problem 1 involves a graphical representation:
In the diagram below, all the "integer" points are marked. Each point in the
grid represents a fraction.

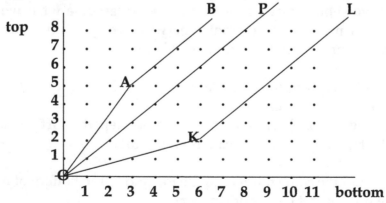

The point **A** represents the fraction **5/3** and the point **K** represents the fraction **2/6**.
The value of the fraction **5/3** is represented by the slope of the line **OA**.
The value of the fraction **2/6** is represented by the slope of the line **OK**.

All points on the line **OP** represent fractions with a value 1, e.g. **2/2 = 3/3 ...**

All points above the line **OP** have a value greater than **1**.
All points below the line **OP** have a value less than **1**.

Now adding **x** to the top and the bottom of "fraction **A**" will send you to some point **A'** where **AA'** is parallel to **OP** and the slope of **OA'** will be less than the slope of **OA**. This applies to any point above **OP**, thus:

Adding x to the top and bottom of any point above OP decreases the value of the fraction.

Adding x to the top and bottom of the "fraction **K**" will send you to some point **K'** where **KK'** is parallel to **OP** and the slope of **OK'** will be greater than the slope of **OK**, thus:

Adding x to the top and bottom of any point below OP increases the value of the fraction.

Combining these two results we deduce that:

If a/b < 1 then (a+x)/(b+x) > a/b and

If a/b>1 then (a+x)/(b+x) < a/b.

(N.B. such a proof that relies on diagrams would not be considered to be a **rigorous proof**)

A more general version of problem 1 is:

Exercise 2:   If $\underline{a} < \underline{x}$ where a, b, x and y are all positive,
                      b   y
Prove that $\underline{a+x}$ lies between $\underline{a}$ and $\underline{x}$
                   b+y   b   y

A rigorous proof would follow the lines of the first proof to problem 1, however, we can also use the graphical demonstration:

In the diagram below, all the "integer" points are marked. Each point in the grid represents a fraction.

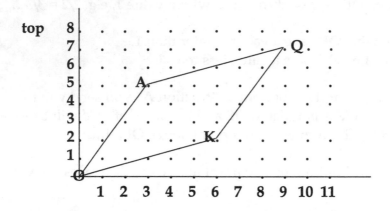

If **a/b** is represented by the point **A** and **x/y** is represented by the point **K**, then **(a+x)/(b+y)** will be represented by the point **Q** where **OAQK** is a parallelogram.

a/b = slope OA , x/y = slope OK and **(a+x)/(b+y) = slope OQ.**

Now the slope of **OQ** is clearly between the slope of **OA** and the slope of **OK** and so we deduce that

$$a/b < (a+x)/(b+y) < x/y$$

A Special Case:

Assuming that we have proved the result of exercise 2, we can deduce problem 1 as a special case:

If $\underline{\text{a}} < \underline{\text{x}}$ where a, b, x and y are all positive,
 b $\phantom{<}$ y

then $\underline{\text{a}}$ < $\underline{\text{a+x}}$ < $\underline{\text{x}}$
 b $\phantom{<}$ b+y $\phantom{<}$ y

Put **x=y** and we have:

If $\underline{\text{a}} < \underline{1}$ where a, b, x and y are all positive,
 b

then $\dfrac{a}{b} < \dfrac{a+x}{b+x} < 1$

thus, if **a/b < 1** then adding x to top and bottom makes the fraction larger.

Put **a=b** and we have:

If $1 < \dfrac{x}{y}$ where a, b, x and y are all positive,

then $1 < \dfrac{a+x}{a+y} < \dfrac{x}{y}$

thus, if **x/y > 1** then adding a to top and bottom makes the fraction smaller.

On Question 2: The length of a roll of paper

I had a friend who set up a small business supplying rolls of paper.
He would buy large rolls of paper and rewind them to the sizes needed by his customers. Among other items, he could supply toilet rolls and rolls for till receipts.
He asked me how to calculate the number of small rolls that he would be able to make from a large roll of a given size. Was there a formula that he could use?

The Problem.

We have a large roll of external radius 20 cm on a core of radius 4 cm. The length of paper on roll is given as 350 metres.
How many small rolls, radius 5 cm could be made if they had a core of radius 2 cm and how long would the small rolls of paper be.

Solution

My first thought was to use the areas of cross section;

Area of large roll = $A = \pi R^2 - \pi r^2$

$$= \pi.400 - \pi.16 = 384\pi$$

Area of small roll = a = $\pi R^2 - \pi r^2$

$= \pi.25 - \pi.4 = 21\pi$

Number of small rolls = A/a = $384\pi / 21\pi$ = about 18 rolls

The length of each roll is 350/18 = about 19 metres

===

Could we find the thickness of the paper?

Imagine the paper unwound so that it makes a long (350 metre) rectangle:

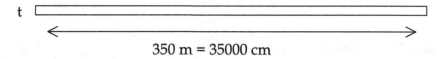

350 m = 35000 cm

We know that the area of the roll is 384π cm^2.

Therefore, 35000t = 384π giving t = 0.03 cm = 1/3 mm.

===

These results seemed satisfactory but thinking of the problem in more detail there was a nagging feeling that things were not quite right after all, the paper does not really form perfect circles, it is actually in the form of a spiral. How does the solution hold up for something which is quite thick?

Diagram for rather thick paper:

We suppose that the strip has the same width along the whole of its length.

Define the length of the strip as the length of the mid-line of the strip (dotted).

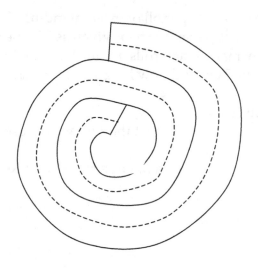

Now examine the area of a circular strip of
external radius R and internal radius r.

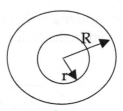

We know, by subtraction,
that the area of the circular ring is $A = \pi R^2 - \pi r^2$

Now the length of this circular strip, by definition, is the length of the mid-
line, which is the length of a circle of radius $r + \frac{1}{2}(R-r) = \frac{1}{2}(R+r)$.

The length of the mid-line is therefore $2\pi . \frac{1}{2}(R+r) = \pi(R+r)$

The width of the strip is (R-r).

Therefore, the area of the circular strip is $\pi(R+r)(R-r) = \pi R^2 - \pi r^2$

This means, that if we multiply the length of the mid-line by its width, we
get the area of the unwound strip:

$$\pi(R-r)$$

Now take just a part of the circle and consider the area of an arc of the strip.
For example, take a strip of width x, radii r and (r+x):

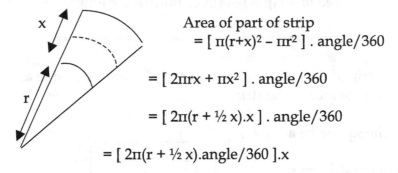

Area of part of strip
$$= [\,\pi(r+x)^2 - \pi r^2\,] . \text{angle}/360$$

$$= [\,2\pi rx + \pi x^2\,] . \text{angle}/360$$

$$= [\,2\pi(r + \frac{1}{2}x).x\,] . \text{angle}/360$$

$$= [\,2\pi(r + \frac{1}{2}x).\text{angle}/360\,].x$$

= length of mid-line . width   (since the radius of the
                                mid-line is $(r + \frac{1}{2}x)$

Hence: Area of part of strip = length of mid-line x width

==================================

Now consider a strip of width x that is wound round an irregular shape as illustrated in the following figure. The mid-line is dotted

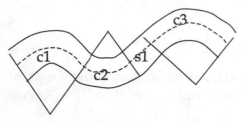

The strip is dissected into portions of circular arc, with maybe a few straight bits.

If the lengths of the pieces of the mid-line are c1, c2, s1 (straight bit), c3 etc then, using the mid-line theorem, we know that the areas of the pieces of the strip are c1.x , c2.x, s1.x, c3.x etc.

Therefore, the area of the strip is c1.x + c2.x + s1.x + c3.x

$$= (c1+c2+s1+c3).x$$

$$= \text{length of mid-line . width.}$$

Therefore, whatever the shape of the core that the strip is wound on, the area of the strip is

Area of strip = length of mid-line x width

==================================

Example   A strip of thickness **x** is wound round a square of side **a**.
              Find the area of the strip.

Let the side of the square be **a**

Let the width of the strip be **x**

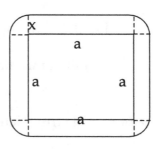

[1] By direct calculation:

area = circle + 4 rectangles

$$= \pi x^2 + 4ax$$

[2] Use the mid-line theorem:

length of mid line = 4a + circle of radius ½x

$$= 4a + 2\pi(½x) \qquad = 4a + \pi x$$

Area = length of mid-line x width

$$= (4a + \pi x).x \; = 4ax + \pi x^2 \; \text{verifying the mid-line theorem}$$

Exercise

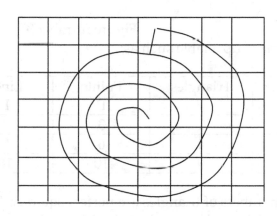

By counting squares, the area of this ancient shellfish was found to be 27 cm²
It has been estimated that these shellfish grow 1.5 cm each year.
How old was the shellfish when it died?
(Hint: use the mid-line theorem)

================================

On question 3

Generalization

The simplest kind of generalization is the kind of problem often asked of
school students in their "investigation project".

Example 1

The student is asked to count the number of lines, points and small triangles in these figures and then suggest formulae for each result;

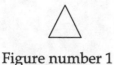

Figure number 1

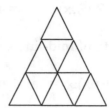

Figure number 2

Figure number 3

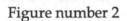

The student might make up a table such as;

Figure number F	Triangles T	Points P	Lines L
1	1	3	3
2	4	6	9
3	9	10	18

We can see that the numbers of triangles are perfect squares: $T = n^2$
The numbers of triangles also increase by the odd numbers:

$$1+3 = 2^2$$

$$1+3+5 = 3^2$$

$$1+3+5+7 = 4^2$$

suggesting that the sum of the first n odd numbers is n^2

The number of points in each row increases by one, giving the value for P as

$$1+2 = 3$$

$$1+2+3 = 6$$

$$1+2+3+4 = 10$$

maybe someone would spot that $2P = (F+1)(F+2)$

The number of lines is easier to spot: $L = T+P-1$

In general, for a figure with n rows, we have

$$T = n^2 \qquad P = 1+2+3+....+n \qquad L = T+P-1$$

======================================

Example 2.   Naturally, the next question would involve squares

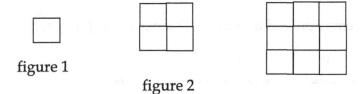

figure 1

figure 2

figure 3

and the student could construct a table such as

Number of rows	Number of squares	Number of points	Number of lines
n	S	P	L
1	1	4	4
2	4	9	12
3	9	16	24
4	16	25	40

In general, if the number of rows is **n**, then we have

$$S = n^2$$

$$P = (n+1)^2$$

$$L = 2n(n+1)$$

We may also note that $P+S = L+1$

Which is a special case of "The Squiggles Theorem" Theorem for plane networks.

Any squiggle is made up of **Points, Regions** and **Links**
Links always cross at a point and links always join two points. For any such squiggle,

Points + regions = links + 1

In this squiggle

In this squiggle, **Points = 6, Regions = 3, Links = 8**

(The Squiggles Theorem is proved in the next chapter, on Topology)

==

Problem 3A: The rolling penny

If I roll one penny round another, what angle will it turn through?

I have two pennies. I hold one of them fixed and roll the other round so that it returns to its starting position.
Through what angle does the rolling penny turn?

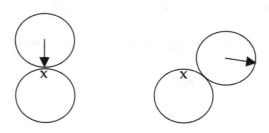

Suppose that initially, the two coins touch at x and that there is an arrow marked on the rolling coin.

As the coin rolls, equal distances are "covered" along each circumference.

When the rolling coin has rolled half of its circumference, the arrow head will touch the fixed circle.

after another 'half circumference' the non tip end of the arrow will touch the fixed circle.

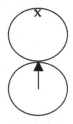

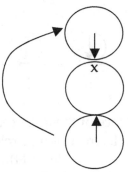

The rolling coin therefore completes two revolutions.

We can get an alternative view if we unroll the circumference of the fixed circle into a straight line of length 2πr, where r is the radius.

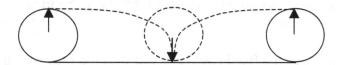

The rolling circle performs one revolution in covering the length 2πr and when we curl the straight line round into a circle, we add another revolution.

Problem 3b
If the rolling circle is only half the radius of the fixed circle through what angle does it now turn?
Let the radius of the small coin be r and the radius of the fixed coin 2r.

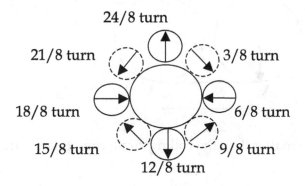

24/8 turn

21/8 turn   3/8 turn

18/8 turn   6/8 turn

15/8 turn   9/8 turn

12/8 turn

The circumference of the small circle is $2\pi r$
The circumference of the large fixed circle is $4\pi r$

If we 'unroll' the fixed circle, then the small circle will roll twice along the straight line:

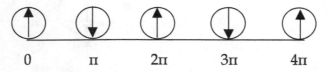

0   π   2π   3π   4π

Since the circumference of the large fixed circle is twice the circumference of the small rolling circle, the rolling circle make two revolutions. Now curl the straight line round into a circle and this adds one more revolution.
The small circle therefore makes three revolutions.

Problem 3c

One of the "jobs" of the mathematician is to generalize problems where possible. The rolling circle is easily generalized if we use the idea of unwrapping the fixed circle.

Suppose that the radius of the fixed circle is **f** and that the radius of the rolling circle is **r**.
If we unroll the circumference of the fixed circle, it will be a line of length **$2\pi f$**.
The circumference of the rolling circle is **$2\pi r$**.

The number of revolutions made by the rolling circle when it rolls along this straight line of length $2\pi f$ will be $\dfrac{2\pi f}{2\pi r} = \dfrac{f}{r}$

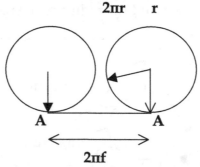

A   A

$2\pi f$

If the fixed circle is smaller than the rolling circle then $\dfrac{f}{r}$ is the fraction of a revolution turned through by the rolling circle. In the previous diagram, we now wrap the line AA round the small fixed circle to add one complete revolution.

The General result:

If a circle of radius **r** rolls once round a circle the fixed circle of radius **f** then the rolling circle turns through $\left(1+\dfrac{f}{r}\right)$ revolutions (for any positive **f** and **r**).

===================================

On questions 4 and 5

The Wobbly Table

An Exercise in Continuity

Suppose that your ideal body weight was 80 kg. You weigh yourself and find that you are 85 kg so you start a daily regime of vigorous exercise. At the end of the week you weigh yourself again and find that you are 75 kg. You conclude that sometime during that week you were at the ideal body weight of 80 kg.

Barring losing an arm or a leg, your body weight varies continuously. It does not jump from one value to another.

If you jumped off of a building 50ft high, you would expect, at some time, to be 10 feet above the ground.

If you draw a continuous curve from one side of a line to the other side then it might seem obvious that your curve must cross the line somewhere.

One of the most important activities for a mathematician is solving equations, for example, finding a value of x for which $f(x) = 0$ where $f(x)$ is some expression involving the unknown number x. If $y=f(x)$ is a continuous curve and we know that $f(a) > 0$ but $f(b) < 0$ then somewhere between a and b we have $f(x)=0$.

These examples illustrate the idea of continuity.

In the real world however, things are sometimes not so simple. Quantum tunnelling is the name given to the path of an electron that skips from one side of a thin insulating layer to the other without going through the layer or going round it!

Problem 4
How can you stop a 4 legged table wobbling on a bumpy floor?

A square table with four equal legs stands on an uneven floor but the table wobbles.
Prove that, if the table is rotated about the centre of the square top then, in less than 90 degrees, a position will be found that stops the wobble.

Solution

First position

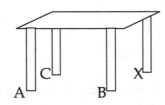

The table stands on three legs A, B and C but leg X is short of the floor because the floor is bumpy. The table legs are all the same length.
Now rotate the table by 90 degrees about the centre of the top trying to keep A, B and C in contact with the floor.
Suppose that we reach the 90 degrees turn.

Second position

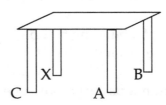

At this position, the table would stand on legs C, A and X and leg B would be short. Suppose that you were to be able to push leg B down until it touched the floor, then leg X would have to be pushed **into** the floor so that X would be below floor level. Now consider X during the rotation: X started above the floor and ended below the floor. Therefore, by continuity, somewhere during the rotation, X must hit the floor. At this position the table would not wobble.   ■

How to Quarter an Area

Problem 5

To prove that you can always find a pair of perpendicular lines to quarter any given area we start with the area **K** and watch a moving line **L**

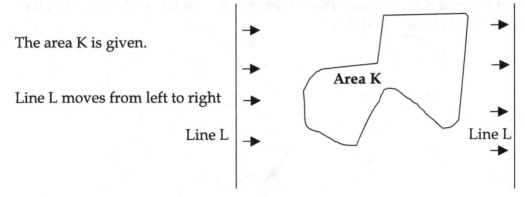

The area K is given.

Line L moves from left to right

Line L

Consider the part of area K to the left of line L.
This area starts with the value zero, and when L hits the area K, the value of the area to the left increases until it reaches the value K.

Since the area to the left of L goes from zero to K, it must pass through the value ½ K. Therefore, there is a position for line L that bisects the area.

Now consider the line M which is at right angles to line L.

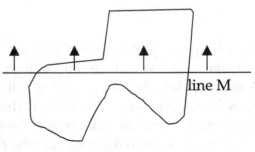

By the same argument, when moving line M from below area K to above area K, the part of area K below line M starts at zero and increases to K.
Therefore, there must be a position where line M bisects area K.

We have shown that there are two lines L and M that bisect the area K (see figure)

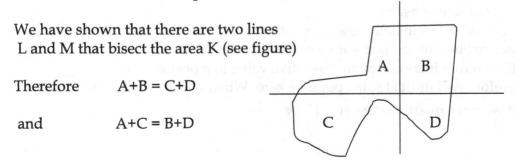

Therefore   A+B = C+D

and   A+C = B+D

Subtracting gives   B-C = C-B   which means that   B=C and A=D.
We want to show that we also have A=B.

Suppose that A<B so that A-B is negative and consider what happens if the lines L and M are rotated anticlockwise, but ensuring that both L and M bisect the area K at all times. The values of A, B, C and D will change but we will always have B=C and A=D.

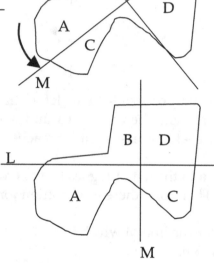

However, when L and M have turned through 90 degrees, we reach the starting figure but with L and M interchanged:

Now think about the value of A-B as L and M rotate.
We want to show that we can reach a position where A=B, but suppose that A starts off less than B so that A-B is initially negative.
Now look at the final value of A-B when L and M have each turned through 90 degrees. The final value of A-B the same as the starting value of C-A. But C is always equal to B throughout the rotation of the lines L and M so the final value of A-B is in fact the starting value of B-A which we have supposed to be positive.
Therefore, if A is initially less than B, the value of A-B will start off being negative but will finished with a positive value.
A-B therefore increases from a negative value to a positive value.
Therefore A-B must at some point be zero. When this happens, A=B=C=D and we have quartered the area K. ∎

On Question 6: Shortest Paths

Extending a Problem

"The shortest distance between two points is a straight line" (Euclid).

[1] Crossing the river

Fred has a plank of wood that is just long enough to reach across the river but where should he put the plank to find the shortest route home?

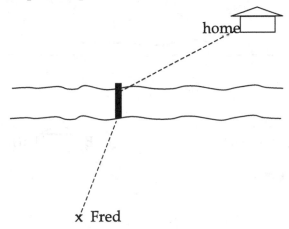

Solution

Fred steps forward a plank length
to point A. The shortest distance from
A to home, is the straight line AB.
His shortest path length is therefore
AB+plank.
The line AB shows ~~him~~
where to put the plank.

Fred's shortest path is
xy+yz+zB

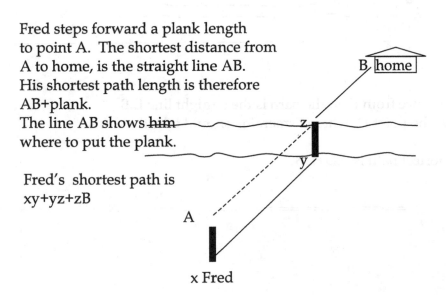

[2] Esmeralda needs a drink

Esmeralda the cow is about to leave the middle of the meadow and head for
the barn. Before going to the barn, Esmeralda wants to take a drink from the
stream that runs along one side of the meadow.
 What is the shortest route?

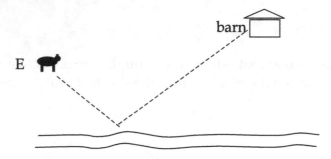

Solution

 Reflect Esmeralda in the
nearside bank of the stream
 to E'.

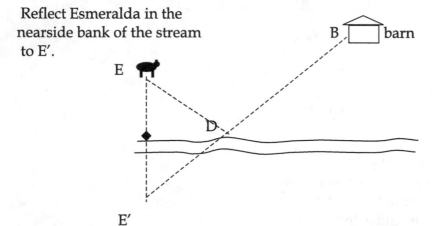

The shortest distance from E' to the barn is the straight line E'B.
This line shows the point D, where Esmeralda should take the drink.

Esmeralda's shortest path is ED+DB

========================

[3] Esmeralda needs a nibble.

There is a hedge running along another side of the meadow and Esmeralda decides to have a nibble from the hedge after the drink.

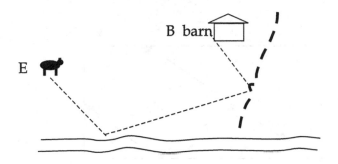

Solution

Suppose that Esmeralda decides to drink at the point D on the river bank and nibble the hedge at the point N.

Reflect Esmeralda in the river bank to E′
and reflect the barn in the hedge to B′.

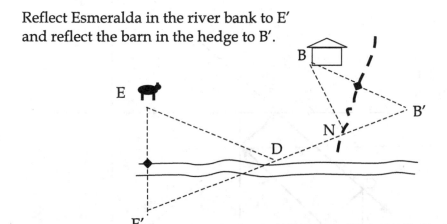

Wherever D is and wherever N is, Esmeralda's path EDNB will always be equal to E′D+DN+NB′.

But the shortest distance from E′ to B′ is the straight line E′B′.
The straight line E′B′ therefore gives the drinking point D and the nibbling point N.

Esmeralda's shortest path is now ED+DN+NB

==================================

[4] Trick Snooker

Show how to play a snooker ball from C off of three cushions so that it returns to its starting position.
(Assume an ideal bounce so that the "angle of incidence equals the angle of reflection")

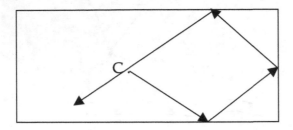

Solution

Reflect C to C1
Reflect C1 to C2
Reflect C2 to C3

Join CC3 to find Z
Join ZC2 to find Y
Join YC1 to find X

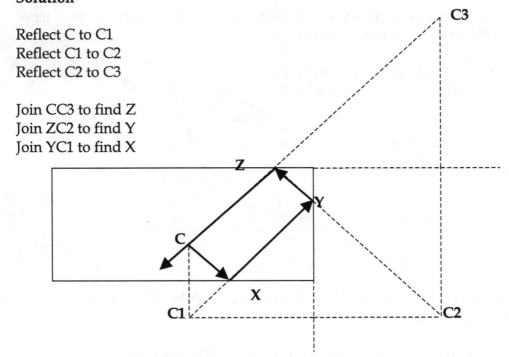

Note that CXYZC = C1XYZC = C2YZC = C3ZC and that C3ZC is the straight line so that CXYZC is the shortest path.

============================

On Question 7

The Buckminster Football

Richard Buckminster Fuller (1895-1983) was an American architect, designer and writer who, during the 1950's, invented, designed and built geodesic domes. Geodesic domes can cover large areas and are self supporting with no need for internal supports.
In 1985 new forms of carbon were discovered that were based on the carbon 60 molecule which has a shape very like Buckminster Fuller's domes.
Carbon molecules can take the form of hollow spheres or long nanotubes which can form strong carbon fibre. These newly discovered carbon molecules were called fullerenes and the spherical carbon 60 molecule was named buckminsterfullerene. The carbon 60 molecule is regular in shape and has hexagonal and pentagonal faces. It is also called a bucky ball.

An old style football is made up of six double panels, each one as long as it is wide.

one double panel

six of these double panels would make the six faces of a cuboid so that one of the old style footballs is in effect, a blown up cuboid.

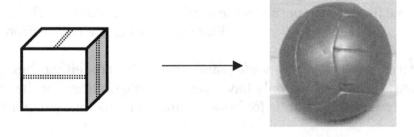

A 1920s soccer ball

Later versions of the blown up cube have an extra panel inserted between the two parts of each double panel

Bending the panels produced a more practical shape but the design is still based on the six faces of a cuboid

The carbon 60 molecule provides us with a different design.
A bucky ball is a carbon molecule in the shape of a polyhedron but with only hexagons and pentagons for the faces.

Designing a Football

Lets try to design a football based on hexagonal and pentagonal faces.

We assume (i) that each vertex of the football has the same structure
and that   (ii) $F+V = E+2$ (Euler's formula for a polyhedron)

Three faces meet at each vertex and the faces are either hexagons or pentagons. Therefore, we only have these four possibilities for the faces that meet at any vertex (H stands for hexagon and P stands for pentagon):

3H … not possible since the vertex would then be flat. This is in fact, the structure of graphite.

3P … this gives a dodecahedron with 12 pentagonal faces so we could make a football with 12 pentagonal panels.

H+2P ... this does not work because each hexagon would have to be surrounded by six pentagons and to keep the same structure at each vertex, the ring of pentagons would have to be surrounded by a ring of hexagons and this would result in a P+2H vertex.

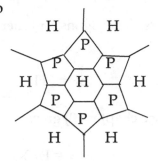

2H+P is therefore the only way of constructing a regular solid with the same structure at each vertex, out of pentagons and hexagons.

Each vertex has the same P+2H structure

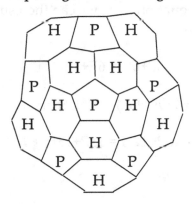

Let   F be the number of faces,
          V be the number of vertices,
          E is the number of edges,
          H is the number of hexagons
          P is the number of pentagons

The faces are either hexagons or pentagons, therefore
$$F = H+P$$

If we count 5 vertices for each pentagon plus 6 vertices for each hexagon then we would have counted each vertex three times, therefore

$$3V = 5P+6H$$

If we count 5 edges for each pentagon plus 6 edges for each hexagon then we would have counted each edge twice, therefore

$$2E = 5P+6H$$

Each pentagon gives 5 hexagons, but then each hexagon would be counted 3 times, therefore

$$5P = 3H$$

Euler's formula (see chapter 10) gives

$$F+V = E+2$$

And now we have 5 equations from which to find F, V, E, H and P. Multiplying this last equation by 6, we can then substitute for F, V and E in terms of H and P, thus

$$6F+6V = 6E+12$$

gives $6(H+P) + 2(5P+6H) = 3(5P+6H)+12$

or $6H+6P + 10P+12H = 15P+18H + 12$

thus $P = 12, H=20$

The Buckminster Football therefore has 12 pentagons and 20 hexagons

The **icosahedron** has 20 triangular faces and 12 vertices. Five triangular faces meet at each vertex. The next figure shows a "top" view of an icosahedron:

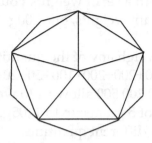

Lopping off the corners of such a solid is called "truncating". If we truncate the icosahedron then we will create 12 new faces and each of these faces will be a pentagon. By carefully choosing how much to lop off, we can create 20 hexagonal faces. All we have to do is remove one third of each edge:

The Buckminster football is therefore a truncated icosahedron, as well as a geodesic dome and a carbon 60 molecule

On Questions 8 and 9

Weighing Things

If you buy a pair of scales with a set of weights you would probably get one 500 gram weight, two 200 gram weights, one 100g, a 50g, two 20g, one 10g and a 5g.

With these weights we can weigh any of the quantities which are multiples of 5g up to the maximum 500+200+200+100+50+20+20+10+5. Thus the weights **5 10 20 20 50 100 200 200 500** allow us to weigh the quantities 5g, 10g, 15g... going up in steps of 5 grams up to 1105g. With the nine given weights we can measure 221 different quantities.

Now if we have 9 weights, there will be $2^9 = 512$ different combinations of the weights so we can see that this given selection of weights is not the most efficient in terms of reducing the number of weights we need. Let us examine the problem from a simple starting point.

Problem:

What is the minimum number of weights needed if we wish to weigh all quantities from 1 gram up to 100 grams?

Notation:

We will use ordinary numbers to represent quantities to be weighed and bold numbers to represent weights that we use. Thus if we weigh a quantity of 95 grams using a fifty, two twenty gram weights and one 5 gram weight we represent this by the equation

$$95 = 50 + 20 + 20 + 5$$

Discussion:

Suppose that we needed a set of weights that would enable us to weigh any whole number weight from 1 to 100, then the simplest solution would be to have one weight for each value. That is, we use 100 weights, one for each quantity:

1, 2, 3, 4, 5, ... **50, 51,** **98, 99, 100**

This set of weights can weigh any value up to 1+2+3=....+99+100 = 5050 since:

With one weight we can weigh

1, 2, 3, ...99, 100

With two weights we can weigh

$$100+1, 100+2, \ldots 100+99$$

With three weights we can weigh

$$100+99+1, 100+99+2, 100+99+3, \ldots 100+99+98$$

with four weights we can weigh

$$100+99+98+1, 100+99+98+2, \ldots 100+99+98+97$$

with five weights we can weight from

$$100+99+98+97+1 \text{ up to } 100+99+98+97+96$$

Continuing with this system we finally end with 100 weights to weigh

$$100+99+98+97+96+\ldots+3+2+1 = 5050$$

All of these weighings except the singleton weighings, have many different solutions, for example we can weigh 200 grams using $200=\textbf{100+99+1} = \textbf{100+98+2} = \textbf{100+97+3} = \textbf{99+98+3} = \textbf{99+97+4} = \textbf{99+96+5} = \ldots\ldots = \textbf{68+67+65}$.

Getting back to the problem of weighing 1,.. 100 grams, if we combine weights in pairs, we can see that we do not need the weights for **51g** up to **100g** since we can combine 51g = **50 + 1**, 52g = **50 + 2** and so on up to 98 = **50 + 48**, 99 = **50 + 49**, 100 = **50 + 49 + 1** so we can still weigh all the values from 1 to 100 but using only the 50 weights

$$\textbf{1, 2, 3, 4, 5} \ldots \ldots \textbf{46, 47, 48, 49, 50}$$

Now can the necessary number of weights be reduced still further and still allow us to weigh from 1g up to 100g?

In the decimal number system, we count by grouping in tens thus if we use the nine weights **1, 2, 3, 4, 5, 6, 7, 8, 9** together with the nine weights **10, 20, 30, 40, 50, 60, 70, 80, 90** then we can make all the values from1 to 99 using just

one or two weights, for example 57 = **50 + 7**, 99 = **90 + 9** and we can make up
the quantity 100 as

$$100 = 90 + 9 + 1$$

Thus we can weigh values from 1 to 100 with just 18 weights:

1 2 3 4 5 6 7 8 9 10 20 30 40 50 60 70 80 90

But this set of weights is clearly inefficient as there are 2^{18} different
combinations that can be made of these eighteen weights.
Can the number of necessary weights be reduced still further by using other
combinations, for example, we do not need the 3 because 3 = **1 + 2** and
similarly, we do not need the 5 gram weight because 5 = **4 +1**.
We certainly need the **1** and the **2**
then we have

	3= **1 + 2**
we need the	**4**
and then	5= **4 + 1**
	6= **4+ 2**
	7= **4 + 2 + 1**
now we need the	**8**
and then	
	9= **8 + 1**

Hence, we could get all the values from 1 to 100 using

	1 2 4 8	for the units digit
and	**10 20 40 80**	for the tens digit

(and we can still make up 100 from 100 = **80 + 20**)

Now we have reduced the number of necessary weights down to eight.
Can we reduce the number still further?
If we have eight different weights then there will be $2^8 = 256$ different
possible combinations.
If we have seven different weights then there will be $2^7 = 128$ different
combinations.
However, if we have just six weights then there will only be $2^6 = 64$ different
combinations and so we would not be able to make up the 100 different
quantities we need.
Clearly, in the most efficient set of weights, there will be no repeated values.
Let use then go through the argument we used above, avoiding the use of
repeated weights.

The Argument

We need a **1** gram weight	**1**
We need a **2** gram weight	**2**
Now we can make up 3 grams	3 = **2** + **1**
but then we need a **4** gram weight	**4**
now we can make up	5 = **4** + **1**
	6 = **4** + **2**
	7 = **4** + **2** + **1**

and at this stage we can get all values from 1 to 7.

So now we need the **8** gram weight	**8**
now we can make up all values from	9 = **8** + **1**
	10 = **8** + **2**
up to	15 = **8** + **4** + **2** + **1**

to go further, we need the	**16** gram weight.
Then we will be able to	
make up all values from	17 = **16** + **1**
up to	31 = **16** + **15**

now we can make up all the quantities from 1 up to 31 grams.
To go any further we will need the   **32** gram weight.
If we now add the **32** gram weight to our collection then we will be able to

make up all the weights from	33 = **32** + **1**
up to	63 = **32** + **31** .

Now we can make up any quantity from 1 gram up to 63 grams.
To go any further, we need the   **64** gram weight.
At this stage we have this set of weights:

$$\textbf{1 } \textbf{2 } \textbf{4 } \textbf{8 } \textbf{16 } \textbf{32 } \textbf{64}$$

and we can make up any quantity from 1 gram up to 127 grams.

The argument has led us to the binary representation for the quantity we wish to weigh, for example, if we wish to weight a quantity of 57 grams, writing this in powers of 2, we have

$$57 = 32 + 16 + 8 + 1$$

If we reverse the order of the weights in our collection we have

$$\textbf{64 } \textbf{32 } \textbf{16 } \textbf{8 } \textbf{4 } \textbf{2 } \textbf{1}$$

and to find which weights to use, we write the quantity we wish to weigh in binary notation. If something weighed 95 grams then we would use

$$95 = 64 + 16 + 8 + 4 + 2 + 1$$

Now 95 decimal = 101111 binary

The binary representation for the quantity we wish to weigh shows us which weights we would use.

Is this the most efficient set of weights?

Remember, with these seven weights, there are $2^7 = 128$ different combinations. If we reduced the number of weights in our set to 6 then there would be only $2^6 = 64$ different combinations and so we would be unable to make up more than 63 quantities (note that the measure of zero grams as one of the combinations). This shows us that we must have at least 7 weights in order to weight the quantities from 1 gram up to 100 grams.
Thus our set of weights **1 2 4 16 32 64** is the best set!

When we buy a set of scales, instead of giving us the set of weights

5g 10g 20g 20g 50g 100g 200g 200g 500g

we could be given the smaller set

5g 10g 20g 40g 80g 160g 320g 640g

and still be able to measure quantities from 5 grams in steps of 5 up to the maximum 1275 grams!

==============================

Base 3?
The binary system gives us the most efficient way of making up various quantities from a given set of weights. As a exercise, lets see what we might

come up with if we used base 3. If we tried a collection of weights suggested by the base 3 representation, we could proceed as follows:

Decimal value
| 0 | 1 | 2 | 3 | 4 | 5 | 6 | 7 | 8 | 9 | 10 | 11 | 12..... |

Base 3
| 000 | 001 | 002 | 010 | 011 | 012 | 020 | 021 | 022 | 100 | 101 | 102 | 110.... |

Two units **1** and **1** would allow us to weigh 1 and 2 grams. Two further weights **3** and **3** will now allow us to weigh from 3 grams up to eight grams, that is from 010_3 up to 022_3
and then we would add two weights 9 an 9 that would allow us to weigh from 022_3 up to 222_3 (which is 26 decimal).
In order to continue to higher values, we now need two weights **27** and **27** that allow us to weigh from 0222_3 up to 2222_3 (=80 decimal).
We now need a weight **81** grams in order to get up to the value 100. Thus, in order to weigh values from 1 gram up to 100 grams, we would use the set of weights

1 1 3 3 9 9 27 27 81

that is, a set of nine weights rather than the set of eight weights which we deduced using the binary system.
With this set of weights, we would be able to weigh

0001_3 0002_3 ....12222_3

In decimal notation this is 1 gram up to 161 grams.
Using the binary system, the first nine weights are **1 2 4 8 16 32 64 128 256**. These nine weights allow us to weigh quantities from 1 gram up to 511 grams.
The larger the number base for the design of the set of weights, the more inefficient the representation becomes.
Using a system based on base 5 numbers, we would require four weights **1** gram, four weights of **5** grams and 4 weights of **25** grams in order to weigh quantities from 1 gram up to 100 grams.
That is a set of twelve weights rather than the set of seven weights **1 2 4 8 16 32 64** based on the binary system.
Using the other Pan

There is one further trick that we could employ, based upon the fact that, using a pair of scales, we can subtract weights as well as add them. Suppose that we had just two weights, a **3** gram weight and a **5** gram weight, then we can certainly weigh 3, 5 and 8 gram quantities but we can also weigh 2 grams by putting the **3** and the **5** on different pans.

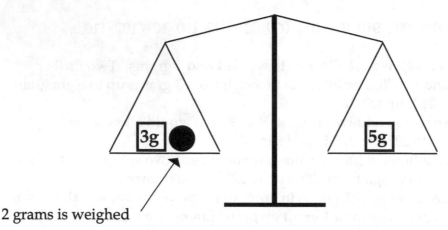

2 grams is weighed

Using the subtraction method now we see what weights are needed to weigh up to 100 grams:

      1 gram will weigh   1 gram

      a **3 gram** weight weighs   3-1, 3 and 3+1

and at this stage we can now weigh all values from 1 to 4.

Note that in base 3 these values are

$$0001_3 \text{ up to } 0011_3$$

We now need a **9 gram** weight to weigh 9-4, 9-3, ...9, 9+1, 9+2,.........9+4 and then, at this stage we can weigh all values from 1 to 13 or , in base 3

$$0001_3 \text{ up to } 0111_3 \,.$$

Now we need the **27 gram** weight so that we can weigh 27-13, 27-12, ...27, 27+1....27+13 and at this stage we can weigh all values from 1 gram up to 40 grams. The next weight we need is the **81 gram** weight. This now allows us to weigh values from 81-40, 81-39....81, 81+1, ...81+40 and at this stage we can weigh all values from 1 up to 121 grams.

We have now found the minimal set of weights which allow us to weigh quantities from 1 gram up to 100 grams:

1 3 9 27 81

Using just one scale pan, the minimal set of weights is given by the base 2 numbers:

1 2 4 8 16 32 64

Using both scale pans, the minimal set of weights is given by the base 3 numbers:

1 3 9 27 81

=========================

Problem 9

 You have 9 balls and a pair of scales. The balls are all the same except for one that is heavier than the others.

[1] How can you find the odd ball in the fewest weighings?

[2] Would it matter if the odd ball was lighter?

[3] What if you did not know if it was heavier or lighter?

===

On Question 10

How to Win at Nim

The word nim probably derives from the 15th century Scandinavian and Old English word for take or steal. The game is usually played with matchsticks or lines drawn with chalk on a blackboard. It is a game for two players who take turns to remove (to nim) any number from just one of the rows. The traditional game starts with four rows containing 1, 3, 5 and 7 matches but any number of matches in any number of rows can form the basis for a game.

The Traditional Game is for two players A and B
The starting position:

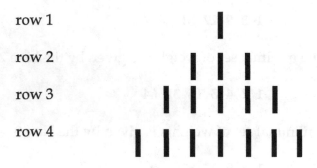

A removes 2 from row2

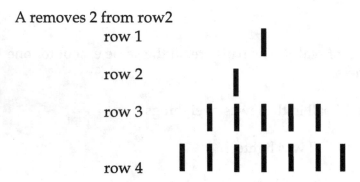

B to move: B removes 2 from row 4

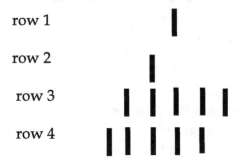

A to move: A removes 3 from row 3

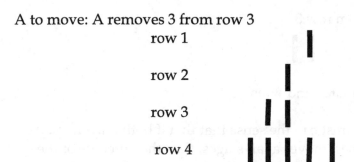

B to move: B removes 3 from row 4

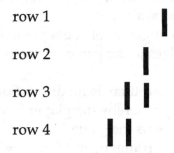

A to move: A removes 1 from row1:

row 2

row 3

row 4

B to move: B removes 1 from row2:

row 3

row 4

A to move: A removes 2 from row 3

row 4   ▌▐

B to move: B removes one match and wins

The game of Nim is deterministic in the sense that that if both players play correctly, that is, use their best moves at all stages, then the outcome of the game is decided from the starting position. This is the same for a game of chess which is also a deterministic game. In order to lose a game of chess, you must make a mistake. In order to win a game of chess, your opponent must make a mistake. In a game of chess, if both players play their best moves at all stages the result must be a draw.

Fortunately, the number of different games of chess possible is such that the human race could devote themselves to the game for the rest of time and still not exhaust all the possibilities.

The game of Nim is deterministic, but there is no draw. Someone must win so now we shall develop a strategy whereby the player from a winning position can ensure that he or she wins the game.

We start with the traditional initial arrangement of rows with 1, 3, 5 and 7 matches but write the numbers in binary:

	decimal value	binary value		
		4	2	1
row 1	1			1
row 2	3		1	1
row 3	5	1	0	1
row 4	7	1	1	1
total number of bits per column		2	2	4

The winning strategy depends on counting the numbers of bits in each of the columns. In this position, the number of 1 bits is 4, the number of 2 bits is 2 and the number of 4 bits is also 2.

Definition:
If the totals for the numbers of bits in each column are all even then the position is called **totally even** otherwise the position is called **an odd position.**

Theorem 1
A totally even position becomes an odd position after one play.
Proof
Any play will alter the bits for just one row in the position.
This will change the totals for some of the columns by one, making the corresponding totals odd so that the position is no longer totally even.

Theorem 2
An odd position can always be restored to a totally even position after a further play.
Proof
Find the leftmost odd column
Choose any row with a 1 bit in this column.
Change the bits of this row wherever the total for any column is odd. This indicates the row and the number of matches to remove from this row.

Example:	1 0 1 1 1	1+2+4+16=23
	0 1 1 1 0	2+4+8=14
	0 0 1 1 1	2+4+8=7
	1 0 1 0 1	1+4+16=21
totals	2 1 4 3 3	

This example shows four rows of matches with 23 in the first row, 14 in the second row, 7 in the third row and 21 matches in the fourth row.
The position is odd because the total for the first column is 3, the total for the second column is 3 and the total for the fourth column is 1. (The other two columns have even totals).
Column 4 is the leftmost odd column.
Choose the second row which has 14 matches.
Change the second row to   0 0 1 0 1 i.e. to 5 matches. So we remove 9 matches from the second row.

We now have the position
　　　　　1 0 1 1 1
　　　　　0 0 1 0 1
　　　　　0 0 1 1 1
　　　　　1 0 1 0 1
which is a totally even position.

These two theorems give us our strategy for keeping a winning position (if one arises).

Once you have an odd position on your move, make a play which leaves the position totally even......**but** care has to be taken at the end game because the aim is to leave a single match and this single match is an odd position!

As the game proceeds, the number of rows with more than 1 match will reduce.

The end game

Keeping your match winning even position, eventually you will reach a position where you can leave singletons in every rows.

As soon as this is possible, break the rule and leave **an odd number of singletons.**

Here is an example of a winning position:

> 1
> 1
> 1
> 22
> 23

If your opponent removes a singleton, then reduce the position to

> 1
> 1
> 22
> 22

If your opponent removes some from one of the larger piles then match the move, for example, if your opponent removes 18 from the third row, you remove 18 from the fourth row, leaving

> 1
> 1
> 4
> 4

If your opponent removes 3 from the third row you now **break the evens rule** and remove all of row 4 leaving the winning position

> 1
> 1
> 1

If you are left with a position which is totally even, as at the start of the traditional game, then you are bound to leave an odd position for your opponent. You must just hope that your opponent leaves an odd position for you also, from which you can maintain a winning position.

If both players play correctly, starting with a totally even position should lose, and starting with an odd position should win.

==============================

On Questions 11 and 12

Symmetry

Symmetry shapes the real world. Mammals need to move efficiently in the forward direction in search of food and so have a bilateral symmetry; left and right are pretty much the same. Left and right are not important for jellyfish and starfish. They have rotational symmetry but because the world above is different from the world below, they have no symmetry from top to bottom. The worm moves forward in search of food and so has no front to back symmetry, but the sun above and the ground below are not important for the worm so it has a significant rotational symmetry.

Symmetry is useful in the abstract world of mathematics.
A equilateral triangle has three axes of reflectional symmetry.
It has rotational symmetry of order 3.

A Triangle of Numbers

The numbers 1,2,3,4,5,6 are placed along side of the triangle so that the three numbers along each side add up to the same total (9).

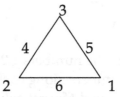

Problem 10

Use the symmetry of the triangle to write five other triangles of numbers with the same property, each side adding up to 9.

Reflections and rotations of a solution do not normally count as being essentially different solutions since we can see the same arrangements of the numbers by going round the back of the diagram or by rotating the page.

The question arises, however, how many essentially different solutions are there?

Duality

If we have a theorem in geometry, about lines and points, we can create another theorem, called the dual theorem, by changing point for line and line for point:

A trivial example:

        Four lines in general give six points.

Dual Theorem

        Four points in general give six lines.

For our triangle if we subtract each of the numbers on the triangle from 7, we do get another solution with the sides adding up to 21-9 = 12. It is called the **dual** of the original.

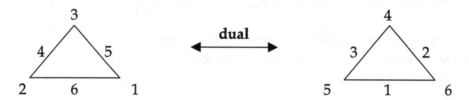

Problem 11

Arrange the numbers 1,2,3,4,5,6 along the sides of the triangle so that the three numbers along each side add up to the same total. How many **essentially different** solutions are there?

Discussion

A good starting point is to decide where the three odd numbers and the three even numbers go:
We can have either

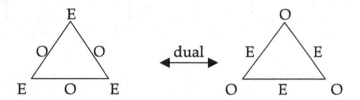

OR

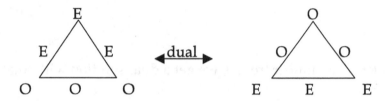

There are four essentially different solutions, with totals 9, 10, 11 and 12.

A Square of Numbers

A square has rotational symmetry of order 4 and has four axes of reflectional symmetry.

The numbers 1,2,3,4,5,6,7,8 are here placed round the sides of the square so that the three numbers along each side add up to the same total.
Here is an example with the total along each side equal to 12.

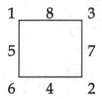

Using rotational and reflectional symmetry, draw 7 more diagrams with the same property:

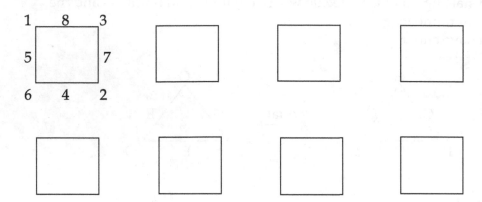

Duality

If we subtract each number from 9, we get a dual solution with total 15:

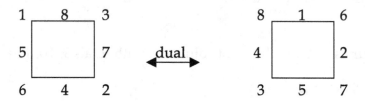

Problem 12 Place the numbers 1..8 along the sides of a square so that three numbers add up to the same total.

Again, one method of attack is to decide where the even numbers could be: We might try for a symmetrical arrangement:

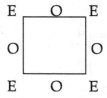

But if we add up the totals , the Evens are counted twice, the Odds once to give

$$2(E+E+E+E) + (O+O+O+O) = 4 \times \text{total}$$

$$2(2+4+6+8) + (1+3+5+7) \quad = 40 + 16 = 56$$

so that the total along each side would have to be 56/4 = 14

but E+O+E must be odd giving a contradiction. Therefore we cannot have this symmetrical arrangement.

A little analysis shows that the only arrangements are

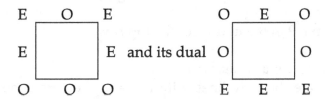

and the only possible totals are 12, 13, 14 and 15.

There is one solution with a total 12, there are two solutions with a total 13, two solutions with a total 14 and one solution with a total 15.

==

On Question 13

Magic Squares

A magic square is a number square that has each row, column and the two diagonals adding up to the same number.

To fix ideas, we fill the cells of the square with numbers the 1,2,3,4,5,6,.....

A 3x3 magic square then uses the numbers 1+2+3+4+5+6+7+8+9 giving a total 45.

Dividing 45 between the three rows gives the sum of each row = 15.

The well known 3x3 magic square is   2 9 4
                                          7 5 3
                                          6 1 8

Of course, we can reflect and rotate this square of numbers to produce seven other but these are not counted as being different.
The question arises, "Are there any other 3x3 magic squares?"

Solution

Consider the 9:
we can only have 9+1+5 and 9+2+4 in the 9 line therefore the 9 cannot be at a corner, for the corner has to have 3 lines adding up to 15.

Therefore, we can put 9 in the middle of the top row.

1 cannot be at a corner for a similar reason:
1+9+5 and 1+8+6 are the only lines with a 1 and the corner needs 3 lines adding up to 15.

Therefore, the middle line must be 9+5+1 or 9+1+5 but 1 cannot be in the middle because there are 4 lines going through the middle number.

Therefore the middle line must be 9+5+1.

It is now easy to show that **there is only one 3x3 magic square.**

Odd number magic squares are easy to construct.
Start with 1 in the middle of the top row,
count right and up;

if out of the top go to the bottom square of the next column;

if out of the right side go up to the start of the next row;

if up right is blocked go to the square underneath:

A 5x5 magic square:

```
17  24   1   8  15
23   5   7  14  16
 4   6  13  20  22
10  12  19  21   3
11  18  25   2   9
```

Magic squares with an even number of sides are more difficult to construct, however, here is a 6x6 magic square:

```
23   1  18  20  36  13     the rows, columns and
34   9  32  30   2   4     diagonals add up to the
22   8   1  25  29  16     magic number 111.
17  31  24  14   6  19
 3  35   5   7  28  33
12  27  21  15  10  26
```

Exercise: Evaluate 12x37,   15x37,   18x37

=================================

Sudoku

The Amercan architect Howard Garns invented this number puzzle in 1979 and called it "Number Place". It became popular in Japan in 1986 where it was renamed Sudoku = Single Number.

The number of possible games is 6 670 903 752 021 072 936 960 (Felgenhaur and Jarvis .. 2005) but this number includes reflections and rotations.
The number of really different games is 5 472 730 538 (Russel and Jarvis ... 2005)

2		9		7		8		6
	8		9	2	4		5	
		5		6		4		
1				9			3	
	7			1			4	
	2			5				1
		7		8		9		
	9		7		5		6	
8		6		4		5		3

The object is to fill in the numbers 1,2,3,4,5,6,7,8,9 in the outlined 3x3 squares in such a way that the numbers 1,2,3,4,5,6,7,8,9 occur in each row, in each column and in the two long diagonals.

Theorem : Every Sudoku square is a Magic Square

Proof Each row, column and the two diagonals adds up to

$$1+2+3+4+5+6+7+8+9$$

Theorem : Not every 9x9 Magic Square is a Sudoku Square

Proof This 9x9 magic square is not a Sudoku square:

47	58	69	80	1	12	23	34	45
57	68	79	9	11	22	33	44	46
67	78	8	10	21	32	43	54	56
77	7	18	20	31	42	53	55	66
6	17	19	30	41	52	63	65	76
16	27	29	40	51	62	64	75	5
26	28	39	50	61	72	74	4	15
36	38	49	60	71	73	3	14	25
37	48	59	70	81	2	13	24	35

We examine here, not how to solve a Sudoku puzzle but how to construct one.

We first choose any 3x3 square of the numbers 1..9 for example: $M =$

1	3	4
7	2	5
9	6	8

Notation:

The operation **r** rotates the rows of the square so that the bottom row becomes the top row and the others shift down.

If **M** is the chosen square then we write

$$M = 1\,3\,4 \qquad\qquad rM = 9\,6\,8 \quad \text{and} \quad r^2M = 7\,2\,5$$
$$7\,2\,5 \qquad\qquad\qquad 1\,3\,4 \qquad\qquad\qquad 9\,6\,8$$
$$9\,6\,8 \qquad\qquad\qquad 7\,2\,5 \qquad\qquad\qquad 1\,3\,4$$

The operation **c** rotates the columns so that the last column becomes the first column and the other two columns shift to the right:

$$cM = \begin{matrix} 4 & 1 & 3 \\ 5 & 7 & 2 \\ 8 & 9 & 6 \end{matrix} \qquad c^2M = \begin{matrix} 3 & 4 & 1 \\ 2 & 5 & 7 \\ 6 & 8 & 9 \end{matrix} \qquad crM = \begin{matrix} 8 & 9 & 6 \\ 4 & 1 & 3 \\ 5 & 7 & 2 \end{matrix}$$

$$cr^2M = \begin{matrix} 5 & 7 & 2 \\ 8 & 9 & 6 \\ 4 & 1 & 3 \end{matrix} \qquad rc^2M = \begin{matrix} 6 & 8 & 9 \\ 3 & 4 & 1 \\ 2 & 5 & 7 \end{matrix} \qquad c^2r^2M = \begin{matrix} 2 & 5 & 7 \\ 6 & 8 & 9 \\ 3 & 4 & 1 \end{matrix}$$

Now we need to note exactly what the left and right diagonals are for each of these squares:

Left diagonal squares					Right diagonal squares				
1 2 8	M	crM	c^2r^2M		4 2 9	M	cr^2M	rc^2M	
9 3 5	rM	c^2M	cr^2M		8 3 7	rM	cM	c^2r^2M	
7 6 4	r^2M	cM	rc^2M		5 6 1	r^2M	c^2M	crM	

To construct a Sudoku square, we choose three of the left diagonal squares, one from each row, to form the left diagonal of the final Sudoku square. Then choose three of the right diagonal squares to form the right diagonal of the final Sudoku square. This ensures that the completed long left diagonal for the Sodoku square will have the numbers 128935764 and that the long right diagonal will have the numbers 429837561.

Complete the Sudoku square by choosing the remaining four squares, while observing two final conditions:

[1] Each row of squares selected must have a term with no r, a term with r and a term with r^2

[2] Each column of the squares selected must have a term with no c, a term with c and a term with c^2.

For example, we might select

$$\begin{matrix} M & rcM & c^2r^2M \\ r^2cM & c^2M & rcM \\ rc^2M & M & r^2M \end{matrix}$$

We have chosen M c^2M and r^2M to give a left diagonal 128359764 and c^2r^2M c^2M r^2M to give a right diagonal 783156942

Each row of the selected squares has just one r and just one r^2 term.

Each column of the selected squares has just one c term and just one c^2 term.

The complete Sudoku square in this case is

```
1 3 4   8 9 6   2 5 7
7 2 5   4 1 3   6 8 9
9 6 8   5 7 2   3 4 1

5 7 2   3 4 1   8 9 6
8 9 6   2 5 7   4 1 3
4 1 3   6 8 9   5 7 2

6 8 9   1 3 4   7 2 5
3 4 1   7 2 5   9 6 8
2 5 7   9 6 8   1 3 4
```

You would now select at least 20 of the numbers to complete the Sudoku puzzle.

The original square M can be selected in 9x8x7x6x5x4x3x2x1 = 362 880 ways so this method only produces a fraction of the complete number of Sudoku squares.

If M was chosen to be a magic square then you would have a cunning ploy for solving the puzzle!

===

Solutions

The shellfish was 18 years old.

Problem 9

[1]   weigh 3 against 3 then you can know where the heavy ball is.
Then weigh 1 against 1 from the group with the heavy ball.

[2]   No

[3]   you need one extra weighing to find out if the odd ball is heavier or
lighter.

Problem 10   Triangles of numbers

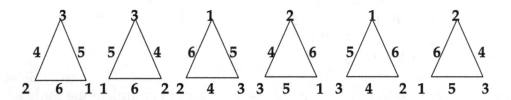

Problem 11

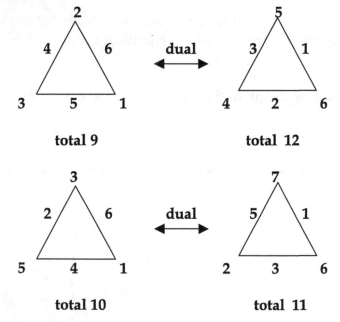

Squares of numbers

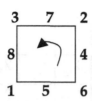

Problem 12

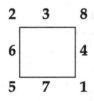

 dual

total 13   total 14

 dual

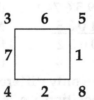

total 13   total 14

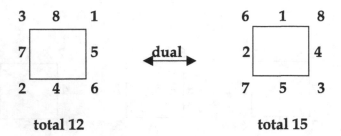

total 12   total 15

Sudoku

2	4	9	5	7	3	8	1	6
6	8	1	9	2	4	3	5	7
7	3	5	8	6	1	4	9	2
1	6	8	4	9	2	7	3	
5	7	3	6	1	8	2	4	9
9	2	4	3	5	7	6	8	1
3	5	7	1	8	6	9	2	4
4	9	2	7	3	5	1	6	8
8	1	6	2	4	9	5	7	3

Problem 14

```
    9567        deduce that
  + 1085        M must be 1
   10652        then S must be 9
                then R must be 8
```

then the only different numbers that give a carry from the units position are 7+5 = 12

[15] The Pineapple Theorem

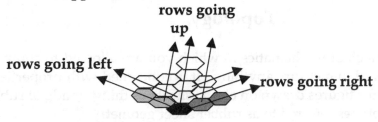

Let **V** be the numbers of rows going up
Let **R** be the numbers of rows going round to the right
Let **L** be the number of rows going round to the left.

Choose one segment (coloured black)
Consider the row going right (dark grey)
And the rows going left (light grey)

The light grey row and the dark grey row must meet at a segment round the back of the pineapple. Let this segment be shaded black and suppose that we have unwrapped the light and dark segments:

L  R

The two black segments to the left and right (marked L and R) coloured black represent the single segment where the light and dark grey rows meet.
The number of segments going up the pineapple is the number of segments between the L and R (including one of them) plus the black at the bottom, which is 7 in this figure, = **light+dark + 2**
But the number of light grey + black is equal to the number of rows going round to the right:
           R = light grey + one black
and the number of dark grey + black segments is the number of rows going round to the left:
           L = dark grey + one black

Now   **(light grey+black) + (dark grey+black) = number of greys + 2**

Therefore   **R + L = V** which proves the pineapple theorem.
           ==========================

CHAPTER 10

Topology

Topology is a branch of mathematics in which you are allowed to count but you are not allowed to measure anything. It is concerned with properties of objects that hold for figures drawn on elastic sheets or things made of rubber. Topology is sometimes referred to as rubber sheet geometry.

If I draw four lines on a sheet of elastic, in general there will always be 3 regions, 6 points and 4 (probably curved) lines. If the elastic is deformed in any way, without tearing, this will still be true.

(Note we ignore such cases of areas with zero area or parallel lines meeting at infinity)

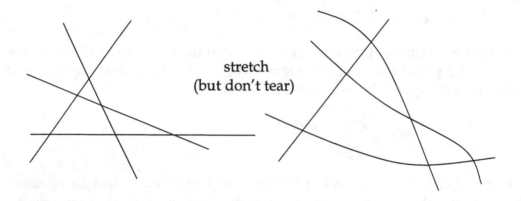

stretch
(but don't tear)

If I made a key out of plasticine, I could mould the key into the shape of a cup:

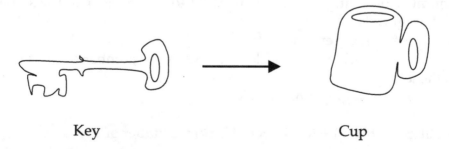

Key   Cup

Basically, I still have a lump of plasticine with a hole in it.
We say, the key and the cup are "topologically equivalent".

Exercise 1

1. Which of the following are topologically equivalent?

A B C D E F G H I

J K L M N O P Q R

S T U V W X Y Z

2 Which of these numerals are topologically equivalent?

0 1 2 3 4 5 6 7 8 9

3. Which of these are topologically equivalent?
   (i) An egg cup,
   (ii) a jug,
   (iii) a rubber tyre,
       (iv) a balloon,
   (v) a hollow tennis ball,
   (vi) a teapot,
       (vii) a mans open front shirt,
   (viii) a tea shirt,
   (ix) a pair of ladies panties,
   (x) a pair of gents pants

Polyhedrons (also **polyhedra**)

A polyhedron is a 3 dimensional shape bounded by flat faces with straight edges and a number of vertices. Vertices are corners where edges meet.

Euler's theorem for polyhedrons requires us to count the faces, edges and vertices.

Examples

		Faces	vertices	edges
Cube		6	8	12
Pyramid		5	5	8
Diamond		8	6	12
Pyramid on a 5 sided base		6	6	10

Euler's theorem states that

$$F + V = E + 2$$

(proof later)

For any polyhedron where **F** is the number of faces, **V** is the number of vertices and **E** is the number of edges.

It is not difficult to prove Euler's Theorem in certain special cases:

Example 1. Prismoids

A prismoid solid has the same cross section along its length.

Proof for a Prismoid solid:

Suppose that the cross section has **n** sides:

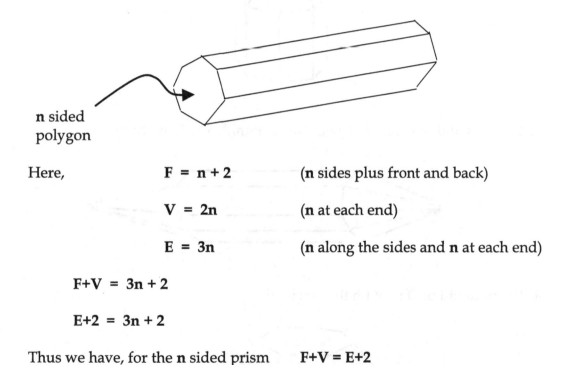

n sided polygon

Here, $F = n + 2$ (**n** sides plus front and back)

$V = 2n$ (**n** at each end)

$E = 3n$ (**n** along the sides and **n** at each end)

$F + V = 3n + 2$

$E + 2 = 3n + 2$

Thus we have, for the **n** sided prism $F + V = E + 2$

================================

Exercises 2

Prove Euler's Theorem for the following solids

1. A pyramid on an **n** sided base

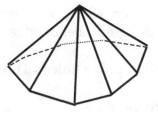

2. The pyramid of Qu 1 stuck on the prismoid of example 1

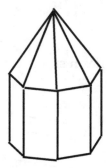

3. The prismoid of Example 1 with two pyramids stuck on the ends

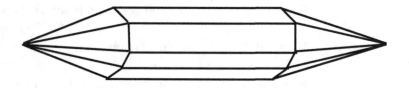

4. The pyramid of Qu 1 with the tip cut off

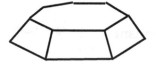

If we have two polyhedra having one face exactly the same shape and size then we can stick the two solids together on this common face, for example:

pyramid

+

cuboid

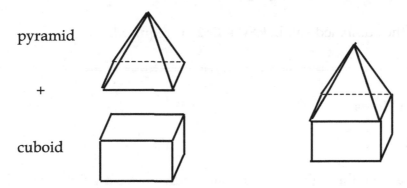

Theorem 1

If Euler's Theorem works for the two separate solids then the Theorem will also work for the combined solid

Proof

For solid 1 we have $\qquad$ $F1 + V1 - E1 = 2$

For solid 2 we have $\qquad$ $F2 + V2 - E2 = 2$

For the combined solid, we lose two faces where they are joined:

$$\text{Faces} = F = F1 + F2 - 2$$

If the common face has **n** edges, we lose **n** vertices at the join

$$\text{Vertices} = V = V1 + V2 - n$$

We also lose **n** edges at the join so that

$$\text{Edges} = E = E1 + E2 - n$$

Therefore, for the combined solid

$$F + V - E = F1+F2 - 2 + V1+V2 - n - (E1+E2 - n)$$

$$= F1 + V1 - E1 + F2 + V2 - E2 - 2$$

$$= 2 + 2 - 2$$

Hence, for the combined solid, **F+V = E+2** as required.

================================

Solids with Holes

Theorem 2

If we bore a hole through an Euler solid and remove a prism shaped piece, then Euler's formula still holds for the remaining shape with the hole.

Proof

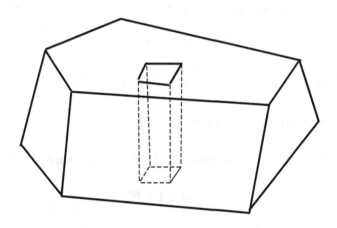

For the complete solid with no hole, $Fc + Vc - Ec = 2$

For the removed piece, $Fr + Vr - Er = 2$

For the solid with the hole, we have $F = Fc + Fr - 2$
because two faces from the piece removed, will not be counted.

Also $V = Vc + Vr$

and $E = Ec + Er$

Hence, $F + V - E = Fc + Fr - 2 + Vc + Vr - (Ec + Er)$

$$= Fc + Vc - Ec + Fr + Vr - Er - 2$$

$$= 2 + 2 - 2$$

$$= 2 \quad \text{as required}$$

Thus, Euler's Theorem still works if we bore a prism shaped hole through from one face to another.

==

Solids with Bits Stuck on

Theorem 3

If we have two solids for which Euler's Theorem holds and we stick one of them in the middle of one of the faces of the other, then for the combined solid

$$F + V \;=\; E + 3$$

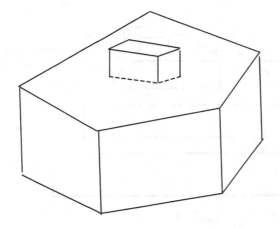

Proof

Suppose there are **n** edges round the "stuck on" face. Then

for solid 1 (big one) $F1 + V1 - E1 = 2$

for solid 2 (stuck on) $F2 + V2 - E2 = 2$

for the combined solid $F = F1 + F2 - 1$ (we have lost the stuck on face)

$V = V1 + V2$ (counting the 'inward' pointing vertices)

$E = E1 + E2$

thus $F + V - E = F1 + F2 - 1 + V1 + V2 - (E1 + E2)$

$= F1 + V1 - E1 + F2 + V2 - E2 - 1$

$= 2 + 2 - 1$

thus $F + V = E + 3$

Corollary

If we continue to add further bits on in this way, not touching or over
lapping any edges, then we get an extra one on the right hand side for each
piece we add, e.g.

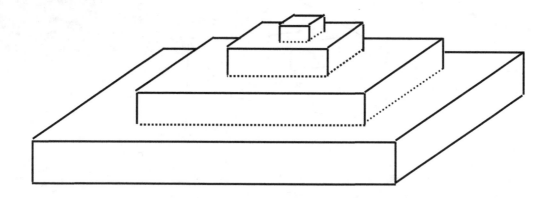

For this step pyramid, with 3 "stuck ons", $F + V = E + 5$

Theorems in two dimensions

Suppose that we take one of the Euler solids, stand it on one face and squash it flat. On the paper, we then have a network of polygons, for example:

The Cube

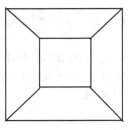

Pyramid on a square base

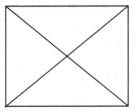

Diamond

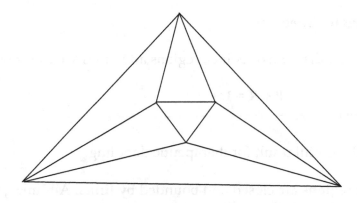

Pyramid on a pentagonal base

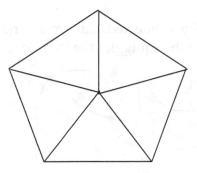

Definitions: Each face **F** has been squashed flat into a region **R**

Each vertex **V** gives a point **P**

Each edge **E** gives one link **L** in the network

Euler's Theorem   **F + V = E + 2** gives a related theorem for the network, but we have lost one face, the face that the solid was standing on, thus **R = F-1.** However, we do have **V=P** and **E=L.**

Thus $F + V = E + 2$

Gives $R+1 + P = L + 2$

or $R + P = L + 1$

regions + points = links + 1

This result is quite general:

The Squiggles Theorem

For any plane diagram of points, regions and links, we can show that

$$P + R = L + 1$$

(proof later)

This is a delightful result for therapeutic doodling.

Note that **regions** are closed and bounded by **links**. All **links** join two **points** or start and finish at the same **point**.

Squiggles must be fully connected.....there are no disjoint parts or isolated floating bits. Applying the squiggles theorem to this squiggle, we have

3+4 = 6+1

Examples

1.

P = 4
R = 2
L = 5   P+R=L+1

2.

P = 8
R = 1
L = 8   P+R=L+1

3.

P = 6
R = 0
L = 5   P+R=L+1

4.

    P=1
    R=2
       L=2

     P=1
    R=1
       L=1

5.

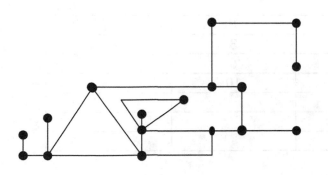

17+4 = 20+1

Exercise 3

1. Complete a table showing regions **R**, points **P** and links **L** for 1x1, 2x2, 3x3 .... size blocks of squares:

size	R	P	L
1x1	1	4	4
2x2	4	9	
3x3	9	16	
4x4			
..			
..			

1x1 square

2x2 square   3x3 square

2. Complete a table showing regions **R**, points **P** and links **L** for 1 row, 2 row, 3 row, 4 row...... networks of triangles:

1 row

2 row

3 row

rows	R	P	L
1	1	3	3
2	4	6	
3	9	10	
4			
..			
..			

3. Complete a table for regions **R,** points **P** and links **L** for hexagonal chessboards of different sizes, verifying that **P+R = L+1**

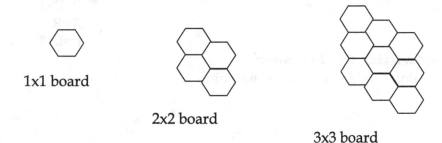

1x1 board

2x2 board

3x3 board

size	R	P	L
1x1	1	6	6
2x2	4	16	19
3x3	9	30	
4x4			

The Squiggles Theorem

For any plane network **P+R=L+1**

Proof

The proof depends on showing that any network can be built up from a simpler network by adding a new point or a new link and that if the Squiggles theorem holds for the simpler network then it must also hold for the larger network.

The different ways that a network can be enlarged, by adding a point or adding a link, are as follows:

(i) **add a point:**
this point can only be added in the middle of a link
e.g.

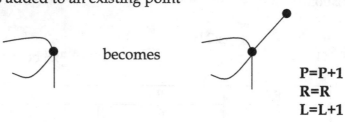

becomes

P=P+1
R=R
L=L+1

(ii) **add a flying link to a point:**
the link is added to an existing point
e.g.

becomes

P=P+1
R=R
L=L+1

(iii) **add a link between two existing points**
this always creates an extra region
e.g.

becomes

P=P
R=R+1
L=L+1

(iv) **add a link starting and ending at the same point**
this also creates an extra region
e.g.

P=P
L=L+1
R=R+1

becomes

or, the loop may enclose some of the network:

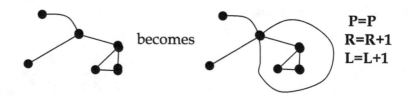

becomes

P=P
R=R+1
L=L+1

In every case, the number of links increases by one and, either the number of points, or the number of regions also increases by one.

Thus, if **P+R=L+1** for the old network, we also have **P+R=L+1** for the new network.

Now the simplest network consists of just one point: ● for which **P+R=L+1**

and any more complex network can be built from this by adding bits using one of (i), (ii), (iii) or (iv) and , as we have seen, each time we add a bit we find that the Squiggles Theorem continues to hold.

=======================================

We are now in a position to prove Euler's theorem for solid figures:

Theorem (Euler's Theorem)

For any solid figure with flat polygonal faces

$$F + V = E + 2$$

Proof

Stand the solid on one face and squash and stretch it down to gives a network of **P** points, **R** regions and **L** links.

Then we have **P=V, R=F-1 and L=E**

We know that **P+R=L+1,** hence **V + F-1 = E+1**

$$\text{Giving} \quad \text{F+V=E+2}$$

=======================================

Plato's Solids

The regular solids are the five polyhedrons that have congruent regular polygons for faces and have exactly similar vertices.
They were discovered by the Greek mathematician Pythagoras, who lived from 582 BC to 507 BC and were described by Euclid (325 BC – 265 BC) in book 13 of his Elements.

Plato, born in 427 BC in Athens, knew of these five regular solids and mentions them in his dialogues written for his Academy that he founded for philosophical discussion in his garden in 385 BC. His academy was the first institution for higher learning in the western world and lasted for 900 years until it was closed down by Emperor Justinian. The five regular solids are called the Platonic solids.

We will use Euler's Theorem to show there are just five regular solids and no more.

The regular polygons that we could consider for the faces of regular solids are:

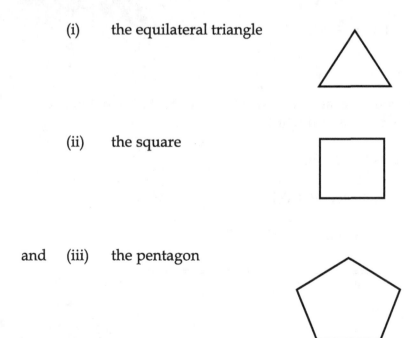

    (i)   the equilateral triangle

    (ii)   the square

and (iii)   the pentagon

Hexagons alone cannot be used because three hexagons at a vertex would make a flat corner since the angle of a regular hexagon is 120°

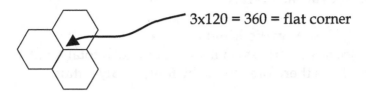

3x120 = 360 = flat corner

At the vertices of a regular solid, we have the following choices:

(i)   3 triangles   (total =3x60=180)
(ii)   4 triangles   (total =4x60=240)
(iii) 5 triangles   (total = 5x60=300)
(iv)   3 squares   (total= 3x90=270)
(v)   3 pentagons (total= 3x108=324)

Each case gives us a different Platonic solid.

Case (i)   3 triangles at each vertex

Let there be **F** faces, **V** vertices and **E** edges.
Since each face has 3 vertices on it we would get a total **V=3F** but each vertex uses 3 faces therefore, divide by three, **V=3F/3** thus

$$V = F$$

Similarly, each face has 3 edges giving a total **E=3F** but each edge is shared by two faces therefore, divide by 2, **E=3F/2** thus

$$E = 3F/2$$

Substitute in Euler's fomula **F+V = E+2** and we get

$$F + F = 3F/2 + 2$$

Hence   **F = 4, V=4, E=6**

Case (i) gives the **tetrahedron**

with 4 faces, 4 vertices and 6 edges.

Case (ii)   4 triangles at each vertex

Let there be **F** faces, **V** vertices and **E** edges.
Since each face has 3 vertices on it we would get a total **V=3F** but each vertex uses 4 faces therefore, divide by four, **V=3F/4** thus

$$V = 3F/4$$

Similarly, each face has 3 edges giving a total **E=3F** but each edge is shared by two faces therefore, divide by 2, **E=3F/2** thus

$$E = 3F/2$$

Substitute in Euler's formula **F+V = E+2** and we get

$$F + 3F/4 = 3F/2 + 2$$

$$4F + 3F = 6F + 8$$

$$F = 8, \quad V = 6, \quad E = 12$$

Case (ii) gives the **octahedron**

With 8 faces, 6 vertices and 12 edges

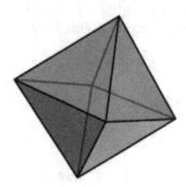

Case (iii)   5 triangles at each vertex

Let there be **F** faces, **V** vertices and **E** edges.
Since each face has 3 vertices we would get a total **V=3F** but each vertex uses 5 faces therefore, divide by five, **V=3F/5** thus

$$V = 3F/5$$

Similarly, each face has 3 edges giving a total **E=3F** but each edge is shared by two faces therefore, divide by 2, **E=3F/2** thus

$$E = 3F/2$$

Substitute in Euler's formula **F+V = E+2** and we get

$$F + 3F/5 = 3F/2 + 2$$

$$10F + 6F = 15F + 20$$

$$F = 20, \quad V = 12, \quad E = 30$$

Case (iii) gives the **icosahedron** with

20 faces, 12 vertices and 30 edges

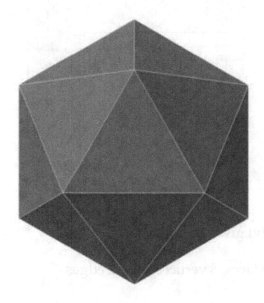

Case (iv)   3 squares at each vertex

Let there be **F** faces, **V** vertices and **E** edges.

Since each face has 4 vertices we would get a total **V=4F** vertices but each vertex uses 3 faces therefore, divide by three, **V=4F/3** thus

$$V = 4F/3$$

Similarly, each face has 4 edges giving a total **E=4F** edges but each edge is shared by two faces therefore, divide by 2, **E=4F/2** thus

$$E = 4F/2$$

Substitute in Euler's formula **F+V = E+2** and we get

$$F + 4F/3 = 4F/2 + 2$$

$$3F + 4F = 6F + 6$$

$$F = 6, \ V = 8, \ E = 12$$

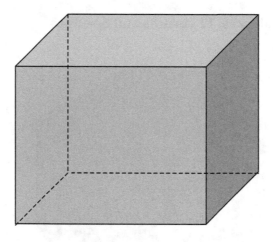

Case (iv) gives the **cube**

With 6 faces, 8 vertices and 12 edges

Case (v)   3 pentagons at each vertex

Let there be **F** faces, **V** vertices and **E** edges.
Since each face has 5 vertices on it we would get a total **V=5F** vertices but each vertex uses 3 faces therefore, divide by three, **V=5F/3** thus

$$V = 5F/3$$

Similarly, each face has 5 edges giving a total of **E=5F** edges but each edge is shared by two faces therefore, divide by 2, **E=5F/2** thus

$$E = 5F/2$$

Substitute in Euler's formula **F+V = E+2** and we get

$$F + 5F/3 = 5F/2 + 2$$

$$6F + 10F = 15F + 12$$

$$F = 12, \quad V = 20, \quad E = 30$$

Case (v) gives the **dodecahedron**

With 12 pentagonal faces, 20 vertices and 30 edges

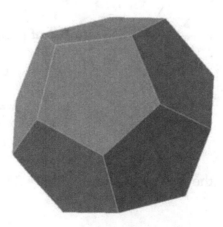

The **tetrahedron, octahedron, icosahedron, cube and dodecahedron** are the only regular solid figures.

Pick's Theorem

Georg Pick (1859-1942) was born in Vienna in 1859 and published the theorem that bears his name in 1899. It is deceptively simple to state and must have formed the basis for many school projects in GCSE maths.
Georg Pick was taken by the Nazis during the second world war. He died in a concentration camp in 1942.

Theorem

If a polygonal shape is drawn on graph paper with its vertices at grid points, with

       (i)   the number of points on the boundary being **n**

      (ii)   the number of points inside the figure being **b**

then the area of the figure is **A = n + ½b − 1**

A Proof of Pick's Theorem will be presented later.

Example 1

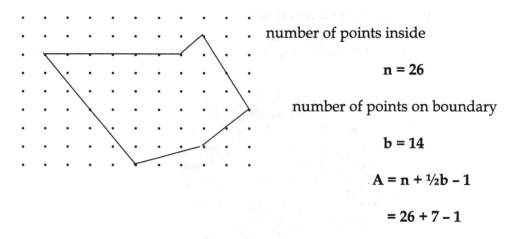

number of points inside

$$n = 26$$

number of points on boundary

$$b = 14$$

$$A = n + \tfrac{1}{2}b - 1$$

$$= 26 + 7 - 1$$

Using Pick's Theorem, the area of this figure is **A = 32**

Example 2
Proof of Pick's Theorem for squares

Suppose that we have an **NxN** square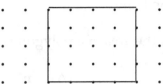

The number of points inside will be $\quad$ **n = (N-2)x(N-2)**
The number of points on the boundary will be $\quad$ **b = 4x(N-1)**

Pick's formula for the area is $\mathbf{n + \tfrac{1}{2}b - 1}$ $= \mathbf{(N-2)(N-2) + 2(N-1) - 1}$

$$= \mathbf{N^2 - 4N + 4 + 2N - 2 - 1}$$

$$= \mathbf{N^2 - 2N + 1}$$

$$= \mathbf{(N-1)^2}$$

Thus Pick's formula gives the correct result for the **NxN** square.

Exercise 4

1. Prove Pick's Theorem for an **MxN** rectangle.

$\quad$ e.g. **M=5, N=4**

(N.B. check for **M < 3 ; N<3**)

2. Prove Pick's Theorem for an isosceles right angle triangle of short side **N**

$\quad$ e.g. **N=5**

(N.B. check for **N<3**)

Pick with Holes

If there is a hole in the polygon then Pick's formula is modified:

Pick with one Hole

Suppose that the area of a polygon is given by Pick as

$$A1 = n1 + \tfrac{1}{2}b1 - 1$$

Suppose that there is one hole in the polygon and that the area of the hole is

$$A2 = n2 + \tfrac{1}{2}b2 - 1$$

Let **N** and **B** be the numbers of points inside and on the boundary for the region with the hole, then we have:

$$N = n1 - n2 - b2$$

and $$B = b1 + b2$$

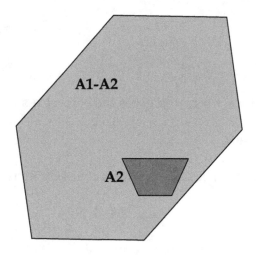

Using Pick, we have $$A1 - A2 = (n1 + \tfrac{1}{2}b1 - 1) - (n2 + \tfrac{1}{2}b2 - 1)$$

$$= n1 - n2 - b2 + \tfrac{1}{2}b1 + \tfrac{1}{2}b2$$

$$= N + \tfrac{1}{2}B$$

Thus the formula for Pick with one hole is **Area(1) = n + ½b**

Pick with two Holes

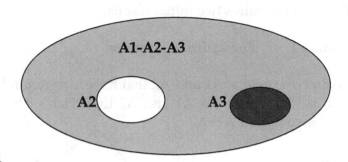

$A_1-A_2-A_3$

A_2 A_3

Using the Pick notation:

$A_1 = n_1 + \frac{1}{2}b_1 - 1$

$A_2 = n_2 + \frac{1}{2}b_2 - 1$

$A_3 = n_3 + \frac{1}{2}b_3 - 1$

For the polygon with two holes, we have

$N = n_1 - n_2 - n_3 - b_2 - b_3$

$B = b_1 + b_2 + b_3$

Thus $A_1 - A_2 - A_3 = (n_1 + \frac{1}{2}b_1 - 1) - (n_2 + \frac{1}{2}b_2 - 1) - (n_3 + \frac{1}{2}b_3 - 1)$

$= n_1 - n_2 - n_3 - b_2 - b_3 + \frac{1}{2}b_1 + \frac{1}{2}b_2 + \frac{1}{2}b_3 + 1$

$= N + \frac{1}{2}B + 1$

Thus we have the following extensions to Pick's Theorem:

Pick with 0 holes $A(0) = n + \frac{1}{2}b - 1$

Pick with one hole $A(1) = n + \frac{1}{2}b$

Pick with two holes $A(2) = n + \frac{1}{2}b + 1$
and similarly,
Pick with three holes $A(3) = n + \frac{1}{2}b + 2$
etc....

Proof of Pick's Theorem

This proof requires two initial results:

Lemma 1   The addition property

Given two regions A1 and A2 that have a common boundary line, we prove that, if Pick works for A1 and A2 then Pick's Theorem also works for the combined region A1+A2

Proof

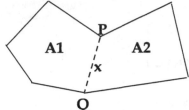

Let there be **x** points along
the common boundary PQ
(not counting P and Q) then

$$A1 = n1 + \tfrac{1}{2}b1 - 1$$

$$A2 = n2 + \tfrac{1}{2}b2 - 1$$

But for the combined region $N = n1 + n2 + x$

$$B = b1 + b2 - 2x - 2 \quad \text{(P and Q were counted}$$

twice)
Hence $$A1+A2 = (n1 + \tfrac{1}{2}b1 - 1) + (n2 + \tfrac{1}{2}b2 - 1)$$

$$= n1 + n2 + x + \tfrac{1}{2}(b1 + b2 - 2x - 2) - 1$$

$$= N + \tfrac{1}{2}B - 1$$

This proves the **addition property**

The extended addition property

Using the above, we can deduce that if Pick's theorem holds for lattice polygons **A1, A2, A3,........** which have common boundaries with each other, then Pick's Theorem will hold for the combined polygon **A1+A2+A3+.....**

Lemma 2   The Subtraction Property

Given two regions B and C, joined together along a common boundary line to form region A, we prove that if Pick's Theorem works for regions A and B, then it will also work for the region C (=A–B).

Let there be x points along the common boundary PQ, (not counting P and Q), then

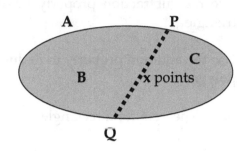

Using an obvious notation

$A = n_a + \frac{1}{2}b_a - 1$

$B = n_b + \frac{1}{2}b_b - 1$

But for region C we have $n_c = n_a - n_b - x$

$$b_c = b_a - b_b + 2x + 2 \text{ (taking account of P and Q)}$$

thus the area C is $C = A - B$

$$= (n_a + \frac{1}{2}b_a - 1) - (n_b + \frac{1}{2}b_b - 1)$$

$$= n_a - n_b - x + \frac{1}{2}(b_a - b_b + 2x + 2) - 1$$

$$= n_c + \frac{1}{2}n_b - 1$$

Which proves the subtraction property

The Extended subtraction property.

If Pick's Theorem holds for lattice polygons **A, B, C, D,....** where **B, C, D,...** are disjoint polygons contained in **A** sharing the boundary of **A**, then Pick's Theorem will hold for the reduced polygon **A-B-C-D-....** Note however, that none of the subtractions must result in a hole. The subtraction property must not be confused with **Pick with Holes**.

Outline of the proof:

(i)   Prove Pick's Theorem for a rectangle

(ii)   Prove Pick's Theorem for a 90° triangle

(iii)   Use the subtraction property to deduce Pick's Theorem for any triangle

(iv)   Use the addition property to deduce Pick's Theorem for any lattice polygon.

(i) Pick's Theorem for a rectangle

Proof (i)
Suppose that the rectangle is **M** points by **N** points
The the number of points inside is $n = (M-2)(N-2)$ for M,N >1

The number of points on the boundary is $b = 2(M-1) + 2(N-1)$

Pick's Formula therefore gives $(M-2)(N-2) + \frac{1}{2}(2(M-1)+2(N-1) - 1$

$$= MN - 2M - 2N + 4 + M - 1 + N - 1 - 1$$

$$= MN - M - N + 1$$

$$= (M-1)(N-1) \text{ as required}$$

(ii) To prove Pick's Theorem for a 90° triangle

Proof (ii)
Complete the smallest
bounding rectangle

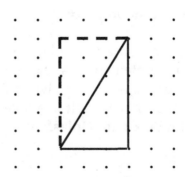

Suppose that there are **x** lattice points on the diagonal.

Let **n** be the number of points inside the triangle, then there will be **(2n + x –2)** points inside the rectangle.

Let there be **b** points on the boundary of the triangle, then there will be **(2b – 2x + 2)** points on the boundary of the rectangle.

If Δ is the area of the triangle, then the area of the square is 2Δ hence, using Pick's Theorem for the rectangle, we have

$$2\Delta = (2n + x - 2) + \tfrac{1}{2}(2b - 2x + 2) - 1$$

$$2\Delta = 2n + \tfrac{1}{2}(2b) - 2$$

$$\Delta = n + \tfrac{1}{2}b - 1$$

which proves Pick's Theorem for the 90° triangle.

(iii) Prove Pick's Theorem for any triangle.

Proof (iii)

Complete the smallest bounding rectangle, then we have a figure corresponding to either

case 1	case 2

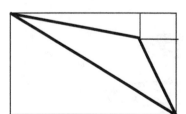 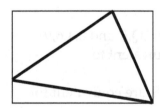

In case 1 we use the extended subtraction principle to deduce Pick's Theorem for the triangle by subtracting three triangles and one rectangle from the bounding rectangle, and in case 2, we just need to subtract three triangles from the bounding rectangle.

This proves Pick's Theorem for triangles of any shape.

(iv) Last part

Proof (iv)

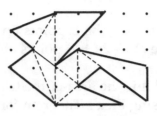

Any polygon drawn on the grid
can be broken down into a set
of triangles:

For example, here, the thick outlined 13agon has been divided up into 10
triangles using the thin dotted lines.
We have proved that Pick's Theorem holds for each of the triangles,
therefore, using the extended addition property, Pick's Theorem will hold for
the complete polygon.

This completes the proof of Pick's Theorem.

==

Solutions

Ex 1

Qu 1 and 2

A,D,O,P,Q,R and 0,4,6,9
are equivalent to

B and 8 are equivalent to

The rest are all equivalent to

Qu 3.

The egg cup and the balloon have no holes, if you take the balloon to be a 3 dimensional object with thickness, then the egg cup and the balloon are equivalent.

If the jug has a complete handle then it has one hole and is equivalent to the tyre. (If the jug does not have a complete handle then it is equivalent to the egg cup)
The tennis ball is not equivalent to any of the other objects. It is the only object that completely encloses a 3 D space.

Assuming that the last five objects are to be taken as 3 dimensional objects with thickness, then the teapot and the open front shirt are equivalent, each having two holes.

The tea shirt and the ladies panties are equivalent, each having 3 holes.

The gents pants stand alone, having four holes.

Ex 2

Qu 1 $F=n+1$, $V=n+1$, $E=2n$ $F+V = E+2$

Qu 2 $F=2n+1$, $V=2n+1$, $E=4n$ $F+V = E+2$

Qu 3 $F=3n$, $V=2n+2$, $E=5n$ $F+V = E+2$

Qu 4 $F=n+2$, $V=2n$, $E=3n$ $F+V = E+2$

Ex 4

Qu 1   MxN rectangle

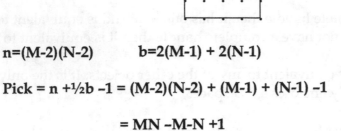

$$n=(M-2)(N-2) \qquad b=2(M-1) + 2(N-1)$$

$$\text{Pick} = n +\tfrac{1}{2}b -1 = (M-2)(N-2) + (M-1) + (N-1) -1$$

$$= MN -M-N +1$$

$$= (M-1)(N-1) \ = \text{area}$$

Qu 2   Isosceles right angled triangle

$$n= 1+2+3+\ldots+(N-3) = \tfrac{1}{2}(N-3)(N-2) \qquad b= 3(N-1)$$

$$\text{Pick} = n+ \tfrac{1}{2}b -1 = \tfrac{1}{2}(N-3)(N-2) + \tfrac{1}{2}.3(N-1) -1$$

$$= \tfrac{1}{2} (N^2 -2N +1)$$

$$= \tfrac{1}{2} (N-1)(N-1) = \text{area}$$

CHAPTER 11

Coordinates

Rene Descartes (1596-1650)

Rene Descartes began his education at the age of eight in a college in Anjou, where he remained for eight years. He was not a strong child and while there his teachers often allowed him to stay in bed till 11 am. So, you teachers, if your student arrives late for the 9 o'clock class do not worry, you may have a budding Descartes! After college he studied law and then travelled throughout Europe and did a two year stint in the Bavarian army. In 1628 he settled in Holland where he began writing on Physics, at the very time when Galileo was condemned to house arrest for suggesting that the Earth went round the Sun.

Descartes's most important mathematical work was "La Geometrie", published in 1637. In it he describes how algebra can be applied to problems in geometry. It became the foundation of what we now call Cartesian coordinate geometry. Descartes died in 1650 while on a visit to Sweden at the invitation of Queen Christina where he was to give her instruction in philosophy and set up an Academy of Sciences.

The bulk of this chapter concerns two dimensional Cartesian coordinates with the emphasis on the straight line and the circle, much as would be found in a first course in mathematics but with the use of column vectors where appropriate. We open, however, with a discussion of coordinates in one dimension.

Lineland

An ant finds himself crawling along a piece of string and has no way of getting off. He can crawl to the left or crawl to the right but cannot get off of the piece of string which extends as far as he knows, an infinite distance in both directions. He is now living in a one dimensional world and finds it most inconvenient because, as soon as he finds an obstacle in his path, he is stuck as there is no way of going around anything in this one dimensional world. In order to get around an obstacle, he would have to come out of the one dimensional world into the world of two dimensions.

Coordinates in one dimension

We a point on the line called the origin of coordinates. The coordinate of a point on the line is the displacement from the origin to the point. We use the word displacement rather than distance because although a distance is always positive, a displacement from the origin in one direction will be taken as positive and a displacement from the origin in the other direction will be taken as negative. In other words, a distance is always positive but a displacement can be either positive or negative depending on the direction from the origin.

The positive direction for an x axis is always chosen to be to the right (but be aware that if you go to the other side of the line, then this would be to the left). Points to the left of the origin have negative coordinates.

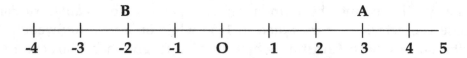

Each point on the axis has a unique x-coordinate. The point A in our diagram is the point with x-coordinate 4, and can be referred to simply as "the point 4". B is the point with x-coordinate –2 and can be referred to as "the point –2". We can also write A(4) and B(-2).

Given two points $A(x_1)$ and $B(x_2)$, we ask the question "How do we get from A to B?"
For example, given A(3) and B(7), the answer would be, move 4 steps to the right.
Given A(5) and B(-2), the answer would be move 7 steps to the left.

The answers to these questions are called the displacements and the displacement from A to B and is written $\overline{AB}$.

Displacements are usually indicated using the overline symbol to distinguish them from distances.

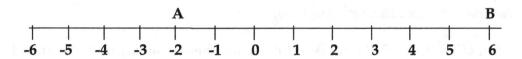

In the above diagram, we have

$$\overline{AB} = 8$$

and
$$\overline{BA} = -8$$

Examples:

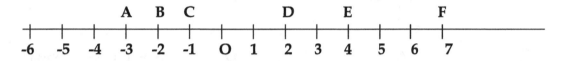

[1] $\overline{DE} = 4 - 2 \quad = 2$

[2] $\overline{FE} = 4 - 7 \quad = -3$

[3] $\overline{DB} = -3 - 2 \quad = -5$

[4] $\overline{AB} = -3 - (-6) = 3$

[5] $\overline{CB} = -3 - (-1) = -2$

Since $\overline{AB} = 1$ and also $\overline{BC} = 1$, we write $\overline{AB} = \overline{BC}$ meaning that the displacement from A to B is the same as the displacement from B to C.

Theorem 1

If the x-coordinate of A is x_1 and the x-coordinate of b is x_2 then
$$\overline{AB} = x_2 - x_1$$

Proof

The following diagrams cover all cases for which A and B are distinct points (i.e. at different positions on the line)

The different cases can be identified by the order of the points **O, A** and **B** on the line. The six different orders are:

OAB, OBA, ABO, BAO, AOB and **BOA** and these cases are each illustrated by a diagram.
The desired result is proved for each separate case.

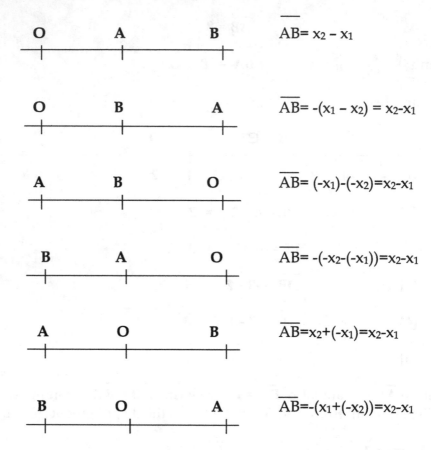

$$\overline{AB} = x_2 - x_1$$

$$\overline{AB} = -(x_1 - x_2) = x_2 - x_1$$

$$\overline{AB} = (-x_1) - (-x_2) = x_2 - x_1$$

$$\overline{AB} = -(-x_2 - (-x_1)) = x_2 - x_1$$

$$\overline{AB} = x_2 + (-x_1) = x_2 - x_1$$

$$\overline{AB} = -(x_1 + (-x_2)) = x_2 - x_1$$

The above diagrams cover all possible cases for which **A** and **B** are distinct points.
If two points or all three points coincide, then the verification of the result is a simple exercise.

Flatland

The two dimensional world is called Flatland after a book by Edwin Abbott (1884). Stephen Hawking notes in "A Brief History of Time" that any inhabitant of Flatland would have to throw up after dinner, because, if the person had an alimentary canal connecting the mouth to the rear end, then the person would fall into two parts.

Flatland is by no means a dull place though, for there are thousands of interesting theorems to be discovered! The only trouble is, that the flatlander would not be able to see them because, to appreciate the theorems he would need to be up above the plane in three space plane and looking down.

Coordinates in Two Dimensions.

The x-y Plane

We normally choose a line Ox for the x-axis and a line Oy at right angles to Ox for our y-axis.

We do not have to choose 90° for the angle (which gives us 'rectangular Cartesian coordinates'), but it is natural and easier to cover our plane with squares rather than parallelograms.

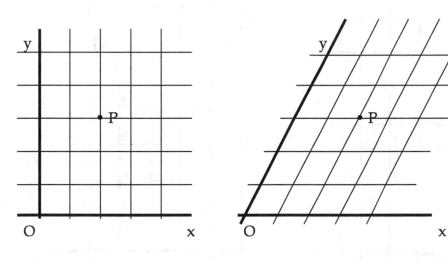

Rectangular axes at 90°   Oblique axes at 60°

The point P has coordinates (2,3) in both sets of axes.

The distance formula for OP² in the rectangular axes is $OP^2 = x^2 + y^2$

The distance formula for OP² in the oblique axes is $OP^2 = x^2 + y^2 + xy$

Using rectangular axes gives the simplest formula for the distance between two points.
In everything that follows we assume rectangular axes with the positive x axis extending to the right and the positive y axis extending upwards.
We refer to this plane with these 90° axes as the Cartesian plane.
Every point P in the x-y plane has a unique pair of real numbers, (x,y), associated with it called its coordinates. We write P(x,y). x is called the x coordinate and y is called the y coordinate.
Every pair of numbers (x,y) gives a single unique point P(x,y). There is a 1→1 correspondence between ordered pairs of numbers (x,y) and points P in the Cartesian plane.

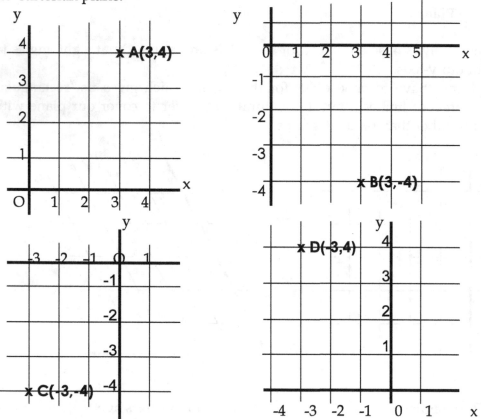

Each of the points A,B,C and D in the above diagrams is at a distance 5 units from the origin O.

Displacements in 2 Dimensions

Again, given two points $A(x_1, y_1)$ and $B(x_2, y_2)$, we ask the question "how do we get from A to B?".

For example, given A(2,3) and B(6,8) as in the following figure, it is clear that we need to go 4 steps in the x direction and then 5 steps in the y direction:

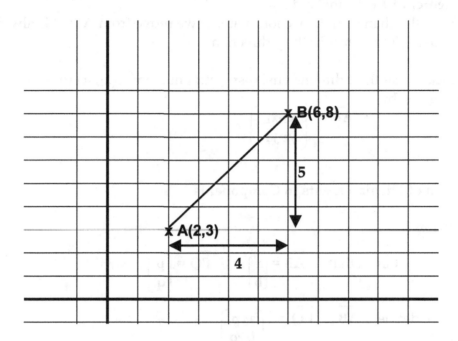

We write this result as

$$\overline{AB} = \begin{pmatrix} 4 \\ 5 \end{pmatrix}$$

We write the 2 dimensional displacement as a **column vector** to avoid confusion with x and y coordinates that are always measured from the origin of the coordinate system. The x displacement is written first and the y displacement written underneath.

Theorem 2

Given A(x_1,y_1) and B(x_2,y_2)

Then $\qquad\qquad \overline{AB} =$ $\qquad \begin{pmatrix} x_2 - x_1 \\ y_2 - y_1 \end{pmatrix}$

Proof

The change in the x coordinate as we move from A to B is always (x_2-x_1) ... by **Theorem 1** from Lineland.

Similarly, the change in the y coordinate, as we move from A to B is always (y_2-y_1), using Theorem 1 in the y direction.

This means that the 2 dimensional displacement from A(x_1,y_1) to B(x_2,y_2) is always given by

$$\overline{AB} = \begin{pmatrix} x_2-x_1 \\ y_2-y_1 \end{pmatrix}$$

Addition of Displacements in Components

Definition 1

$$\text{Given that} \quad \overline{AB} = \begin{pmatrix} a \\ b \end{pmatrix} \qquad \overline{PQ} = \begin{pmatrix} p \\ q \end{pmatrix}$$

$$\text{we define} \quad \overline{AB} + \overline{PQ} = \begin{pmatrix} a+p \\ b+q \end{pmatrix}$$

Definition of Equality

Given $\overline{AB} = \begin{pmatrix} a \\ b \end{pmatrix}$ and $\overline{PQ} = \begin{pmatrix} p \\ q \end{pmatrix}$, $\overline{AB} = \overline{PQ}$ means that a=b and p=q

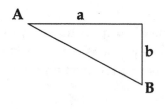

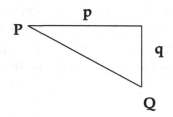

From the above figure, it is clear that the two triangles are congruent and that
AB = PQ. Further, the directions of AB and PQ are the same. That is, AB//PQ.
This means that AB and PQ are opposite sides of a parallelogram.

If two displacements are equal, then their line segments are equal and parallel.

Adding Displacements

Suppose that we have two displacements $\overline{AB} = \begin{pmatrix} a \\ b \end{pmatrix}$ and $\overline{EF} = \begin{pmatrix} e \\ f \end{pmatrix}$

then, by definition 1 $\overline{AB} + \overline{EF} = \begin{pmatrix} a+e \\ b+f \end{pmatrix}$

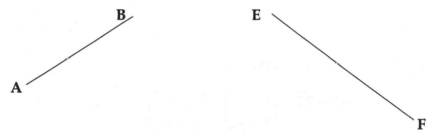

Figure 1

Now draw BD equal and parallel to EF. Then we know that $\overline{BD} = \overline{EF}$

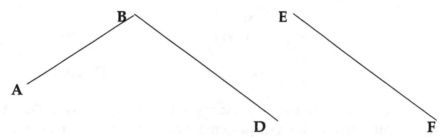

Figure 2

To add AB + EF we therefore draw BD equal and parallel to EF and write

$$\overline{AB} + \overline{EF} = \overline{AB} + \overline{BD}$$

Theorem 3 $\overline{AB} + \overline{BC} = \overline{AC}$

It may seem that Theorem 3 is stating the obvious, i.e. that going from A to B and then going from B to C is the same as going from A to C, but we are interested here in using the component coordinates.
 For completeness we give separate proofs for 1 dimension and for 2 dimensions however, we could derive the 1D theorem from the 2D theorem simply by assigning all the y coordinates to zero. However,

Proof for 1D

Let $A(x_1)$, $B(x_2)$ and $C(x_3)$

Then $\overline{AB} + \overline{BC} = (x_2 - x_1) + (x_3 - x_2)$ by Theorem 1

                            $= x_3 - x_{11}$

                            $= \overline{AC}$ by Theorem 1

Proof for 2D

Let $A(x_1, y_1)$, $B(x_2, y_2)$ and $C(x_3, y_3)$

then $\overline{AB} + \overline{BC} = \begin{pmatrix} x_2 - x_1 \\ y_2 - y_1 \end{pmatrix} + \begin{pmatrix} x_3 - x_2 \\ y_3 - y_2 \end{pmatrix}$ Theorem 2

              $= \begin{pmatrix} x_2 - x_1 + x_3 - x_2 \\ y_2 - y_1 + y_3 - y_2 \end{pmatrix}$ Definition 11.7

              $= \begin{pmatrix} x_3 - x_1 \\ y_3 - y_1 \end{pmatrix}$

              $= \overline{AC}$ Theorem 2

For convenience, we will follow what many books do, that is, we omit to use the overline symbol for a displacement but type displacements using bold face.

 Thus, instead of $\overline{AB}$, we may write AB

Thus in figure 2 above, we can deduce that **AB + EF = AB + BD** because the components of **EF** are the same as the components of **BD**.

Thus **AB + EF = AB + BD = AD** (Theorem 3)

This gives us the "nose to tail" addition rule for displacements.
(This is often referred to as the **Triangle Rule** of addition)

Definition 2 **Adding more than two displacements**

By **AB + EF + PQ, we mean (AB+EF) + PQ**

Theorem 4 **Nose to tail addition**

$$AB + BC + CD = AD$$

Proof

$$AB + BC + CD = (AB+BC) + CD \quad \text{(definition 2)}$$

$$= AC + CD \quad \text{(theorem 3)}$$

$$= AD \quad \text{(theorem 3)}$$

Definition 3 **The zero displacement.**

We write **0** for the zero displacement $\begin{bmatrix} 0 \\ 0 \end{bmatrix}$

Theorem 5 For any three points A, B and C

$$AB + BC + CA = 0$$

Proof
Let $A(x1,y1)$, $B(x2,y2)$, $C(x3,y3)$

then **AB+BC+CA = AA** (theorem 4)

$$= 0 \quad \text{(definition 3)}$$

Definition 4   **Position Vectors**

Given P(x,y), the displacement **OP** is called the position vector of P with respect to origin O.

In order to simplify some notations, we often adopt the convention that points are denoted by capital letters and their position vectors are denoted by the corresponding lower case. Thus the position vector of P(x,y) is **OP=p**

Definition 5   **The negative of a displacement:**

Given $\mathbf{AB} = \begin{bmatrix} x \\ y \end{bmatrix}$ we define $-\mathbf{AB} = \begin{bmatrix} -x \\ -y \end{bmatrix}$

Definition 6   **Subtraction**

$\quad\quad$ **AB – PQ**   means **AB + (- PQ)**

Definition 7   **Multiplication by a number**

$\quad\quad\quad\quad$ k**AB** means $\begin{bmatrix} kx \\ ky \end{bmatrix}$

Rules:

From the basic rule of addition we see that rules of arithmetic are obeyed by displacements. Thus

Commutative rule   **AB + PQ = PQ + AB**

Associative rule   **(AB+CD)+EF = AB+(CD+EF)**

Distributive rule   **k(AB+PQ) = kAB + kPQ**

We do not go into great detail here, but will accept these rules, which follow from the basic definitions, for example, given $\mathbf{AB} = \begin{bmatrix} a \\ b \end{bmatrix}$ and $\mathbf{PQ} = \begin{bmatrix} p \\ q \end{bmatrix}$

then $\mathbf{AB+PQ} = \begin{bmatrix} a+p \\ b+q \end{bmatrix} = \begin{bmatrix} p+a \\ q+b \end{bmatrix} = \mathbf{PQ+AB}$

Theorem 6: $AB = b\text{-}a$

Proof

     $AB = AO+OB = OB+AO = OB\text{-}OA = b\text{-}a$

(Using theorem 3, the commutative rule and the definition of the negative and subtraction)

Theorem 7   **The Mid point theorem**

     If M is the mid-point of AB then $m = \tfrac{1}{2}(a+b)$

Proof

Since M is the mid point of AB $AM = MB$

therefore $m\text{-}a = b\text{-}m$ (Theorem 7)

                             $2m = b+a$

                             $m = \tfrac{1}{2}(a+b)$ ∎

In terms of coordinates this result shows that

the **mid point of A(x_1,y_1) B(x_2,y_2) is M** $\left(\dfrac{x_1+x_2}{2} , \dfrac{y_1+y_2}{2} \right)$

Exercise 1

1.   Given A(a,b), B(a+x,b+y), P(p,q) and Q(p+a,q+b)

     Prove that the mid point of AQ is the same as the mid point of BP.

     Why is this true for all values of a,b,x,y,p and q?

2.   Given A(a,b), C(c,d), E(e,f) and G(g,h)
     P is the mid point of AC, Q is the mid poit of CE, R is the mid point of EG and S is the mid point of GA. Show that the mid point of PR is the same as the mid point of QS.

Theorem 8   **Point of Division**

If P divides **AB** internally in the ratio m:n

then $p = \dfrac{mb + na}{m+n}$

Proof

since P divides AB in the ratio m:n

$$nAP = mPB$$

therefore $n(p-a) = m(b-p)$

$$(m+n)p = mb + na$$

$$p = \frac{mb + na}{m+n}$$

In terms of coordinates, given $A(x_1,y_1)$ and $B(x_2,y_2)$

then P is $\left(\dfrac{nx_1+mx_2}{m+n} \ , \ \dfrac{ny_1+my_2}{m+n} \right)$

Exercise 2

1.   Find the point that divides A(4,7) and B(9,2) in the ratio 2:3

2.   Find the point that divides P(9,3) and Q(2,10) in the ration 3:4

3.   Given A(a,b), B(b,c) and C(c,a)

Write down the coordinates of P the mid point of AB and Q the mid point of BC.
Find the point that divides CP in the ration 2:1
Find the point that divides AQ in the ration 2:1

Theorem 9

The centre of mass of m_1 at A_1 and m_2 at A_2 is given by $\mathbf{g} = \dfrac{m_1 a_1 + m_2 a_2}{m_1 + m_2}$

Proof

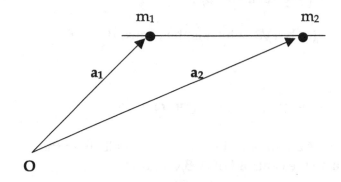

If G is the centre of mass then G divides $A_1 A_2$ in the ratio $m_2 : m_1$

Hence, by theorem 8 $\mathbf{g} = \dfrac{m_1 a_1 + m_2 a_1}{m_1 + m_2}$

Definition 8

The centre of mass of m_1 at A_1, m_2 at A_2, m_3 at A_3 ...... m_n at A_n

is at the point $\mathbf{g} = \dfrac{m_1 a_1 + m_2 a_2 + m_3 a_3 + \ ...... + m_n a_n}{m_1 + m_2 + m_3 + ... + m_n}$

Definition 9   **The centroid of a set of points is the centre of mass of equal masses placed at those points.**

Example

If A′, B′, C′ are the mid points of BC, CA, AB prove that the centroid of ABC is the same as the centroid of A′B′C′

Solution

$$a' = \tfrac{1}{2}(b+c), \; b' = \tfrac{1}{2}(c+a), \; c' = \tfrac{1}{2}(a+b)$$

The centroid of A'B'C' is $(a'+b'+c')/3$

$$= \tfrac{1}{2}(b+c+ c+a + a+b)/3$$

$$= (a+b+c)/3 \text{ which is the centroid of ABC.}$$

Exercise 3

1.   If G is the centroid of A, B, C then   **GA+GB+GC = 0**

2.   If M is the mid point of AB and N is the mid point of CD then the mid point of M and N is also the centroid of A,B,C and D.

3.   If P, A and B are any three points and M is the mid point of AB then **2PM = PA + PB**

4.   If **GA+GB+GC = 0** and P is any point then **PA+PB+PC = 3PG**

Theorem 10   **Pythagoras in Rectangular Cartesian axes**

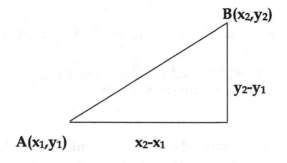

We know from Lineland, that the displacement from $A(x_1,y_1)$ to $B(x_2,y_2)$ in the x direction is always (x_2-x_1) and that the displacement from A to B in the y direction is (y_2-y_1).

Since we are using rectangular ($90°$) axes we can use Pythagoras's Theorem to give

$$AB^2 = (x_2-x_1)^2 + (y_2-y_1)^2$$

Exercise 4

1.   Calculate the distance AB for each of the following pairs of points:

      (i)   A(2,4)   B(5,8)

      (ii)   A(6,8)   B(3,12)

      (iii)   A(12,12)   B(6,4)

      (iv)   A(10,8)   B(5,-4)

      (v)   A(3,4)   B(8,-8)

      (vi)   A(-8,-4)   B(16,3)

      (vii) A(-5,-5)   B(10,3)

      (viii) A(24,-20)   B(-24,-6)

2. Labelling both the x axis and the y axis from –7 to +7, plot the following points on graph paper: (3,6), (2,5), (4,7), (-2,1), (-4,-1), (-6,-3)
Check that all the points lie on a straight line.

We say that **y=x+3** is the equation of this line.

We note also, that all points above the line have coordinates that satisfy the inequality **y>x+3**

y>x+3

all points below the line have coordinates that satisfy the inequality   **y<x+3**

y<x+3

Equations of Curves

Any equation such as y=x+3 determines a set of points with (x,y) coordinates that satisfy the equation. That set of points is often a continuous curve in which case we call the equation "the equation of the curve". The set of points determined by the equation could be empty, for example there are no points with coordinates satisfying $x^2 = -1$. There is only one point for the curve $x^2 + y^2 = 0$. We shall see that $x^2 + y^2 = 4$ represent the set of points on a circle centre (0,0) and readius 2.

The Equation   y = mx

The origin (0,0) and the point M(1,m) have coordinates that satisfy the equation **y = mx.**

Suppose that P(X,Y) also has coordinates that satisfy the equation i.e. Y = mX

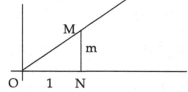

Now enlarge triangle OMN from centre O with scale factor X to give triangle OHK

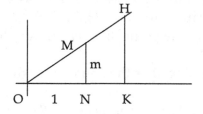

Then since the scale factor is X,   OK = X.ON = X
                              and   KH = X.NM = mX

Therefore H is the point (X,mX) = (X,Y) in other words, H is the point P.

If X>1 then P is on OM produced
If 0<X<1 then P is between O and M
If X<0 then P is on MO produced backwards.

Therefore, for any value X, if the coordinates of the point P(X,Y) satisfy the equation y=mx, then P lies on the straight line OM.

The value m represents the rise in the value of y when moving one step horizontally and then moving up or down onto the line.
m is called the **gradient** or **slope** of the line

conversely, suppose that P(X,Y) lies on the line OM

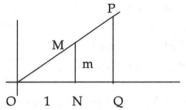

The triangle OPQ (see fig.)
is an enlargement of OMN
with scale factor X since the
triangles are similar and OQ=X.ON

Therefore, QP = X.NM = mX
Therefore the coordinates of P(X,Y) satisfy the equation y = mx.

Thus   y = mx **is the equation of a straight line**
through the origin with gradient m.

==

The reader should consider the diagram for the case when **m** is negative (or zero) to complete the theorem:

The origin (0,0) and the point M(1,m) have coordinates that satisfy the equation **y = mx.**

Suppose that P(X,Y) also has
coordinates that satisfy the
equation i.e. Y = mX

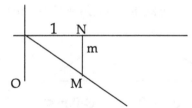

Here, the value of **m** is negative

==

Consider what happens if we shift the line **y = mx** (line 1) vertically by a distance **c** (plus or minus) to give another line called line 2. Any point **P(x,y)** will be shifted to some point **Q(x, y+c)**.

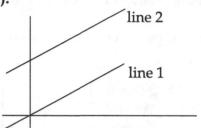

Suppose that **Q(X,Y)** has coordinates
That satisfy **Y = mX + c**
then   **Y- c = mX**
so that **P(x, Y-c)** is on line 1
and therefore **Q(X,Y)** is on line 2.

Suppose that **Q(X,Y)** is a point on line 2.
Then **P(X. Y- c)** will be a point on line 1.
Therefore **Y- c = mX**
and so **Y = mX + c**

> Therefore any point **(X,Y)** with coordinates satisfying
> **Y = mX + c** lies on line 2.
> Also, any point on line 2 has coordinates that satisfy
> the equation **y = mx + c,** therefore

Theorem 11

y = mx+c is the equation of a straight line with gradient m that passes through (0,c).

The reader may like to consider these four different cases:

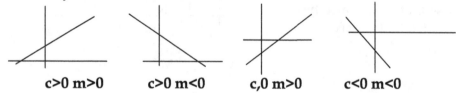

c>0 m>0   c>0 m<0   c,0 m>0   c<0 m<0

c is called the intercept on the y axis.

The value of m in the equation y=mx+c represents the number of steps you need to go up (or down, if m is negative), to get back on the line if you take one step along in the x direction. This is called the **gradient** of the line.

=================================

Example

Thus the straight line y=3x – 4 has a gradient 4, and if you start at any point on the line and move 1 step in the x direction, you will have to then go 3 steps up, to get back on the line.

The line y=3x-4
has a gradient 3

3 steps up

one step along

Theorem 12:   **Linear equations**

Any equation of the form **ax+by=c** is called a linear equation because it represents a set of points lying on a straight line.

If b=0 the equation becomes x=c/a which is the set of points on the line through (c/a, 0) and parallel to the y axis.
If b≠0 then the equation can be rewritten

$$y = -\frac{a}{b}x + \frac{c}{b}$$

which represents the straight line with gradient -a/b passing through (0,c/b).

==============================

Example

A straight line has the equation   **6x + 3y = 4**

Rewrite the equation as   y = -2x + 4/3

The gradient of the line is m=-2 which tells us that we go

one step along, 2 steps down

1

2

Exercise 5

[1]   Show that the lines y = 2x+2, 2y = -x+9

y = 2x –3 and 2y = -x + 4   form a square.

Find the vertices of the square.

[2]   Show that the line 2x – 3y = 3 passes through (3,1), (6, 3), (9, 5)… and that the line 3x + 2y = 11 passes through (3,1), (5,-2), (7, -5) .. and that they are perpendicular.

===================================

Linear mappings and Straight Line Graphs

Question :   What is term number 200 of this sequence?

5, 7, 9, 11, 13, 15, 17, 19, 21, 23, 25, …….

Solution There is, of course, no general way of finding a formula for the nth term of a sequence. Indeed, most sequences do not have a formula for the nth term, as for example, this one: 2, 3, 5, 7, 11, 13, 17, 19, 23, 29, 31, …., the sequence of prime numbers.
However, in many cases, a useful formula does exist.

We notice that in this example, the terms increase by 2 each time. The sequence is in fact, an arithmetic sequence.

To find a formula for an arithmetic sequence, we use the common difference as something like a scale factor, in other words, we multiply the term number by the common difference.

We therefore try **2n + constant** for the general term.

To find term number 1, we try 2x1 and adjust by adding 3   2x1+3 = 5
To find term number 2, we try 2x2 and adjust by adding 3   2x2+3 = 7
similarly, term number 3 is   2x3+3 = 9

With this clue, for term number 1, multiply the term number by 2, giving 2 and then we have to add 3 to get five.

In general the mapping from the term number to the number in the sequence is

$$n \longrightarrow 2n + 3$$

this tells us that term number 200 is $2 \times 200 + 3 = 403$.

We can relate this mapping to the straight line equation $y = 2x + 3$ and if we plot the graph of the line, we can see the terms of the sequence sit at integer points of the grid:

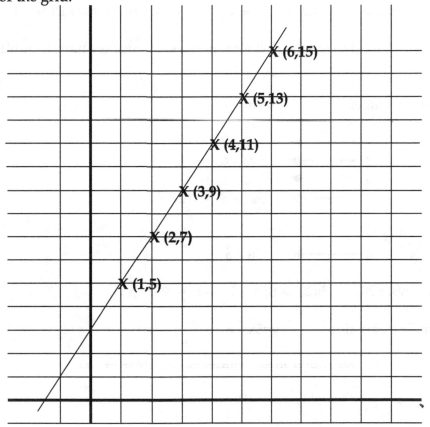

The graph shows the terms of the sequence 5,7,9,11,13,.....

Example 2

Given the sequence -3, 2, 7, 12 ....

Find the mapping **n** ➝ ?

Find term number 200

Find the relevant straight line equation

Solution

The terms of this sequence increase by 5, so this is our scale factor:

Now 5x1 = 5, and to get the first term in the sequence we have to subtract 8 hence:

Term number   term

1 ➝   5x1 – 8 = -3
2 ➝   5x2 – 8 =   2
3 ➝   5x3 – 8 =   7
4 ➝   5x4 – 8 = 12

The mapping is
        n ➝   **5n – 8**

Term number 200 = 5x200 – 8 = 1000 – 8 = **992**

Straight line equation   **y=5x – 8**

==================================

Exercise 6

Find (i) the mapping (ii) term number 200 (iii) the equation for the straight line graph

[1] 8, 10, 12, 14,.......

[2] –4, -1, 2, 5,.....

[3] –8, -6, -4, -2,.....

[4] 9, 10, 11, 12, ...

[5] 12, 8, 4, 0,.....

[6] 12, 7, 2, -3,....

Parallel Lines

In the standard equation for a line in the form $y = mx + c$ the **m** is the gradient of the line and is equal to the tangent of the angle that the line makes with the x-axis (see chapter 19)

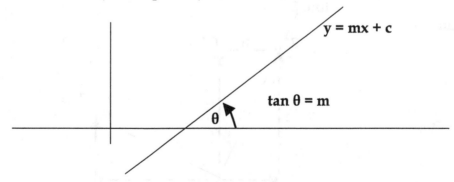

Any two lines that are parallel will have the same gradient.

If $y = m_1x + c_1$ is parallel to $y = m_2x + c_2$ then $m_1 = m_2$

This means any two equations that represent parallel lines can have the same x and y terms thus

Theorem 13

$$ax+by = c \text{ is parallel to } ax+by = d$$

Example

Write down the line passing through the point (2,3) that is parallel to the line
$3x+4y = 5$

Solution
An line parallel to $3x+4y = 5$ will have an equation of the from

$$3x + 4y = k \quad \text{for some value of k.}$$

To find the value of k, substitute (2,3) to get

$$3x2 + 4x3 = k \quad \text{giving} \quad k = 6+12 = 18$$

the required equation is therefore $3x+4y = 18$

Perpendicular Lines
In figure 1, line 1 is rotated by 90 degrees about P to give line 2.
The gradient of line 1 is m_1 and the gradient of line 2 is m_2.

Figure 1

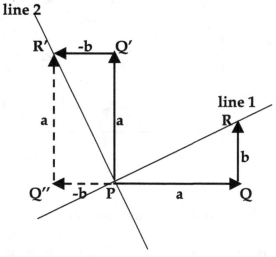

The gradient of line 1 is given by $m_1 = \dfrac{QR}{PQ} = \dfrac{b}{a}$ where **a** and **b** are positive.

The gradient of line 2 is given by $m_2 = \dfrac{Q''R'}{-PQ''} = \dfrac{a}{-b}$ note that here, **b** "points" in the negative direction

The product of the two gradients is $m_1.m_2 = \dfrac{b}{a} \times \dfrac{a}{-b} = -1$

Figure 2

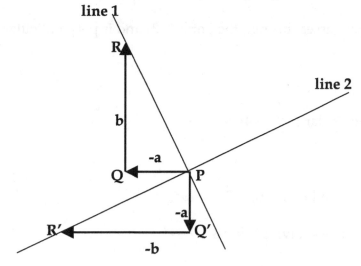

In figure 2 we again have line 1 rotated about P to give line 2.

Here, the gradient of line 1 is given by $\dfrac{QR}{-PQ} = \dfrac{b}{-a}$

The gradient of line 2 is given by $\dfrac{-PQ'}{-R'Q'} = \dfrac{-a}{-b}$ where **a** and **b** are positive.

The product of the two gradients is $m_1.m_2 = \dfrac{b}{a} \times \dfrac{a}{-b} = -1$

thus we have

Theorem 14

In general, if $y = m_1x + c_1$ **is perpendicular to** $y = m_2x + c_2$ **then** $m_1.m_2 = -1$

In all cases except when the two lines are parallel to the axes.

===================

Alternatively,
it is useful to note that the line $ax + by = c_1$
is perpendicular to the line $bx - ay = c_2$ for any values of c_1 and c_2

Example

Find the line that passes through the point (2,3) and is perpendicular to the
line $3x+4y = 5$

Solution

Any line perpendicular to $3x + 4y = 5$

is of the form $4x - 3y = k$

Substitute (2,3) to find **k giving**

$$4 \times 2 - 3 \times 3 = k \quad \text{giving} \quad k = 8 - 9 = -1$$

the required line is therefore $4x - 3y = -1$

Theorem 15   **The Intercept Form for a Line**

$$\frac{x}{a} + \frac{y}{b} = 1$$

is the straight line that cuts off intercepts **a** and **b** from the x and y axes
respectively.

Proof

The equation is a linear equation in x and y and therefore, by sections
theorem 12 represents a straight line.

The point A(a,0) satisfies the equation and therefore lies on the line.
The point B(0,b) satisfies the equation and therefore lies on the line.

The equation therefore represents the straight line AB as required.

============================

The form   L1 + λ L2 = 0

By Theorem 12, we know that **ax+by+c=0** represents a straight line in **2d** Cartesian coordinates and we therefore refer to an expression such as ax+by+c as a **linear form.**

Suppose that **L1** and **L2** are two such linear forms.

We could have for example **L1=ax+by+c** and **L2=px+qy+r.**

Then we know that **L1=0 and L2=0** are the equations of two straight lines.

Then **L1 + λ L2 = 0** where λ is some number, is also a linear equation in x and y and therefore by theorem 12,also represents a straight line.

Case 1 L1=0 and L2=0 intersect.

Let (h,k) be the point of intersection then the coordinates (h,k) make both **L1=0** and **L2=0** at the same time. Therefore, the coordinates (h,k) will also make the expression **L1 + λ L2 = 0** since both terms will be zero.
Therefore, the point of intersection of **L1=0 and L2=0** also lies on the line **L1 + λ L2 = 0**

Conclusion:

Theorem 16a   L1 + λ L2 = 0 is the equation of a straight line that passes through the intersection of **L1=0** and **L2=0** if these two lines are not parallel

Case 2: L1=0 and L2=0 are parallel

Suppose that **L1=ax+by+c** and **L2=px+qy+r**

Then the gradient of **L1=0** will be **-a/b = m** (say)
And the gradient of **L2=0** will be **-p/q = m** (since the lines are parallel)

Therefore **a= - bm** and **p = -qm**

Now the gradient of **L1 + λ L2 = 0** will be

$$-(a+\lambda p)/(b+\lambda q) = (bm+\lambda qm)/(b+\lambda q) = m(b+\lambda q)/(b+\lambda q) = m$$

Therefore:

Thereom 16b   If **L1=0** and **L2=0 are parallel then L1 + λ L2 = 0** is also
                         a line parallel to **L1=0** and **L2=0**

Theorem 17   **L1=0, L2=0** and **L3=0** represent three **straight** lines.
                         If we can find **non-zero** numbers **a, b and c** such that

$$aL1 + bL2 + cL3 = 0$$

then (i) if none of the lines are parallel, the three lines are concurrent.
    (ii) if two of the lines are parallel then all three are parallel

Proof
(i)
If **L1=0** and **L2=0** intersect at (X,Y) then when we substitute these coordinates
into

$$aL1 + bL2 + cL3 = 0$$

the first two terms will be zero. Which means that **cL3=0** for the point (X,Y).

This then means that the point of intersection of **L1=0** and **L2=0** also lies on
the line **L3=0.** Therefore, the three lines are concurrent.

(ii) Suppose that **L1=0** and **L2=0** are parallel, then by theorem 16b
                L1 + λ L2 = 0 will also be parallel to them,
therefore   **L1+(b/a)L2 = 0**   is parallel to them,
therefore   **aL1+bL2=0**   is parallel to them,
therefore   **-cL3 = 0**   is parallel to them
therefore   **L3 = 0**   is also parallel to them

Example

A triangle is formed by the lines $2x + 3y = 4,$ $3x - 4y = 12,$ $y - 2x = 2$.
Find the equations of the altitudes of the triangle and verify that the three altitudes are concurrent.

Solution

Let **L1=0**   be $2x + 3y - 4 = 0$ **L2=0** is $3x - 4y - 12 = 0$ **L3=0** is $y - 2x - 2 = 0$

L1 is BC, L2 is CA, L3 is AB.

Then AD is of the form **L2 + a.L3 = 0** for some number **a**.

AD is $3x - 4y - 12 + a(y - 2x - 2) = 0$

or $x(3 - 2a) + y(-4 + a) - 12 - 2a = 0$

Now AD is perpendicular to **L1** and is therefore by theorem 14 of the form

$3x - 2y = \text{const}$

Therefore $\dfrac{-4 + a}{3 - 2a} = \dfrac{-2}{3}$

giving $-12 + 3a = -6 + 4a$

$a = -6$

AD is $x(3 + 12) + y(-4 - 6) - 12 + 12 = 0$

AD is $15x - 10y = 0$

or $3x - 2y = 0$

===================

BE is of the form L3 + bL1 = 0 for some number **b**.

Exercise 7   **Show that the line BE is $8x + 6y - 7 = 0$**

===================

CF is of the form L1 + c.L2 = 0 for some number **c**.

Exercise 8   **Show that CF is the line $17x + 34y - 18 = 0$**

===================

The altitudes are:

AD $\qquad\qquad 3x - 2y\ = 0$

BE $\qquad\qquad 8x + 6y - 7\ = 0$

CF $\qquad\qquad 17x + 34y - 28 = 0$

To verify that these three lines are concurrent, by theorem 17 we need to find numbers **a, b and c** so that **a.AD + b.BE + c.CF = 0** (using the obvious notation)

Equating the x terms will give $\qquad 3a + 8b + 17c\ = 0 \qquad$ **(1)**

Equating the y terms gives $\qquad -2a + 6b + 34c\ = 0 \qquad$ **(2)**

Equating the numbers gives $\qquad\qquad -7b - 28c\ = 0 \qquad$ **(3)**

From (3) we have $\ \mathbf{b = -4c}$. $\qquad$ Try c=1 and b=-4

To give, from equation (1): $\qquad\qquad 3a - 32 + 17\ = 0$

$$3a = 15, \ \ a=5$$

Thus, if we take **a=5, b=-4, and c=1**

We get $\qquad\qquad\qquad\qquad$ **5.AD - 4.BE + CF = 0**

So that, by Theorem 17, $\quad$ **AD, BE and CF** are concurrent.

Theorem 18 The line through (x_1, y_1) with gradient m is $y - y_1 = m(x - x_1)$

Proof

> **The line is of the form** $y = mx + c$
>
> **Substitute (x_1, y_1) to find** c:
>
> **Then we have** $c = y_1 - m.x_1$
>
> The equation is therefore $y = mx + y_1 - m.x_1$
>
> or $y - y_1 = m(x - x_1)$

Example

Find the equations for the altitudes of the triangle formed by $A(2,3)$, $B(6,10)$ and $C(4,5)$ **and verify that they are concurrent.**

Solution

(note: we will use the shortened form grad(AB) for the gradient of the line AB)

> Given **B(6,10) and C(4,5)**
>
> We have $\text{grad(BC)} = \dfrac{5-10}{4-6} = \dfrac{-5}{-2} = \dfrac{5}{2}$
>
> therefore $\text{grad(AD)} = \dfrac{-2}{5}$
>
> **A** is the point **(2,3)** therefore using Theorem 18
>
> **AD** is $y - 3 = \dfrac{-2}{5}(x-2)$
>
> or $5y - 15 = -2x + 4$
> **AD** is $5y + 2x - 19 = 0$

Exercise 9   Show that the altitude **BE** is $y+x-16 = 0$

======================================

Exercise 10   Show that the altitude **CF** is $7y+4x-51 = 0$

======================================

The results are:

AD is $5y+2x-19 = 0$

BE is $y+x-16 = 0$

CF is $7y+4x-51 = 0$

We note here that **CF = AD + 2BE**

That is $7y+4x-51 = 5y+2x-19 + 2(y+x-16)$

Hence, by Theorem 17 **AD, BE and CF** are concurrent.

These altitudes in fact, meet at the point **H(61/3, -13/3)**

Exercise 11

Find the altitudes of the triangle **A(18,9), B(27,45), C(63,27)** and show that the altitudes meet at the point **H(31,35)**.

==============================

Theorem 19   **To find the line joining $A(x_1,y_1)$ to $B(x_2,y_2)$**

Solution

The gradient of the line AB is $\frac{y_2 - y_1}{x_2 - x_1}$

therefore, Theorem 18 , the equation of the line AB is

$$y - y_1 = \frac{(y_2 - y_1)}{(x_2 - x_1)} (x - x_1) \qquad \blacksquare$$

===============================

Theorem 20

The medians of the triangle A(x_1,y_1), B(x_2,y_2), C(x_3,y_3) are concurrent.

Proof

We use the conventional notation, A', B', C', for the mid points of BC, CA, AB.

Using Theorem 7, A' is the point A'($\frac{1}{2}(x_2+x_3)$, $\frac{1}{2}(y_2+y_3)$)

Since A is the point A(x_1,y_1), using Theorem 19

AA' is the line $\qquad y - y_1 = \frac{(\frac{1}{2}(y_2+y_3) - y_1)}{(\frac{1}{2}(x_2+x_3) - x_1)} (x-x_1)$

$$y - y_1 = \frac{(y_2+y_3 - 2y_1)}{(x_2+x_3 - 2x_1)} (x-x_1)$$

$$y(x_2+x_3-2x_1) - y_1(x_2+x_3 - 2x_1) = x(y_2+y_3 - 2y_1) - x_1(y_2+y_3 - 2y_1)$$

AA' is the line

$$y(x_2+x_3-2x_1) - x(y_2+y_3 - 2y_1) - y_1(x_2+x_3-2x_1) + x_1(y_2+y_3 - 2y_1) = 0$$

by changing the suffices, we can write down the equations of the lines **BB'** and **CC'**.

Thus we have:

AA' is $\qquad y(x_2+x_3-2x_1) - x(y_2+y_3 - 2y_1) - y_1(x_2+x_3-2x_1) + x_1(y_2+y_3 - 2y_1) = 0$

Change suffices $1\to2, 2\to3, 3\to1$

BB' is $\quad y(x_3+x_1-2x_2) - x(y_3+y_1 - 2y_2) - y_2(x_3+x_1-2x_2) + x_2(y_3+y_1 - 2y_2) = 0$

Now change $1\to2, 2\to3, 3\to1$ in this equation

CC' is $\quad y(x_1+x_2-2x_3) - x(y_1+y_2 - 2y_3) - y_3(x_1+x_2-2x_3) + x_3(y_1+y_2 - 2y_3) = 0$

Adding these three equations we see that

$$AA' + BB' + CC' = 0$$

Thus the medians **AA', BB', CC'** are concurrent by Theorem 17.

==

(**Note:** a shorter method is to find the point that divides AA' in the ratio 2:1.

By Theorem 8 this point is $\dfrac{a + 2a'}{3}$

i.e. $\left(\dfrac{x_1 + 2.\frac{1}{2}(x_2+x_3)}{3} , \dfrac{y_1 + 2.\frac{1}{2}(y_2+y_3)}{3} \right)$

or $\left(\dfrac{x_1+x_2+x_3}{3} , \dfrac{y_1+y_2+y_3}{3} \right)$

by the symmetry of this result, we know that we will find the same point if we divide BB' in the ratio 2:1, and also if we divide CC' in the ratio 2:1.

This point therefore lies on all three medians and they are therefore concurrent.)

====================================

Exercise 12

Use the method of the example above to prove that the altitudes of the triangle $A(x_1,y_1)$, $B(x_2,y_2)$, $C(x_3,y_3)$ are concurrent.

Hint: derive the equations for the altitudes and show that **AD+BE+CF = 0**

==========================

Theorem 21

The equation for a circle

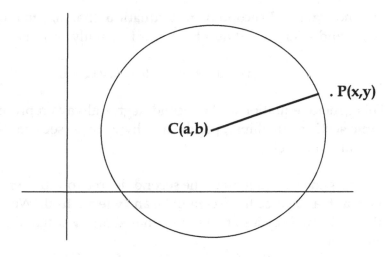

Given a circle, centre **C(a,b)** and radius **r**, we suppose that a point **P(x,y)** moves round the circle. Then **PC²** is always equal to the square of the radius **r** and using the Pythagoras distance formula for **PC²** we have

$$(x-a)^2 + (y-b)^2 = r^2$$

This equation is satisfied by all points on the circle centre **C(a,b)** and radius **r**.
====================

For any points inside the circle we have $(x-a)^2 + (y-b)^2 < r^2$

For any points outside the circle we have $(x-a)^2 + (y-b)^2 > r^2$

Theorem 22:   **The general equation for a Circle**

Any circle has a centre and a radius and therefore has an equation of the form

$$(x-a)^2 + (y-b)^2 = r^2$$

Expanding, we have

$$x^2 - 2ax + a^2 + y^2 - 2by + b^2 - r^2 = 0$$

or $x^2 + y^2 - 2ax - 2by +$ **number terms** $= 0$

The most **general second degree equation**, that is, equation with only x^2, y^2, xy , x and y terms and number terms is usually written

$$ax^2 + by^2 + 2hxy + 2gx + 2fy + c = 0$$

This general equation of the second degree always represents some kind of **conic section**, an ellipse, parabola or hyperbola (see chapter 28). A circle is a special kind of ellipse.

For the general equation of the second degree to represent a circle we must have **a=b** and since there cannot be an **xy** term, **h=0**. We can always then divide through by **a** so that the **x** squared and **y** squared terms have a coefficient 1.
The general equation for a circle is therefore of the form

$$x^2 + y^2 + 2gx + 2fy + c = 0$$

Completing the square terms we have

$$x^2 + 2gx + g^2 + y^2 + 2fy + f^2 = g^2 + f^2 - c$$

$$(x+g)^2 + (y+f)^2 = g^2 + f^2 - c$$

This shows that, if $g^2 + f^2 > c$ **the equation**

$$x^2 + y^2 + 2gx + 2fy + c = 0$$

represents a circle with centre **(-g, -f)** and radius $\sqrt{(g^2 + f^2 - c)}$

Finding a Tangent to a Circle

Example 1

Find the equation of the tangent to the circle

$$(x-9)^2+(y-12)^2 = 25 \quad \text{at the point P(5,9)}$$

Solution

The centre of the circle is C(9,12)

The gradient of CP is $\dfrac{9-12}{5-9} = 3/4$

Since the tangent is perpendicular to the radius, the gradient of the tangent is $-4/3$

By Theorem 18 the equation of the tangent is therefore

$$y - 9 = -4/3(x-5)$$

or
$$3y - 27 = -4(x-5)$$

$$3y - 27 = -4x + 20$$

$$3y + 4x - 47 = 0$$

==================================

The general form of the tangent

There is a neat way of writing down the equation for the tangent to any second degree curve (i.e. some kind of conic section) at the point (x_1,y_1). The trick is to replace

$$x^2 \quad \text{by} \quad x.x_1$$
$$y^2 \quad \text{by} \quad y.y_1$$
$$2xy \quad \text{by} \quad (x.y_1+y.x_1)$$
$$2x \quad \text{by} \quad (x+x_1)$$
$$\text{and} \quad 2y \quad \text{by} \quad (y+y_1)$$

Given the general equation for a curve of the second degree in the form

$$ax^2+by^2+2hxy+2gx+2fy+c = 0$$

Then, the equation of the tangent to this curve at the point (x_1,y_1) is

$$ax.x_1+by.y_1+h(x.y_1+y.x_1)+g(x+x_1)+f(y+y_1)+c = 0$$

================================

Theorem 23   **The Tangent to a Circle**

Using the above general result, the equation for the tangent to the circle

$$x^2+y^2+2gx+2fy+c = 0$$

at the point (x_1,y_1) is

$$x.x_1+y.y_1+g(x+x_1)+f(y+y_1)+c = 0$$

==============================

Example: Find the equation for the tangent to the circle

$$x^2+y^2-4x-6y-12 = 0 \text{ at the point } (5,7).$$

Solution

Using the above general result, the equation for the tangent is

$$5x+7y-2(x+5)-3(y+7)-12 = 0$$

$$5x+7y-2x-10-3y-21-12 = 0$$

$$3x+4y-43 = 0$$

================================

Exercise 13

[1]   Find the centre and radius of the circle $x^2+y^2-6x-8y-75=0$
and find the equation of the tangent at (9,12)

[2]   Find the centre and radius of the circle $x^2+y^2-10x-10y=119$
and find the equation of the tangents at the two points where
the circle cuts the y axis.

[3]   Prove that the circles
$$x^2+y^2-2x-2y-11=0$$
and
$$x^2+y^2-6x-8y-75=0$$
touch and find the equation of the common tangent at the point
of contact.

Common Chords

Theorem 24   If $C1 = 0$ and $C2 = 0$ are two circles, with equal x^2 coefficients,
that intersect then the equation of their common chord is $C1 - C2 = 0$.

Proof
In a similar notation to that used for linear forms in Theorem 16 we will use a
single symbol, say **C,** for a second degree form that represents a circle. We
will assume that the coefficients of the **x^2 and y^2** terms are one.

Let C1 **represent the form** $x^2+y^2+2g_1x+2f_1y+c_1$

and C2 **represent the form** $x^2+y^2+2g_2x+2f_2y+c_2$

Consider the equation $C1 - C2 = 0$

This is a linear equation because the squared terms cancel out, and therefore
represents a straight line. If **C1** intersects **C2** at a point **(X,Y)**, then **C1 - C2 = 0**
will be satisfied by the point **(X,Y).**
C1 - C2=0 therefore represents a straight line that passes through any points
where **C1=0** intersects **C2=0.**
Therefore, if **C1=0** intersects **C2=0** then **C1 - C2=0** is the equation of the
common.

■

If **C1** touches **C2** then **C1 - C2=0** is the equation of the common tangent where they touch.

Example :   Two circles, C1=0 and C2=0 intersect at two points A and B. Find the equation of the line AB if

$$\text{C1:} \qquad x^2+y^2-4x-6y+12 = 0$$

$$\text{C2:} \qquad x^2+y^2-10x-8y+32 = 0$$

Solution:

Any common chord has the equation C1-C2 = 0   (Theorem 23)
which gives the equation

$$-4x+10x -6y+8y +12 - 32 = 0$$

The common chord
is therefore

$$6x + 2y - 20 = 0$$

or $$y = -3x + 10 \qquad \blacksquare$$

Note: if the two circles do not intersect, then **C1 - C2 = 0** gives the equation of a straight line called the radical axis of the two circles

================================

Exercise 14   Given three circles C1=0, C2=0 and C3=0 prove that the three radical axes of the circles, taken in pairs, meet at a point.

(Hint: use Theorem 17)

================================

Three Dimensional Space

Many of the results that we have developed for one and two dimensional Cartesian space are readily extended to three dimensions.

Simple equations in x,y and z coordinates, however, represent surfaces in three dimensional x,y,z axes. A linear equation in x,y and z represents a plane in 3D Cartesian axes. The straight line can only be represented as the intersection of two planes and thus needs two x,y,z equations. In 3D, the straight line is best dealt with using vector methods that are investigated in chapter 24.

Coordinates in Three Dimensions

In three dimensional coordinate geometry (sometimes called **solid geometry**) we fix the position of a point in the x-y plane as usual, using Ox and Oy axes. For the third dimension, we choose the Oz axis perpendicular to the Oxy plane but so that a turn from Ox to Oy would send a right handed screw along the z axis. We then have a **right handed set of coordinate axes.**

Handedness

Handedness in Lineland

The inhabitants of line land might have concept of in front or behind but there could not be any understanding of left and right.

Given a rod AB in lineland, to the linelander, if end A is in front of end B then B would always remain behind A. There would be no way that B could get in front of A.

However, a one dimensional Albert Einstein living in lineland might suggest that B could get in front of A if the rod AB were to come out into 2 dimensional space, turn over and then go back to lineland:

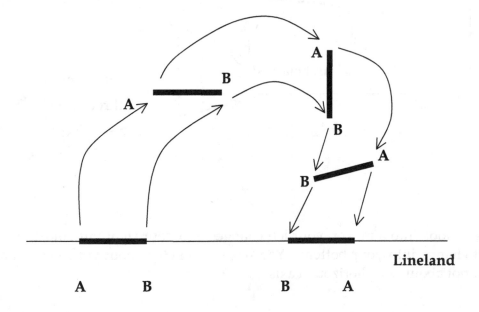

Handedness in Flatland

The inhabitants of flatland might have a right hand and a left hand, but they would not be able to clap hands because they would never be able to place their right hand on top of the left hand and perform a clap.

A two dimensional Einstein living in flatland, however, would see that flatlanders could clap hands if they used the third dimension. The flatlander could send his right hand into the third dimension, turn it over and then clap onto his left hand.

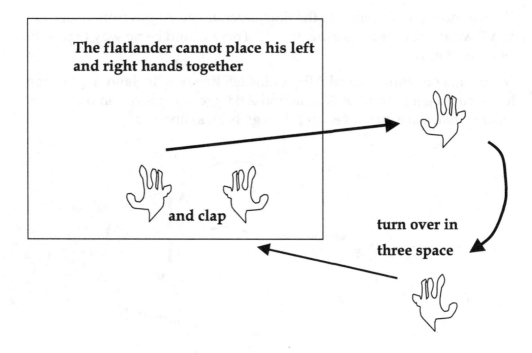

If you look into a mirror, your left changes into your right but your top does not change into your bottom. You have symmetry about your vertical axis but not about your horizontal axis.

Handedness in Three Space

Left and Right handed dice

In the gambling world, it is required that a fair six sided dice has its opposite faces adding up to seven.

However, this requirement is not enough to specify a unique six sided dice.

The following diagrams show that there are two kinds of dice in three space: a **right handed dice** and a **left handed dice**:

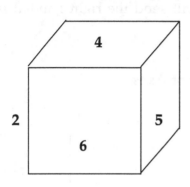
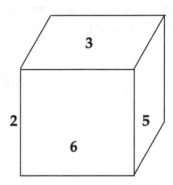

Exploded views:

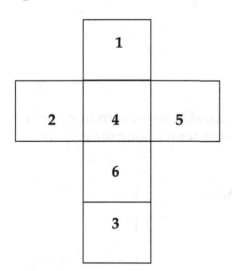

left handed dice   right handed dice

To place the left handed dice onto the right handed dice, so that the numbers match, pick up the left handed dice in 4 dimensional space, turn it over and then you can place it back in 3 dimension as a right handed dice.

Right Handed and Left Handed Axes

In 3 dimensions, a **right handed set of axes** is such that a right handed screw turn from the x axis round to the y axis will send the right handed screw along the positive z axis.

Right Handed Sets of Axes

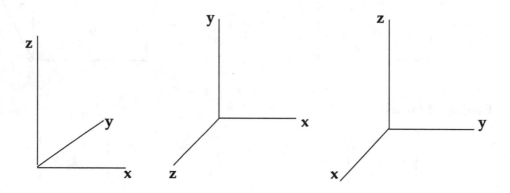

A **left handed set of axes** is such that a right handed screw turn from the x axis round to the y axis will send a right handed screw along the negative z axis.

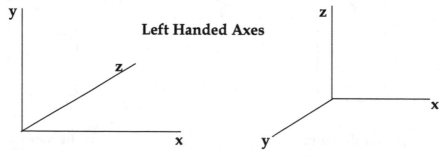

Left Handed Axes

In general, we avoid using Left Handed Axes.

Many of the theorems and results that we have discussed in one and two dimensions readily extend to three dimensional x,y,z axes.

Theorem 25 **Displacements in three dimensions**

Given $A(x1,y1,z1)$ and $B(x2,y2,z2)$ we have

$$AB = \begin{pmatrix} x2\text{-}x1 \\ y2\text{-}y1 \\ z2\text{-}z1 \end{pmatrix}$$

Addition

Given $\quad AB = \begin{pmatrix} a \\ b \\ c \end{pmatrix} \qquad PQ = \begin{pmatrix} p \\ q \\ r \end{pmatrix}$

then $\quad AB + PQ = \begin{pmatrix} a+p \\ b+q \\ c+r \end{pmatrix}$

In general, for any displacements in three dimensions,

$$AB + BC = AC$$

$$AB + BC + CD = AD$$

$$AB + BC + CA = 0$$

$$AB + PQ = PQ + AB$$

$$(AB+CD) + EF = AB + (CD+EF)$$

$$k(AB + PQ) = kAB + kPQ$$

We remind ourselves of the convention that the position vector of the point $P(x,y,z)$ is denoted by $\mathbf{p} = \mathbf{OP}$

If M is the mid point of AB then $\quad \mathbf{m} = \frac{1}{2}(\mathbf{a}+\mathbf{b})$

If P divides AB in the ratio $\mathbf{m}:\mathbf{n}$ then $\mathbf{p} = \dfrac{\mathbf{mb} + \mathbf{na}}{\mathbf{m+n}}$

The formulae for centres of mass and centroids in Theorems 9 and 10 are valid for x,y,z coordinates.

Exercise 1 is equally valid in three dimensions.

Pythagoras's Theorem in three dimensions takes the form:

$OP^2 = ON^2 + NP^2$

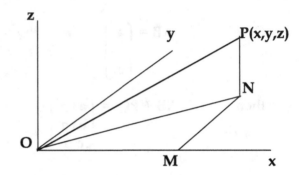

$\quad = OM^2 + MN^2 + NP^2$

$\quad = x^2 + y^2 + z^2$

Theorem 26   If $A(x_1,y_1,z_1)$ and $B(x_2,y_2,z_2)$

then $\quad AB^2 = (x_2-x_1)^2 + (y_2-y_1)^2 + (z_2-z_1)^2$

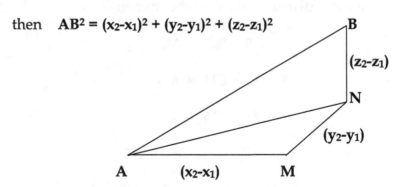

Proof

$$AB^2 = AN^2 + NB^2$$
$$= AM^2 + MN^2 + NB^2$$
$$AB^2 = (x_2-x_1)^2 + (y_2-y_1)^2 + (z_2-z_1)^2 \qquad \blacksquare$$

Exercise 15

[1] Given triangle **A(4,2,3), B(4,6,5), C(1,6,5)**
Find AB, BC and CA and deduce that the triangle is right angled.

[2] Given triangle **A(7,4,5), B(3,8,5), C(3,4,9)**.
Find AB, BC, CA and deduce that triangle ABC is equilateral.

[3] Calculate the distance **AB** for each of the following pairs of points:

[a] **A(5,6,20)** and **B(2,2,8)** [b] **A(10,5,6)** and **B(2,-4,-6)**

[4] Show that the following points lie on a straight line:

A(2,3,4), B(4,4,7), C(6,5,10), D(8,6,13), E(10,7,16)

Check that **2y=x+4,** and **2z = 3x+2**

[5] Given P is the mid point of AB, Q is the mid point of BC, R is the mid point of CD and S is the mid point of DA, prove that the mid point of PR is the same of the mid point of QS.

[6] Prove that **A(3,5,6), B(5,8,10), C(9,14,18)** lie on a straight line.

Linear Equations in 3D

In two dimensional Cartesian coordinates, an equation in x and y generally represents a curve of some kind, that is , a 1 dimensional figure. The equation represents a restriction on all points of the 2D plane so that the dimension is reduced from 2 to 1.

In three dimensions, an equation in x, y and z represents some kind of surface that is a 2 dimensional object. The equation represents a restriction on the points of 3 D space, reducing the dimension of the set of points that satisfy the equation from 3 to 2.

In particular, a linear equation in 2 dimensional (x,y) coordinates represents a one dimensional straight line whereas a linear equation in 3 dimensional (x,y,z) coordinates represents a 2 dimensional plane.

Thus for example, $3x-4y+z = 12$ represents the plane that cuts the coordinate axes at the points (4,0,0), (0,-3,0) and (0,0,12).

Straight line geometry in three dimensions is best dealt with by vector methods and these are studied in the chapter 23.

A comparison of some Objects in Two and Three Dimensions:

In 2D	In 3D
x=2	x=2
is a line // Oy	is a plane // to the Oyz plane

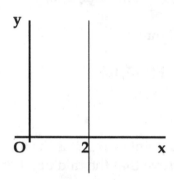 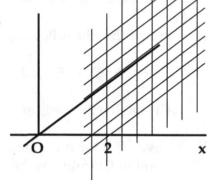

in 2D	in 3D
y=x	y=x
is a line at 45° to the axes	is a plane containing Oz and bisecting yOx

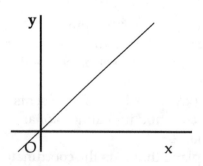

$x^2+y^2=4$

**circle radius 2
centre O**

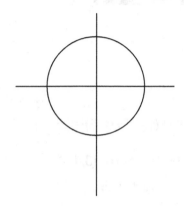

$x^2+y^2=4$

**cylinder, radius 2.
axis is Oz**

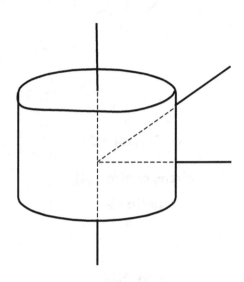

$x^2+y^2=1$

circle radius 1

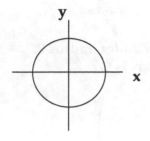

$x^2+y^2+z^2=1$

sphere radius 1

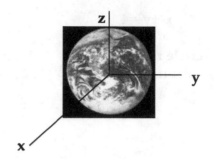

$(x-3)^2+(y-4)^2= 25$

a circle, centre (3,4)
radius 5

$(x-3)^2+(y-4)^2+(z-5)^2=25$

a sphere, centre (3,4,5)
radius 5

$2x+3y = 12$

the linear equation
is a line

$2x+3y+4z = 12$

the linear equation
is a plane

==

Solutions and Answers

Exercise 1   [1] the diagonals of the parallelogram bisect each other

[2] both are $\{ (a+c+e+g)/4 , (b+d+f+h)/4 \}$

Exercise 2   [1] (6,5) [2] (6,6) [3] both are $\{ (a+b+c)/3, (a+b+c)/3 \}$

Exercise 3   [1]   **GA+GB+GC = a-g + b-g + c-g = a+b+c – 3g = 0**
since **g = (a+b+c)/3**

[2]   **m=(a+b)/2 and n=(c+d)/2.** mid point of MN is
(a+b+c+d)/4 which is the centroid of A,B,C,D

[3]   **m=(a+b)/2, 2PM = 2(m-p) = 2(a+b)/2 – 2p**
= a-p + b-p =PA+PB

[4]   **PA+PB+PC = PG+GA + PG+GB + PG+GC**
= 3PG + GA+GB+GC
= 3PG   since GA+GB+GC=0

Exercise 4   5, 5, 10, 13, 13, 25, 17, 50

Exercise 5   The vertices of the square are (1,4), (0,2), (3,3) and (2,1)

Exercise 6   [1] 2n+6, 406, y=2x+6;   [2] 3n-7, 593, y=3x-7;
[3] 2n-10, 390, y=2x-10;   [4] n+8, 208, y=x+8;
[5] 16-4n, -784, y=-4x+16; [6] 17-5n; -983, y=-5x+17

Exercise 11   AD is y = 2x –27, BE is 2y+5x = 225, CF is 4y+x = 171

Exercise 13   [1] Centre (3,4) radius 10, 6x + 8y = 150

[2] Centre (5,5) radius 13

tangents are 12y – 5x = 204 and 5x + 12y + 84 = 0

[3] The radii are $\sqrt{13}$ and $2\sqrt{13}$. The distance between the
centres is $3\sqrt{13}$. The point of contact can be found by dividing
the line of centres in the ratio 1:2. This gives (4,3).

The common tangent is perpendicular to the line of centres and is the line $3x + 2y - 18 = 0$.
Alternatively, subtract the two equations to get $C1 - C2 = 0$ as in the following section.

Exercise 14   The radical axes are $L1 = C1-C2 = 0$
$$L2 = C2-C3 = 0 \text{ and } L3 = C3-C1 = 0$$
Adding we get $L1+L2+L3 = 0$ and therefore, by Theorem 17, the lines are concurrent

Exercise 15   [1] $AB=\sqrt{20}, \; BC=3, \; CA=\sqrt{29}$

                   $AB^2+BC^2 = CA^2 \; \therefore \; \angle B = 90$ converse of Pythagoras

            [2] $AB=BC=CA = 4\sqrt{2}$

            [3]   (a) $AB=13$, (b) $AB=17$

            [4] ·   They all lie on the intersection of the planes $2y = x+4$ and $2z = 3x+2$

            [5]   The mid points of both PR and QS are **(a+b+c+d)/4**

            [6] $\mathbf{AB} = \begin{pmatrix} 2 \\ 3 \\ 4 \end{pmatrix} \quad \mathbf{BC} = \begin{pmatrix} 4 \\ 6 \\ 8 \end{pmatrix} \quad \therefore \mathbf{BC} = 2\mathbf{AB}$

                                                       $\therefore \mathbf{BC} \,/\!/\, \mathbf{AB}$

===

CHAPTER 12

Transformations and Matrices

Equations for Solving Problems

Our first encounter with algebra and equations in school, usually involves problems that ask you to "find the number", for example:

"Dad gave me £5. I spent some on a bag of crisps and twice as much on chocolate. I still had £2.60 left. How much was a bag of crisps?"

Solution
Let x pence be the answer then I spent x pence on a bag of crisps and 2x pence on chocolate.
Therefore
$$x + 2x = 500 - 260$$
$$3x = 240$$
$$x = 80$$

A bag of crisps was therefore 80 pence ∎

and we are soon expected to be able to solve equations of various types:

Exercise 1

1 If $\qquad$ $2x + 3 = 24 - 5x$ $\qquad$ find x.

2 Solve $x^2 - 9x + 20 = 0$

3 Solve $\qquad 3x - 2y = 4$
$\qquad\qquad 4x - 3y = 3$

4 Fred has twice as many marbles as Tash. Fred gives Tash 3 marbles and then finds that he still has 5 more marbles than Tash.
How many marbles does Fred now have?

5 I have a rectangular piece of paper. One side is twice as long as the other. If I fold it in two, I find that one side is six inches longer than the other.
How long is my piece of paper?

Equations for Curves

Around the same time, we also learn about Cartesian coordinates and plotting graphs. We find that an equation can represent a line or a curve:

Example 1   Plot y = 2x + 4

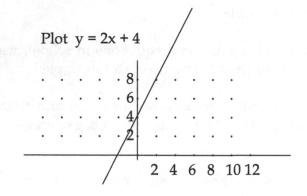

All the points (x,y) satisfying the equation lie on a line that cuts the y axis at (0,4) and has a gradient 2.

or

Example 2.

Complete the following table of values for y = x + 1/x for values of x from 0.2 to 3 using the values shown in the table:

X	0.2	0.4	0.6	0.8	1	2	3
1/x							
Y							

Plotting y against x
gives a hyperbola
something like this, that
has a suspicious minimum
that you may have to
find (its 2 when x=1)

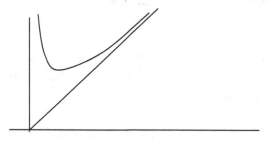

and that there are lines called asymptotes.

Example 3.

You soon learn that $x^2 + y^2 = 9$ is satisfied by points on a circle radius 3.

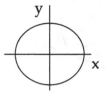

It has become popular in modern syllabi to include inequalities as well as equalities, so that the student may be required to recognize that

$$y > 2x + 4$$

is satisfied by all points in the Cartesian x-y plane that are above the line y=2x+4 and

$$y < 2x + 4$$

is satisfied by all points in the Cartesian x-y plane that are below the line y=2x+4 while

$$y=2x+4$$

is satisfied only by points on the line.

$x^2 + y^2 < 9$　is satisfied by the coordinates of points inside the circle

$x^2 + y^2 > 9$　is satisfied by the coordinates of points outside the circle

Thus we have two distinctly different uses for equations in algebra.

1. We can solve equations to find solutions.

2. We can plot x-y equations to see their shape.

Equations for transformations

In this chapter we investigate another use for equations, and we show how they can be used to change the shapes of given figures that are plotted on x-y Cartesian axes.

In other words, we find that we can use equations to represent transformations. These various transformations include:

translations

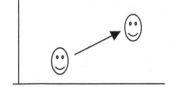

enlargements

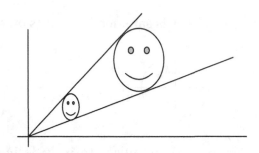

shear

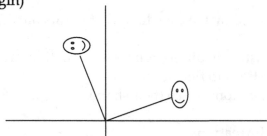

rotation
(90º about the origin)

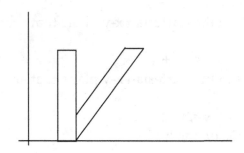

reflection
(in the y axis)

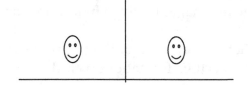

Translations

Example: The diagram shows the following points joined up to form an **object** figure:

A(1,1), B(1,5), C(3,5), D(3,1), E(3,3), F(4,3) and G(3,4)

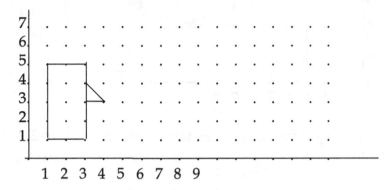

Using the equations $x' = x + 4$
                        $y' = y + 2$

calculate new points (x',y') to form the **image figure** from the given points (x,y).
Join up the appropriate points in the image.

solution

using the equations we calculate

A(1,1)→A'(5,3),
B(1,5)→B'(5,7),
C(3,5)→C'(7,7),
D(3,1)→D'(7,3),
E(3,3)→E'(7,5),
F(4,3)→F'(8,5),
G(3,4)→G'(7,6)

Where the arrows can be interpreted by "goes to"

If we plot the new points on the same diagram we have

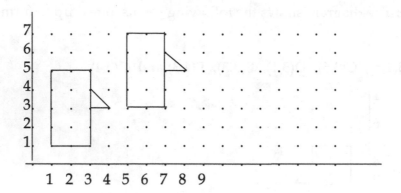

Each point of the object figure has been moved along 4 and up 2 as described by the equations

$$x' = x + 4$$
$$y' = y + 2$$

The object has been translated by the vector $\begin{bmatrix} 4 \\ 2 \end{bmatrix}$

Exercise 2

1. Draw the same object figure and transform the figure using the equations

$$x' = x - 5$$
$$y' = y + 2$$

Identify the translation vector $\begin{bmatrix} -5 \\ 2 \end{bmatrix}$ on your diagram.

2. As for question 1 but use the transformation

$$x' = x - 5$$
$$y' = y - 3$$

Identify the translation vector $\begin{bmatrix} -5 \\ -3 \end{bmatrix}$ on your diagram.

3. On your diagram for question 1, translate the image that you have drawn by the vector $\begin{bmatrix} 4 \\ -3 \end{bmatrix}$

Hence show that the translation $\begin{bmatrix} 4 \\ 2 \end{bmatrix}$ followed by the translation $\begin{bmatrix} 4 \\ -3 \end{bmatrix}$

is equivalent to the translation $\begin{bmatrix} 8 \\ -1 \end{bmatrix}$

Note that
$$\begin{bmatrix} 4 \\ 2 \end{bmatrix} + \begin{bmatrix} 4 \\ -3 \end{bmatrix} = \begin{bmatrix} 8 \\ -1 \end{bmatrix}$$

Enlargements.

Example

The points (1,1), (2,1), (2,2) and (1,3) are joined up to form an object figure.

Plot the image of this figure under the transformation described by the equations

$$x' = 3x$$
$$y' = 3y$$

Solution

Using the transformation equations we have

$(1,1) \rightarrow (3,3)$, $(2,1) \rightarrow (6,3)$, $(2,2) \rightarrow (6,6)$ and $(1,3) \rightarrow (3,9)$

Plotting these points gives the following figure:

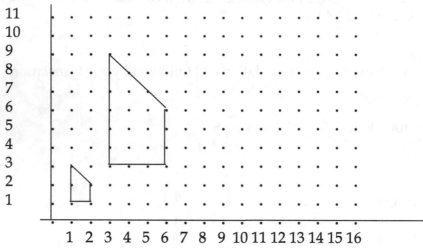

The equations

$$x'=3x$$
$$y'=3y$$

represent an enlargement from the origin by a scale factor 3.

Further examples of simple transformations:

A negative enlargement, scale factor -2:

$$x' = -2x$$
$$y' = -2y$$

$(1,1)\rightarrow(-2,-2)$, $(2,1)\rightarrow(-4,-2)$, $(2,2)\rightarrow(-4,-4)$, $(1,3)\rightarrow(-2,-6)$

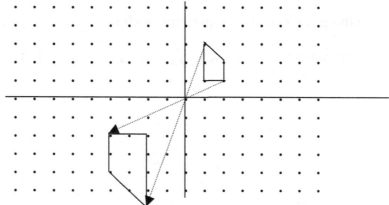

A stretch parallel to Ox, stretch factor 3:

$$x' = 3x$$
$$y' = y$$

$(1,1) \rightarrow (3,1)$, $(2,1) \rightarrow (6,1)$, $(2,2) \rightarrow (6,2)$, $(1,3) \rightarrow (3,3)$

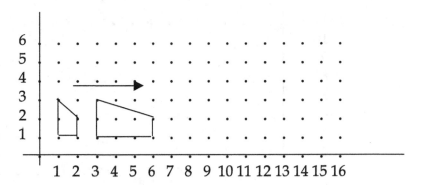

A stretch parallel to Oy by a factor 3:

$$x' = x$$
$$y' = 3y$$

$(1,1) \rightarrow (1,3)$, $(2,1) \rightarrow (2,3)$, $(2,2) \rightarrow (2,6)$, $(1,3) \rightarrow (1,9)$

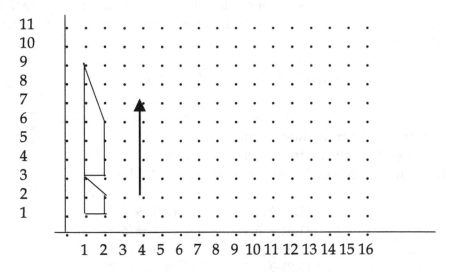

A two way stretch by 3 in the x direction and 2 in the y direction:

$$x' = 3x$$
$$y' = 2y$$

$(1,1) \rightarrow (3,2)$, $(2,1) \rightarrow (6,2)$, $(2,2) \rightarrow (6,4)$, $(1,3) \rightarrow (3,6)$

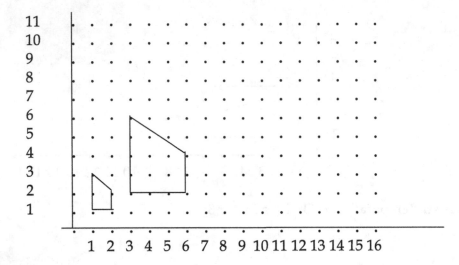

Shear Transformations

Shear transformations are often described by pushing over a stack of books. The diagram below shows a stack of four boxes with corners at (1,1), (1,5), (2,1) and (2,5).

The stack is operated on by the transformation
$$x' = x+y$$
$$y' = y$$
which is a shear, parallel to the x axis.
The object points are transformed as follows:
(they are written in this way to match up with the object figure)

$(1,5) \rightarrow (6,5)$	$(2,5) \rightarrow (7,5)$
$(1,4) \rightarrow (5,4)$	$(2,4) \rightarrow (6,4)$
$(1,3) \rightarrow (4,3)$	$(2,3) \rightarrow (5,3)$
$(1,2) \rightarrow (3,2)$	$(2,2) \rightarrow (4,2)$
$(1,1) \rightarrow (2,1)$	$(2,1) \rightarrow (3,1)$

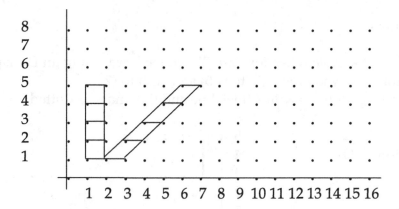

The distance moved by the point (1,1) is called the "shear factor" so that this shear transformation has a shear factor of 1

Shear transformations can be parallel to the y axis. The equations

$$x' = x$$
$$y' = 2x+y$$

push the given set of boxes upwards.

If the corners are (01,), (02), (4,1) and (4,2) then the corners of the image are

(01)→(0,1), (0,2)→(0,2), (4,1)→(4,9) and (4,2)→(4,10)

The other lines in the image can be put in without needing any calculation (we will see why, later on in the chapter)

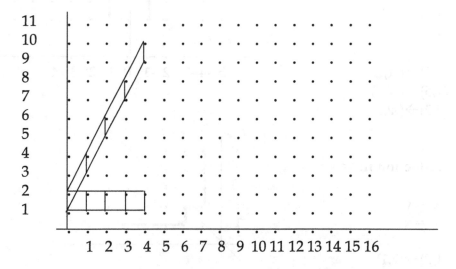

The point (1,1) moves 2 places so the shear factor for this transformation is 2.

Reflections

A number of simple reflections are illustrated here, using an L shaped object figure, formed by joining (1,2) to (1,5) and (1,2) to (2,2).
The object figure is draw in a thick line and the image is dotted.

Reflection in Ox

$x' = x$
$y' = -y$

(1,2)→(1,-2)

(1,5)→(1,-5)

(2,2)→(2,-2)

Reflection in Oy

$x' = -x$
$y' = y$

(1,2)→(-1,2)
(1,5)→(-1,5)
(2,2)→(-2,2)

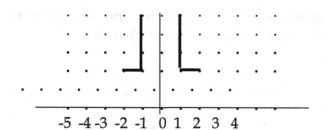

Reflection in y=x

$x' = y$
$y' = x$

(1,2)→(2,1)
(1,5)→(5,1)
(2,2)→(2,2)

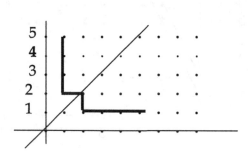

Reflection in y = -x

$x' = -y$
$y' = -x$

$(1,2) \rightarrow (-2,-1)$
$(1,5) \rightarrow (-5,-1)$
$(2,2) \rightarrow (-2,-2)$

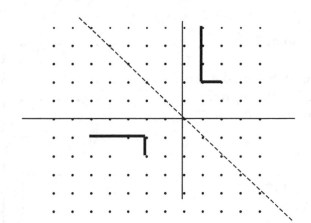

Rotations of 90°, 180°, 270°

Rotation of 90°

$x' = -y$
$y' = x$

$(1,2) \rightarrow (-2,1)$
$(1,5) \rightarrow (-5,1)$
$(2,2) \rightarrow (-2,2)$

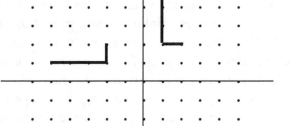

Rotation of 270°
(or –90°)
$x' = y$
$y' = -x$

$(1,2) \rightarrow (2,-1)$
$(1,5) \rightarrow (5,-1)$
$(2,2) \rightarrow (2,-2)$

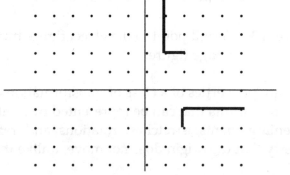

Rotation of 180°

(sometimes called a
reflection through the
origin)
   $x' = -x$
   $y' = -y$

(1,2)→(-1,-2)
(1,5)→(-1,-5)
(2,2)→(-2,-2)

The Matrix of a Transformation

In the above examples, when we construct the image figure for a particular object, we have been making some grand assumptions that have not been proved. We have been mathematically naughty!
In order to draw the image figure, we have been plotting three or four points and then joining them up to complete the figure of the image. What justification is there for constructing the image in this way?
Later in the chapter we will prove three theorems that justify this approach:

Theorem 1   A straight line transforms into a straight line

Theorem 2   Parallel lines in the object transform into parallel lines in the
                image.

Theorem 3   A mid point in the object figure transforms into a mid point in
the   image figure.

These are properties of a class of transformations called **Linear Transformations** that can be represented by a set of linear equations.
The enlargements, stretches, rotations and reflections all have a further property that the origin does not move, unlike the translations.

Linear transformations in the x-y Cartesian plane that leave the origin fixed all have transform equations of the form

$$x' = ax + by$$
$$y' = cx + dy$$

where a,b,c and d are fixed numbers. The nature of the transformation and its effect on any object figure depends only on the values of the numbers a,b,c and d. It is natural, therefore, to group these numbers together, in a rectangular array called the matrix of the transformation.
Thus the equations

$$x' = ax + by$$
$$y' = cx + dy$$

describe the transformation with matrix $\begin{pmatrix} a, & b \\ c, & d \end{pmatrix}$

and we write

$$\begin{bmatrix} x' \\ y' \end{bmatrix} = \begin{pmatrix} a, & b \\ c, & d \end{pmatrix} \begin{bmatrix} x \\ y \end{bmatrix}$$

This means that the transformation determined by the numbers $\begin{pmatrix} a, & b \\ c, & d \end{pmatrix}$

sends the point with position vector $\begin{bmatrix} x \\ y \end{bmatrix}$ to the point $\begin{bmatrix} x' \\ y' \end{bmatrix}$

Example
Find the image of the square A(2,1), B(5,1), C(5,4), D(2,4) under the transformation

$$\begin{pmatrix} 2, & 3 \\ 4, & 1 \end{pmatrix}$$

Solution: The transformation as a matrix equation is

$$\begin{bmatrix} x' \\ y' \end{bmatrix} = \begin{pmatrix} 2, & 3 \\ 4, & 1 \end{pmatrix} \begin{bmatrix} x \\ y \end{bmatrix}$$

which means $\qquad\qquad x' = 2x + 3y$
and $\qquad\qquad\qquad y' = 4x + 1y$
Under this transformation the point A(2,1) goes to A′(7,9),
$\qquad\qquad\qquad\qquad\qquad$ B(5,1) goes to B′(13,21)
$\qquad\qquad\qquad\qquad\qquad$ C(5,4) goes to C′(22,24)
$\qquad\qquad$ and $\qquad\qquad$ D(2,4) goes to D′(16,12)

we can write these calculations in matrix form as

$$\begin{pmatrix} 2, 3 \\ 4, 1 \end{pmatrix}\begin{pmatrix} 2 \\ 1 \end{pmatrix} = \begin{pmatrix} 2\times2 + 3\times1 \\ 4\times2 + 1\times1 \end{pmatrix} = \begin{pmatrix} 7 \\ 9 \end{pmatrix} \qquad \begin{pmatrix} 2, 3 \\ 4, 1 \end{pmatrix}\begin{pmatrix} 5 \\ 1 \end{pmatrix} = \begin{pmatrix} 2\times5 + 3\times1 \\ 4\times5 + 1\times1 \end{pmatrix} = \begin{pmatrix} 13 \\ 21 \end{pmatrix}$$

$$\begin{pmatrix} 2, 3 \\ 4, 1 \end{pmatrix}\begin{pmatrix} 5 \\ 4 \end{pmatrix} = \begin{pmatrix} 2\times5 + 3\times4 \\ 4\times5 + 1\times4 \end{pmatrix} = \begin{pmatrix} 22 \\ 24 \end{pmatrix} \qquad \begin{pmatrix} 2, 3 \\ 4, 1 \end{pmatrix}\begin{pmatrix} 2 \\ 4 \end{pmatrix} = \begin{pmatrix} 2\times2 + 3\times4 \\ 4\times2 + 1\times4 \end{pmatrix} = \begin{pmatrix} 16 \\ 12 \end{pmatrix}$$

In a diagram, we see that the square ABCD is transformed into a parallelogram A′B′C′D′

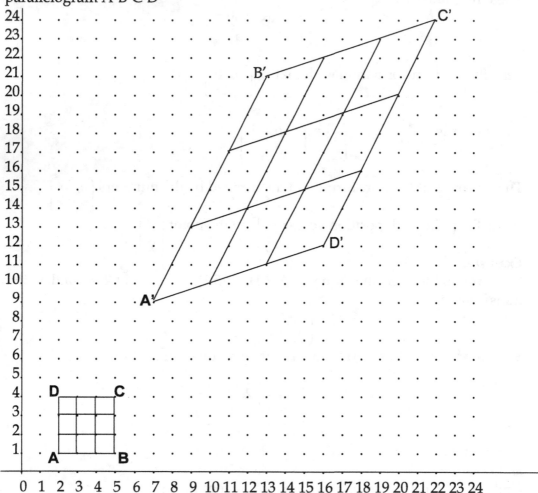

Notice that in this case, the 'orientation' of the figure has been changed. A→B→C→D is anticlockwise in the object figure, but A'→B'→C'→D' is clockwise in the image. The orientation of the figure is changed when the value of ad-bc is negative. In this case, ad-bc = -10.

The value of ad-bc is the scale factor for area. In this case ad-bc = -10 and the area of the object has been magnified by a scale factor 10 but its orientation has been changed.

Finding the Coordinates of the Image Figure

Instead of writing down separate calculations for the coordinates of each point of the image figure, we can use the row x column multiplication to find the coordinates of all points we need in one go:

$$\begin{pmatrix} 2,3 \\ 4,1 \end{pmatrix} \overset{A\ B\ C\ D}{\begin{pmatrix} 2,\ 5,\ 5,\ 2 \\ 1,\ 1,\ 4,\ 4 \end{pmatrix}} = \overset{A'\ B'\ C'\ D'}{\begin{pmatrix} 7,\ 13,\ 22,\ 16 \\ 9,\ 21,\ 24,\ 12 \end{pmatrix}}$$

The x-coordinates of the image are given by 2xtop row + 3x bottom row.

The y-coordinates of the image are given by 4xtop row + bottom row.

Example

An object figure is defined by O(0,0), A(1,0), B(0,1), C(1,1), D(0,2), E(1,2), F(0,3), G(1,3).

Calculate the coordinates of the image figure, under the transformation given by

$$\begin{bmatrix} x' \\ y' \end{bmatrix} = \begin{pmatrix} 2,\ 1 \\ 1,\ 3 \end{pmatrix} \begin{bmatrix} x \\ y \end{bmatrix}$$

Solution

$$\begin{pmatrix} 2 & 1 \\ 1 & 3 \end{pmatrix} \overset{O\ A\ B\ C\ D\ E\ F\ G}{\begin{pmatrix} 0 & 1 & 0 & 1 & 0 & 1 & 0 & 1 \\ 0 & 0 & 1 & 1 & 2 & 2 & 3 & 3 \end{pmatrix}} = \overset{O\ A'\ B'\ C'\ D'\ E'\ F'\ G'}{\begin{pmatrix} 0 & 2 & 1 & 3 & 2 & 4 & 3 & 5 \\ 0 & 1 & 3 & 4 & 6 & 7 & 9 & 10 \end{pmatrix}}$$

Note that it is quicker to find all the x coordinates first: 2xtop + bottom and then all the y coordinates: top + 3xbottom

Diagram showing the image points OA′B′C′D′E′F′G′

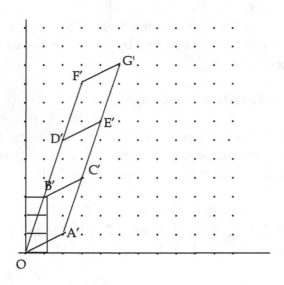

Exercise 3

1. Evaluate

$$\begin{pmatrix} 3 & 4 \\ 1 & 2 \end{pmatrix}\begin{pmatrix} 2 \\ 3 \end{pmatrix}$$

$$\begin{pmatrix} 3 & -4 \\ 1 & 2 \end{pmatrix}\begin{pmatrix} 2 \\ 3 \end{pmatrix}$$

$$\begin{pmatrix} 3 & -4 \\ -1 & 2 \end{pmatrix}\begin{pmatrix} 2 \\ 3 \end{pmatrix}$$

$$\begin{pmatrix} -3 & -4 \\ -1 & 2 \end{pmatrix}\begin{pmatrix} 2 \\ 3 \end{pmatrix}$$

$$\begin{pmatrix} 3 & -4 \\ -1 & 2 \end{pmatrix}\begin{pmatrix} -2 \\ -3 \end{pmatrix}$$

2. Evaluate

$$\begin{pmatrix} 3 & 1 \\ 1 & 2 \end{pmatrix} \begin{pmatrix} 1 & 1 & 1 & 2 \\ 4 & 3 & 2 & 2 \end{pmatrix}$$

and hence write down the coordinates of the image of the L shaped figure A-B-C-D where A(1,4), B(1,3), C(1,2), D(2,2) under the transformation.

$$x' = 3x + y$$
$$y' = x + 2y$$

Plot the object ABCD and its image A′B′C′D′.

Simple Transformation Matrices

There follows a list of matrices for various simple reflections:

The object figure is light and the image dark.

Reflect in the x-axis
$$\begin{pmatrix} -1 & 0 \\ 0 & 1 \end{pmatrix}$$
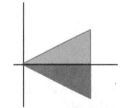

Reflect in the y-axis
$$\begin{pmatrix} 1 & 0 \\ 0 & -1 \end{pmatrix}$$
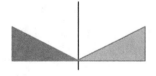

Reflect in the origin
(=rotate 180°)
$$\begin{pmatrix} -1 & 0 \\ 0 & -1 \end{pmatrix}$$
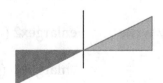

Reflect in the line y = x $\begin{pmatrix} 0 & 1 \\ 1 & 0 \end{pmatrix}$

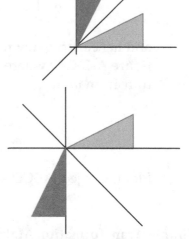

Reflect in the line y = -x $\begin{pmatrix} 0 & -1 \\ -1 & 0 \end{pmatrix}$

You may notice that although each of these transformations will send the light triangle to the dark triangle, the same transformation also sends the dark triangle to the light triangle: each of these transformations is "self inverse".

Thus we may write

reflect(light triangle) = dark triangle

and also   reflect(dark triangle) = light triangle

Reflections are always self inverse: if you reflect the image then you always get back to the object.

Enlargements and translations are not self inverse:

for example:

[1] Enlarge by 2

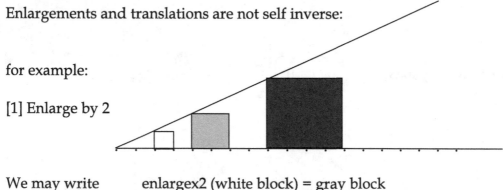

We may write   enlargex2 (white block) = gray block

but   enlargex2 (gray block) = black block

not the white block as it would need to be if the enlargement was self inverse. Enlargements are not self inverse unless the scale factor is +1 or -1.

[2] Translate by $\begin{bmatrix} 2 \\ 1 \end{bmatrix}$

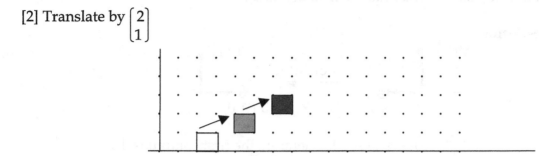

translate(white block) = gray block

But   translate(gray block) = black block,

The translation does not send the gray block back to the white block. Translations are not self inverse.

Finding the Matrix of a Transformation.

How would we, for example, find the matrix for a rotation about the origin by 90 degrees?

Note the following

$$\begin{pmatrix} a & b \\ c & d \end{pmatrix} \begin{bmatrix} 1 \\ 0 \end{bmatrix} = \begin{bmatrix} a \\ c \end{bmatrix}$$

the image of the point (1,0) is the first column of the transformation matrix.

$$\begin{pmatrix} a & b \\ c & d \end{pmatrix} \begin{bmatrix} 0 \\ 1 \end{bmatrix} = \begin{bmatrix} b \\ d \end{bmatrix}$$

the image of the point (0,1) is the second column of the transformation matrix.

This means that, if we can work out where the transformation sends the two points A(1,0) and B(0,1) then we can write down the matrix of the transformation.

(note: positive rotations are anti-clockwise, negative rotations are clockwise)

Example (i)

Find the matrix for a rotation of 90 degrees.

Solution

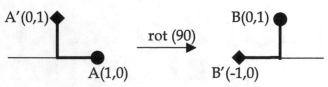

The point A $\begin{bmatrix} 1 \\ 0 \end{bmatrix}$ rotates to A' $\begin{bmatrix} 0 \\ 1 \end{bmatrix}$ therefore the first column is $\begin{bmatrix} 0 \\ 1 \end{bmatrix}$

The point B $\begin{bmatrix} 0 \\ 1 \end{bmatrix}$ rotates to B' $\begin{bmatrix} -1 \\ 0 \end{bmatrix}$ therefore the last column is $\begin{bmatrix} -1 \\ 0 \end{bmatrix}$

The matrix for a rotation of 90 degrees is therefore $\begin{pmatrix} 0 & -1 \\ 1 & 0 \end{pmatrix}$

Example (ii)

Find the matrix for a rotation of 270 degrees

Solution

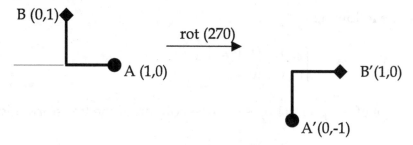

The point A(1,0) rotates to A'(0,-1) therefore the first column is $\begin{bmatrix} 0 \\ -1 \end{bmatrix}$

The point B(0,1) rotates to B'(1,0) therefore the last column is $\begin{bmatrix} 1 \\ 0 \end{bmatrix}$

The matrix for a rotation of 270 degrees is therefore $\begin{pmatrix} 0 & 1 \\ -1 & 0 \end{pmatrix}$

Exercise 4   Use this method to check the matrices for

> 1. reflection in Ox
> 2. reflection in Oy
> 3. reflection through the origin
> 4. reflection in y=x
> 5. reflection in y=-x

We have at this point, discovered all possible transformation matrices made out of two zeros and two ones. There are eight of them and they describe the symmetries of a square. We will see in the next chapter that the eight matrices form a group under matrix multiplication…but more of that later. Here is the list of matrices formed from two zeros and two ones that do not collapse objects down onto a line showing their effect on a square ABCD centred on the origin.

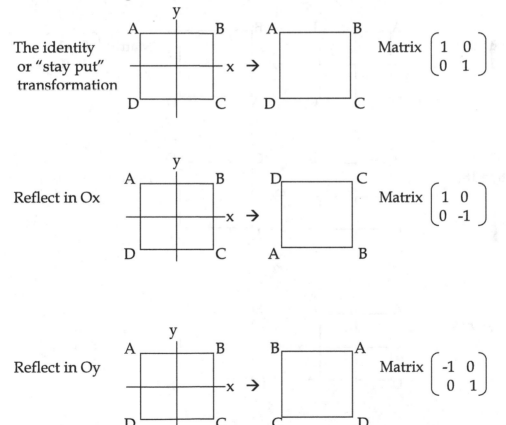

The identity or "stay put" transformation   Matrix $\begin{pmatrix} 1 & 0 \\ 0 & 1 \end{pmatrix}$

Reflect in Ox   Matrix $\begin{pmatrix} 1 & 0 \\ 0 & -1 \end{pmatrix}$

Reflect in Oy   Matrix $\begin{pmatrix} -1 & 0 \\ 0 & 1 \end{pmatrix}$

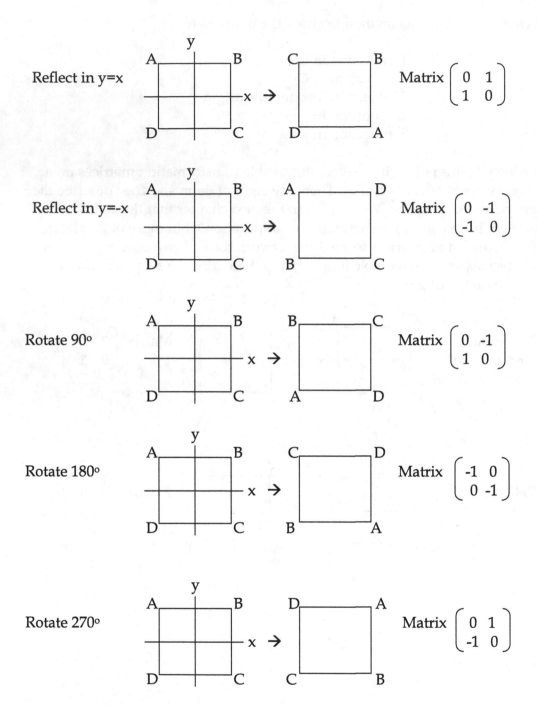

Reflect in y=x

Matrix $\begin{pmatrix} 0 & 1 \\ 1 & 0 \end{pmatrix}$

Reflect in y=-x

Matrix $\begin{pmatrix} 0 & -1 \\ -1 & 0 \end{pmatrix}$

Rotate 90°

Matrix $\begin{pmatrix} 0 & -1 \\ 1 & 0 \end{pmatrix}$

Rotate 180°

Matrix $\begin{pmatrix} -1 & 0 \\ 0 & -1 \end{pmatrix}$

Rotate 270°

Matrix $\begin{pmatrix} 0 & 1 \\ -1 & 0 \end{pmatrix}$

The Area Scale Factor

The following figures show the unit square and its image under $\begin{pmatrix} a & b \\ c & d \end{pmatrix}$

The coordinates of the image of the unit square are given by

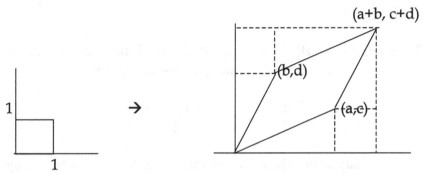

$$\begin{pmatrix} a & b \\ c & d \end{pmatrix} \begin{pmatrix} 0 & 1 & 0 & 1 \\ 0 & 0 & 1 & 1 \end{pmatrix} = \begin{pmatrix} 0 & a & b & a+b \\ 0 & c & d & c+d \end{pmatrix}$$

The area of the outside rectangle bounding the image is $(a+b)(c+d)$

The areas of the four triangles are $ac/2$ (two of these) and $bd/2$ (two of these)

The area of each of the two corner rectangles is bc

To get the area of the parallelogram, we subtract the bits from the total giving:

$$\text{area of parallelogram} = (a+b)(c+d) - ac - bd - 2bc$$

$$= ad - bc$$

This is the area scale factor for any image and it is called the **determinant** of the matrix.

Handedness

If the determinant is negative, then the "handedness" of the image is opposite to the handedness of the image, i.e. if ABC goes clockwise in the object, then A'B'C' goes anticlockwise in the image.

The determinant of a reflection is therefore always negative.
The determinant of a rotation is always positive.

Exercise 5

[1] Find the image of the unit square under each of the following transformations:

(i) $\begin{pmatrix} 2 & 1 \\ 4 & 3 \end{pmatrix}$ (ii) $\begin{pmatrix} -2 & -5 \\ 2 & 4 \end{pmatrix}$ (iii) $\begin{pmatrix} -3 & 3 \\ 2 & -3 \end{pmatrix}$

In each case, draw the parallelogram for the image and state the area of this parallelogram.

[2] Given A(1,1), B(1,2), C(2,2), D(2,1), find the image of the square ABCD under these transformations and plot the image in each case

(i) $\begin{pmatrix} 5 & 2 \\ 3 & 2 \end{pmatrix}$ (ii) $\begin{pmatrix} 2 & 4 \\ 3 & 4 \end{pmatrix}$

What you say about the area and the handedness of these images?

[3] By finding the images of the nine marked points, complete the image of this figure under the transformation $\begin{pmatrix} -1 & 2 \\ 2 & 1 \end{pmatrix}$

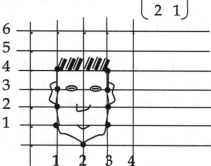

In order to complete question 3, we have to assume that the image of the figure is going to be "well behaved" and not spring any surprises or develop any unexpected holes. To justify this we need a few theorems.

We continue with proofs of three theorems that justify the name of 'linear transformation' for any non-degenerate transformation that has a transformation matrix. These theorems allow us to complete the drawing of the image without needing to find the coordinates of every single image point.

Theorem 1

The transformation $\begin{pmatrix} a & b \\ c & d \end{pmatrix}$ transforms straight lines into straight lines.

Proof

The equations for the transformation are

$$x' = ax + by$$
$$y' = cx + dy$$

Suppose that the object figure is the straight line $y = mx + k$.

Choose some point (x,y) on this line.

The equations give
$$dx' = adx + bdy$$
$$by' = bcx + cdy$$

subtract:
$$x(ad-bc) = dx' - by'$$
similarly:
$$y(ad-bc) = ay' - cx'$$

Provided that $ad-bc \neq 0$ (in which case the transformation collapses 2 space onto a line or a point), we can write

$$x = (dx' - by')/(ad-bc)$$

and $$y = (ay' - cx')/(ad-bc)$$

Substituting for x and y in $y=mx+k$, we get a **linear** equation in x' and y' which shows that (x',y') lies on some fixed straight line.
Whichever point (x,y) we choose, on $y=mx+k$, the image point (x',y') will be on this same straight line.

Thus, the points on $y=mx+k$ are mapped onto points on a straight line.

==

Corollary:

Parallel lines are mapped onto parallel lines

Proof

$$x = (dx'-by')/(ad-bc)$$

and $\qquad y = (ay'-cx')/(ad-bc)$

thus y=mx+k transforms into the straight line

$$(ay'-cx')/(ad-bc) = m(dx'-by')/(ad-bc) + k$$

we do not need to expand and rearrange this equation, but just observe that the gradient of this line only depends on a,b,c,d and m.

Thus parallel lines of the form y=mx + k (for different values of k but the same m) transform into lines with the same gradient, i.e. parallel lines.

================================

Theorem 2   Parallel lines transform into parallel lines (second proof)

Proof consider the line joining two points P(p,q) and R(r,s)

Case 1:   r<>p so that the line PR is not vertical

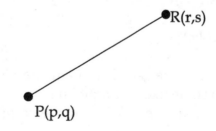

$$R(r,s)$$

$$P(p,q)$$

Let grad(PR) = $\dfrac{s-q}{r-p}$ = m $\qquad$ (r<>p else PR is vertical)

Now P(p,q) → P'(ap+bq, cp+dq)

and R(r,s) → R'(ar+bs, cr+ds)

So that $\text{Grad}(P'R') = \dfrac{cr+ds-(cp+dq)}{ar+bs-(ap+bq)}$

$$= \dfrac{c(r-p)+d(s-q)}{a(r-p)+b(s-q)}$$

$$= \dfrac{c(r-p)+dm(r-p)}{a(r-p)+bm(r-p)} \qquad \text{since s-q = m(r-p)}$$

$$= \dfrac{c+dm}{a+bm}$$

Therefore, any line with gradient m is transformed into a line of gradient
$\dfrac{c+dm}{a+bm}$

so, two parallel lines with gradient m are transformed into two parallel lines
of gradient $\dfrac{c+dm}{a+bm}$

Case 2: if p=r then the line PR is vertical and we have

$\text{Grad } P'R' = \dfrac{cr+ds-(cr+dq)}{ar+bs-(ar+bq)} = \dfrac{ds-dq}{bs-bq} = d/b$

Thus any vertical lines are transformed into lines with gradient d/b and will
thus be parallel

==

Theorem 3   Mid points transform into midpoints

Proof

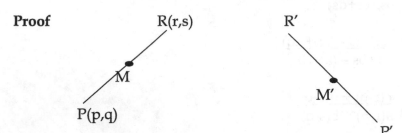

Let P(p,q) and R(r,s), then the mid point of PR is M$\left(\dfrac{p+r}{2} \quad \dfrac{q+s}{2}\right)$

P(p,q) → P'(ap+bq, cp+dq)

R(r,s) → R'(ar+bs, cr+ds)

$$M\left(\frac{p+r}{2}, \frac{q+s}{2}\right) \rightarrow M'\left(\frac{a(p+r)}{2} + \frac{b(q+s)}{2}, \quad \frac{c(p+r)}{2} + \frac{d(q+s)}{2}\right)$$

$$= M'\left(\frac{ap+bq+ar+bs}{2}, \quad \frac{cp+dq+cr+ds}{2}\right)$$

which is the mid point of P'R'

==

Combining Transformations

Let **q** represent a quarter turn (anticlockwise) about the origin. Then we could write

$$\mathbf{q} = \begin{pmatrix} 0 & -1 \\ 1 & 0 \end{pmatrix}$$

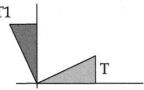

The transformation **q** sends triangle T to triangle T1
and so write

$$\mathbf{q}(T) = T1$$

Let **r** represent a reflection in the line y=x. Then we could write

$$\mathbf{r} = \begin{pmatrix} 0 & 1 \\ 1 & 0 \end{pmatrix} \qquad \text{T1}$$

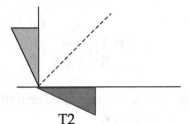

the transformation **r** sends T1 to triangle T2

We can therefore write

$$\mathbf{r}(T1) = T2$$

Combining transformation **q** with transformation **r**, we can write

$$\mathbf{r}(T1) = \mathbf{r}(\mathbf{q}(T)) = T2$$

which we can abbreviate to
$$\mathbf{rq}(T) = T2$$

Suppose that the coordinates of T are (0,0), A(2,0) and B(2,1), then the coordinates of T1 are (0,0), A1(0,2) and B1(-1,2).

So that we can write **q**(T) = T1 in matrix form as

$$\begin{pmatrix} 0 & -1 \\ 1 & 0 \end{pmatrix} \begin{matrix} \text{O A B} \\ \begin{pmatrix} 0 & 2 & 2 \\ 0 & 0 & 1 \end{pmatrix} \end{matrix} = \begin{matrix} \text{O A1 B1} \\ \begin{pmatrix} 0 & 0 & -1 \\ 0 & 2 & 2 \end{pmatrix} \end{matrix}$$

Now reflect T1 in y=x, to get the coordinates of T2: (0,0), A2(2,0) and B2(2,-1) so that we can write **r**(T1)=T2 in matrix form as

$$\begin{pmatrix} 0 & 1 \\ 1 & 0 \end{pmatrix} \begin{matrix} \text{O A1 B1} \\ \begin{pmatrix} 0 & 0 & -1 \\ 0 & 2 & 2 \end{pmatrix} \end{matrix} = \begin{matrix} \text{O A2 B2} \\ \begin{pmatrix} 0 & 2 & 2 \\ 0 & 0 & -1 \end{pmatrix} \end{matrix}$$

which is the matrix form of

$$\mathbf{r}\,(\mathbf{q}(T)) = T2$$

Combining these two matrix equations we have

$$\overset{\mathbf{r}}{\begin{pmatrix} 0 & 1 \\ 1 & 0 \end{pmatrix}} \overset{\mathbf{q}}{\begin{pmatrix} 0 & -1 \\ 1 & 0 \end{pmatrix}} \overset{\text{O A B}}{\begin{pmatrix} 0 & 0 & -1 \\ 0 & 0 & 1 \end{pmatrix}} = \overset{\text{O A2 B2}}{\begin{pmatrix} 0 & 2 & 2 \\ 0 & 0 & -1 \end{pmatrix}}$$

where we note that transformation **q** is applied first, followed by transformation **r**.

Now we ask, "what is the transformation" T → T2 ?

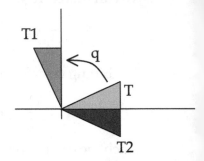

We recognize that T → T2 is simply a reflection in the x axis:

If we write **x** to represent the reflection in Ox then we have

$$\mathbf{x} = \begin{pmatrix} 1 & 0 \\ 0 & -1 \end{pmatrix}$$

we observe that the matrix product of **r.q** gives the matrix **x** :

$$\begin{pmatrix} 0 & -1 \\ 1 & 0 \end{pmatrix} \begin{pmatrix} 0 & 1 \\ 1 & 0 \end{pmatrix} = \begin{pmatrix} 1 & 0 \\ 0 & -1 \end{pmatrix}$$

Thus the matrix of the combined transformation is the product of the matrices of the two transformations.
Note that the first transformation is on the **right** thus **rq** represents a quarter turn followed by a reflection in y=x.

In Theorem 4 at the end of this chapter, we prove in general that the matrix for a combination of two transformations is the matrix product of the matrices of the two transformations.

It is satisfying when we find that the rule for a matrix product that we used for finding the coordinates of an image figure is precisely the same rule that we use for finding the matrix of the combination of two transformations.

Exercise 6

[1]

> i is the identity "stay put" transformation
> x is a reflection in the x axis
> y is a reflection in the y axis
> h is a half turn

Complete the following table:

Right hand (= first)

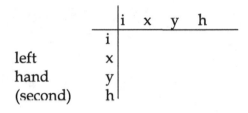

left
hand
(second)

[2]

> i is the identity "stay put" transformation
> q is a quarter turn
> h is a half turn
> t is a three quarter turn

Complete the following table:

Right hand (= first)

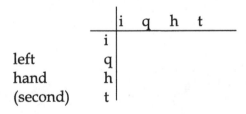

left
hand
(second)

[3]
     a is a reflection in the line y=x
     b is a reflection in the line y=-x

Using the notation introduced in [1] and [2], complete the following table for combining transformations:

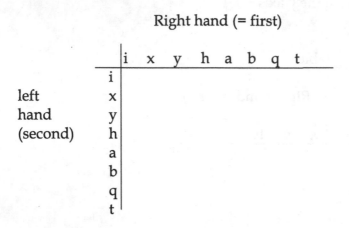

Right hand (= first)

		i	x	y	h	a	b	q	t
	i								
left	x								
hand	y								
(second)	h								
	a								
	b								
	q								
	t								

Exercise 6 illustrates the fact that transformations are grouped together as in these exercises, they form an algebraic structure called a **group.** Groups are studied in the next chapter.
We end chapter 12 with a proof of a fundamental theorem:

Theorem 4   **The matrix for a combination of two transformations is the matrix product of the matrices of the two transformations.**

proof

Let transformation $A = \begin{pmatrix} a & b \\ c & d \end{pmatrix}$

Let transformation $P = \begin{pmatrix} p & q \\ r & s \end{pmatrix}$

Suppose that $X(x,y) \rightarrow X1(x1,y1)$ under transformation P

and that $X1(x1,y1) \rightarrow X2(x2,y2)$ under transformation A

Then the combined transformation AP sends $X(x,y) \rightarrow X2(x2,y2)$

We have
$$\begin{pmatrix} x1 \\ y1 \end{pmatrix} = \begin{pmatrix} p & q \\ r & s \end{pmatrix} \begin{pmatrix} x \\ y \end{pmatrix} = \begin{pmatrix} px+qy \\ rx+sy \end{pmatrix}$$

so that
$$\begin{pmatrix} x2 \\ y2 \end{pmatrix} = \begin{pmatrix} a & b \\ c & d \end{pmatrix} \begin{pmatrix} px+qy \\ rx+sy \end{pmatrix}$$

$$= \begin{pmatrix} apx+aqy+brx+bsy \\ cpx+cqy+drx+dsy \end{pmatrix}$$

$$= \begin{pmatrix} (ap+br)x + (aq+bs)y \\ (cp+dr)x + (cq+ds)y \end{pmatrix}$$

$$= \begin{pmatrix} ap+br, & aq+bs \\ cp+dr, & cq+ds \end{pmatrix} \begin{pmatrix} x \\ y \end{pmatrix}$$

but $AP = \begin{pmatrix} a & b \\ c & d \end{pmatrix} \begin{pmatrix} p & q \\ r & s \end{pmatrix} = \begin{pmatrix} ap+br, & aq+bs \\ cp+dr, & cq+ds \end{pmatrix}$

Therefore the matrix for the combined transformation is AP

===

Exercise 7

Evaluate (i) $\begin{pmatrix} 2 & 3 \\ 4 & 5 \end{pmatrix}\begin{pmatrix} 1 & 1 \\ 1 & 1 \end{pmatrix}$ (ii) $\begin{pmatrix} 1 & 1 \\ 1 & 1 \end{pmatrix}\begin{pmatrix} 1 & 1 \\ 1 & 1 \end{pmatrix}$ (iii) $\begin{pmatrix} 2 & 2 \\ 2 & 2 \end{pmatrix}\begin{pmatrix} 2 & 2 \\ 2 & 2 \end{pmatrix}$

(iv) $\begin{pmatrix} 3 & 1 \\ 1 & 1 \end{pmatrix}\begin{pmatrix} 2 & 1 \\ 1 & 1 \end{pmatrix}$ (v) $\begin{pmatrix} 2 & 1 \\ 1 & 1 \end{pmatrix}\begin{pmatrix} 3 & 1 \\ 1 & 1 \end{pmatrix}$ (vi) $\begin{pmatrix} 2 & 5 \\ 1 & 3 \end{pmatrix}\begin{pmatrix} 3 & -5 \\ -1 & 2 \end{pmatrix}$

The Inverse Transformation

In the last example of exercise 6, we observe that $\begin{pmatrix} 2 & 5 \\ 1 & 3 \end{pmatrix}\begin{pmatrix} 3 & -5 \\ -1 & 2 \end{pmatrix} = \begin{pmatrix} 1 & 0 \\ 0 & 1 \end{pmatrix}$

This means that the combined transformation represented by the product of the two matrices, is the "**stay put**" transformation.

Let **T2** be the transformation $\begin{pmatrix} 2 & 5 \\ 1 & 3 \end{pmatrix}$ and **T1** be $\begin{pmatrix} 3 & -5 \\ -1 & 2 \end{pmatrix}$ then **T2.T1** $= \begin{pmatrix} 1 & 0 \\ 0 & 1 \end{pmatrix}$

thus, if **T1** sends the point **A** to **P** then **T2** sends **P** back to **A**.

We say that **T2** is the inverse of the transformation **T1**.

The matrix $\begin{pmatrix} 2 & 5 \\ 1 & 3 \end{pmatrix}$ is the inverse of the matrix $\begin{pmatrix} 3 & -5 \\ -1 & 2 \end{pmatrix}$

We can easily verify that **T2** undoes the work of **T1** for, if **A** is the point **(a,b)** and **A(a,b)** is sent to the point **P**, then **P** is determined by

$\begin{pmatrix} 3 & -5 \\ -1 & 2 \end{pmatrix}\begin{pmatrix} a \\ b \end{pmatrix} = \begin{pmatrix} 3a - 5b \\ -a + 2b \end{pmatrix}$ so that we have P(3a-5b, -a+2b)

Now apply **T2** to **P** and we have

$\begin{pmatrix} 2 & 5 \\ 1 & 3 \end{pmatrix}\begin{pmatrix} 3a-5b \\ -a+2b \end{pmatrix} = \begin{pmatrix} 6a-10b - 5a+10b \\ 3a-5b - 3a+6b \end{pmatrix} = \begin{pmatrix} a \\ b \end{pmatrix}$

which we can abbreviate to   **T2.T1(A) = T2(P) = A.**

In general, if **T1** is the matrix $\begin{pmatrix} a & b \\ c & d \end{pmatrix}$ then the inverse of **T1** is the matrix

$T2 = \begin{pmatrix} p & q \\ r & s \end{pmatrix}$ (if it exists)   such that $\begin{pmatrix} p & q \\ r & s \end{pmatrix}\begin{pmatrix} a & b \\ c & d \end{pmatrix} = \begin{pmatrix} 1 & 0 \\ 0 & 1 \end{pmatrix}$

Finding the inverse

To find the matrix for **T2,** we note that $pb+qd = 0$ giving $p/q = -d/b$

$$\text{and } ra+sc = 0 \text{ giving } r/s = -c/a$$

So we can let $p=-dh$, $q=bh$ and $r=-ck$ and $s=ak$ where h and k are to be found.

Now check the 1 terms $pa+qc = 1$ gives $-adh+bch = 1$

thus $h = -1/(ad-bc)$

and $rb+sd = 1$ gives $-bck+adk = 1$

thus $k = 1/(ad-bc)$

now we remember that $(ad-bc)$ is the area scale factor (call it **sf**)

and we have the solution $p= d/sf$, $q= -b/sf$, $r=-c/sf$, $s=a/sf$

The inverse matrix is therefore $\dfrac{1}{sf} \begin{pmatrix} d & -b \\ -c & a \end{pmatrix}$ **where sf is the scale factor for area: sf = ad-bc**

=========================

Example

Find the image of the line $2x+3y = 4$ under the transformation $\begin{pmatrix} 2 & 6 \\ 1 & 4 \end{pmatrix}$

Solution
Call the given line **L1** and the image we are trying to find, **L2.**

Let (x,y) on **L1** be transformed to (x', y') on **L2**

Then $\begin{pmatrix} 2 & 6 \\ 1 & 4 \end{pmatrix}\begin{pmatrix} x \\ y \end{pmatrix} = \begin{pmatrix} x' \\ y' \end{pmatrix}$

If we use the inverse matrix, we will "undo" the transformation and get back from (x', y') to (x,y). Let **sf** be the scale factor for the transformation then

$$\begin{pmatrix} x \\ y \end{pmatrix} = \frac{1}{sf} \begin{pmatrix} 4 & -6 \\ -1 & 2 \end{pmatrix} \begin{pmatrix} x' \\ y' \end{pmatrix}$$

therefore $x = (4x' - 6y')/sf$
$y = (-x' + 2y')/sf$

Substitute in **L1** to get $2(4x' - 6y')/sf + 3(-x' + 2y')/sf = 4$
$8x' - 12y' - 3x' + 6y' = 4sf$

Now the scale factor for area is $sf = 2 \times 4 - 1 \times 6 = 2$

Therefore we have $5x' - 6y' = 8$

This holds for all image points (x',y') of any (x,y) on **L1**

The equation of the line L2 is therefore $5x - 6y = 8$

============================

note: an alternative method is to find the images of the two points where **L1** meets the axes. These are $(2,0)$ and $(0, 4/3)$ and the images points are $(4,2)$ and $(8, 16/3)$. Then use $y-y1 = (y2-y1)/(x2-x1)(x-x1)$ ... but this method is a little boring.

Exercise 8

[1]   Find the image of the line $2x+y = 3$ under the transformation $\begin{pmatrix} 2 & 1 \\ 1 & 2 \end{pmatrix}$

[2]   Find the image of the line $4x+3y = 12$ under the transformation $\begin{pmatrix} 1 & -1 \\ 1 & 1 \end{pmatrix}$

[3]   Find the image of the circle $x^2+y^2 = 1$ under $\begin{pmatrix} 2 & 1 \\ 1 & 2 \end{pmatrix}$

[4]   Find the image of the circle $x^2+y^2 = 1$ under $\begin{pmatrix} 4 & -3 \\ 3 & 4 \end{pmatrix}$

Solutions

Exercise 1   (Answers are 3, 4, 5, 6 ,7, 19, 8)

Exercise 3

[1] $\begin{bmatrix} 18 \\ 8 \end{bmatrix}$ $\begin{bmatrix} -6 \\ 8 \end{bmatrix}$ $\begin{bmatrix} -6 \\ 4 \end{bmatrix}$ $\begin{bmatrix} -18 \\ -4 \end{bmatrix}$ $\begin{bmatrix} 6 \\ -4 \end{bmatrix}$

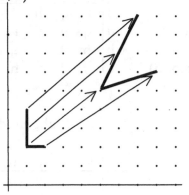

[2]   The image figure is $\begin{pmatrix} 7 & 6 & 5 & 8 \\ 9 & 7 & 5 & 6 \end{pmatrix}$

Exercise 5

[1]   (i) $\begin{pmatrix} 0 & 2 & 1 & 3 \\ 0 & 4 & 3 & 7 \end{pmatrix}$ (ii) $\begin{pmatrix} 0 & -2 & -5 & -7 \\ 0 & 2 & 4 & 6 \end{pmatrix}$ (iii) $\begin{pmatrix} 0 & -3 & 3 & 0 \\ 0 & 2 & -3 & -1 \end{pmatrix}$

             A=2   A=2   A=3

[2] $\begin{bmatrix} 7 & 9 & 14 & 12 \\ 5 & 7 & 10 & 8 \end{bmatrix}$ A=4   same $\begin{pmatrix} 6 & 10 & 12 & 8 \\ 7 & 11 & 14 & 10 \end{pmatrix}$ A=4   opposite

Exercise 6

[1]   Right hand (= first)

		i	x	y	h
	i	i	x	y	h
left	x	x	i	h	y
hand	y	y	h	i	x
(second)	h	h	y	x	i

[2]   Right hand (= first)

	i	q	h	t
i	i	q	h	t
q	q	h	t	i
h	h	t	i	q
t	t	i	q	h

left hand (second)

[3]   Right hand (= first)

	i	x	y	h	a	b	q	t
i	i	x	y	h	a	b	q	t
x	x	i	h	y	t	q	b	a
y	y	h	i	x	q	t	a	b
h	h	y	x	i	b	a	t	q
a	a	q	t	b	i	h	x	y
b	b	t	q	a	h	i	y	x
q	q	a	b	t	y	x	h	i
t	t	b	a	q	x	y	i	h

left hand (second)

Exercise 7   (i) $\begin{pmatrix} 5 & 5 \\ 9 & 9 \end{pmatrix}$ (ii) $\begin{pmatrix} 2 & 2 \\ 2 & 2 \end{pmatrix}$ (iii) $\begin{pmatrix} 8 & 8 \\ 8 & 8 \end{pmatrix}$

(iv) $\begin{pmatrix} 7 & 4 \\ 3 & 2 \end{pmatrix}$ (v) $\begin{pmatrix} 7 & 3 \\ 4 & 2 \end{pmatrix}$ (vi) $\begin{pmatrix} 1 & 0 \\ 1 & 0 \end{pmatrix}$

Exercise 8

[1] $x = 3$

[2] $x + 7y = 24$

[3] $5x^2 + 5y^2 - 8xy = 9$

[4] $x^2 + y^2 = 25$

=================================

CHAPTER 13

Groups

The 30th of May, 1832 and dawn is breaking over the park at Gentilly in Paris. Near the pond, three figures appear in quiet conversation and then step apart. Suddenly the peace is broken by the sharp crack of a single shot and the young man crumples to the ground. The two other figures quietly leave the scene…honour has been satisfied.

Some time later, a passer by helps the wounded man to the Cochin Hospital but the following morning Galois is dead. He had been shot in the stomach. He was twenty years old.

Evariste Galois was a mathematical genius, born in Bourg-La-Reine, Gentilly, now a suburb of Paris, in 1811. While in his late teens he solved the problem of why it was so hard to find a formula for the solution of a fifth degree polynomial equation. In fact he proved that it is impossible to find a general formula and in doing so he laid the foundations of an important branch of abstract algebra – Galois group theory.

Algebraic Structures

An algebraic structure is a set of distinguishable objects or elements, which may be combined by various rules or operations to give other elements that may or may not be elements of the set. These objects could be, numbers, sets, vectors, matrices, statements etc. or may be undefined. The elements are represented by symbols, for example, **a, b, c, x, y, u, v, A, B,** etc. When, for example, we write **a = b** we mean that the symbols **a** and **b** represent the same element. Each algebra has its own rules for combination. We illustrate this by appeal to the algebra of numbers, called **arithmetic.**

The Rules of Arithmetic

The Associative rule for Addition

$$a + (b+c) = (a+b) + c$$

for example $3+(4+5) = 3+9 = 12$
$\qquad\qquad\quad (3+4)+5 = 7+5 = 12$

The Associative rule for Multiplication

$$a(bc) = (ab)c$$

e.g. $3x(4x5) = 3x20 = 60$
               $(3x4)x5 = 12x5 = 60$

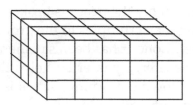

The number of blocks is the same
whether we calculate 3 layers of 20
blocks each, or 5 walls of 12 blocks.

Division does not obey the associative rule:

$$12÷(6÷2) = 12÷3 = 4$$

$$(12÷6)÷2 = 2÷2 = 1$$

When we write 12÷6÷2, without brackets, we are supposed to evaluate from
the left, thus

               $12÷6÷2$ means $(12÷6)÷2$ which is 1.

In general where the same operation is used a number of times, we should
evaluate from the left:
               $36÷3÷2÷3÷2 = 12÷2÷3÷2 = 6÷3÷2 = 2÷2 = 1$

One exception to this rule could be the power operation:

If we want to evaluate the tower of numbers 2^{3^2}

it would appear as $2^\wedge3^\wedge2 = 8^\wedge2 = 64$ or $2↑3↑2 = 8↑2 = 64$

however, when writing towers of numbers by hand, the numbers are written
smaller and smaller as we ascend the tower so that

               2^{3^2} would be intended as $2^9 = 512$

The Commutative Rules

For addition $\quad\quad$ 3+4 = 4+3

For multiplication $\quad$ 3x4 = 4x3

Subtraction and division do not obey the commutative rule

$\quad\quad$ 3-4 ≠ 4-3 $\quad\quad\quad\quad$ 3÷4 ≠ 4÷3

The Distributive Rule

Multiplication distributes over addition

$$3(4+5) \; = \; 3x4 + 3x5$$

Identities

Zero is the identity for addition

$\quad\quad$ n + 0 = 0 + n = n $\quad$ for any number n

One is the identity for multiplication

$\quad\quad$ nx1 = 1xn = n $\quad\quad$ for any number n

The Definition of a Group

A group is the simplest kind of algebraic structure. There is only one rule of combination (or operation), and there are only four requirements:

1. a group is closed under its rule of combination

2. the rule is associative

3. there is an identity (later, we will see that we only need a **left** identity)

4. each element has an inverse (later we will see that we only need a **left** inverse)

Closure

The set of whole numbers is closed under addition.

Let N={1,2,3,4,....}, then for all elements a and b belonging to N, a+b is also a member of N.

In short, for all a,b ε N,   a+b ε N which we can abbreviate still further to

$$\forall \, a,b \, \varepsilon \, N, \quad a+b \, \varepsilon \, N$$

N is also closed under multiplication:

$$\forall \, a,b \, \varepsilon \, N, \quad a.b \, \varepsilon \, N$$

N is not closed under subtraction or division.
We only need one counter example to destroy closure:

for example 2-3 $\notin$ N   therefore N is not closed under subtraction

also 2÷3 $\notin$ N therefore N is not closed under division

The Associative Rule

N is associative under +

$$\forall \, a,b,c \in N \ \text{ we have } \ a+(b+c) = (a+b) + c$$

N is associative under x

$$\forall \, a,b,c \in N \ \text{ we have } \ ax(bxc) = (axb)xc$$

Identities

Let Z be the set of integers, Z = {0, ±1, ±2, ±3, ±4 .....}

Then the identity for addition is 0: $\forall \, a \in Z, \ a+0 = 0+a = a$

There is an identity for multiplication; $\forall \, a \in Z, \ a \times (+1) = (+1) \times a = a$

Inverses

Every element in Z has an inverse under addition. If n∈Z, then the inverse of n under addition is (-n) and we have

$$n + (-n) = (-n) + n = 0$$

The only numbers in Z that have inverses under x are +1 and –1. These two integers are in fact "self inverse":

$$(+1)x(+1) = +1 \quad \text{and} \quad (-1)x(-1) = +1$$

An example of a Group

The integers **Z = { 0, ±1, ±2, ±3, ... }** form a group under addition

Proof

Z is closed under +   ∀ a,b ε Z, a+b ε Z

Z is associative under +   ∀ a,b,c ε Z, a+(b+c) = (a+b)+c

The identity for + is 0   ∀ n ε Z, n+0 = 0+n = n

The positive number +n has an inverse -n
The negative number –n has an inverse +n   (+n)+(-n) = (-n)+(+n) = 0
The inverse of 0 is 0   0 + 0 = 0

Therefore Z is a group under the binary operation +

===================================

The operation "followed by"

Example 1

In this example we have three operations that change the position of an object figure. Half represents a half turn, Vert means flip vertical and Hor means flip horizontal.
These are illustrated by the following:

Half $\curvearrowright$ = $\curvearrowleft$ a half turn

Hor $\curvearrowright$ = $\curvearrowleft$ flip horizontally

Vert $\curvearrowright$ = $\curvearrowright$ flip vertically

We can combine these operations by performing one and then doing the next operation on the result. This gives the following relations:

Half Hor $\curvearrowright$ = Half $\curvearrowleft$ = $\curvearrowright$ = vert $\curvearrowright$

Half Vert $\curvearrowright$ = Half $\curvearrowright$ = $\curvearrowleft$ = hor $\curvearrowright$

Hor Vert $\curvearrowright$ = Hor $\curvearrowright$ = $\curvearrowleft$ = half $\curvearrowright$

Comparing these results with the original definitions, we have for any object:

$$\text{Half Hor} = \text{Vert}$$

$$\text{Half Vert} = \text{Hor}$$

$$\text{Hor Vert} = \text{Half}$$

Now see what happens if we combine three of these together. We have a choice of which two operations to combine first. Suppose we do the second two operations first:

To evaluate Half (Hor Vert $\curvearrowright$) we first evaluate (Hor Vert $\curvearrowright$) = Half $\curvearrowright$ = $\curvearrowleft$

so that we have   Half (Hor Vert $\curvearrowright$) = Half $\curvearrowleft$ = $\curvearrowright$

Now suppose that we combine the first two operations together first

To evaluate (Half Hor) Vert $\curvearrowright$ we first evaluate Half Hor = Vert

and In this case we have (Half Hor) Vert $\curvearrowright$ = Vert Vert $\curvearrowright$ = $\curvearrowright$

and we note that Half(Hor Vert) gives the same result as (Half Hor) Vert

Should we be surprised at this? Well no! When we perform Half (Hor Vert) we perform (Hor Vert) first which requires Vert and then Hor and then we follow with Half. So we do Vert then Hor and then Half.

When we perform (Half Hor) Vert, we perform Vert first and then (Half Hor), but performing Half Hor requires Hor first and then Half. Thus when we perform (Half Hor) Vert we perform Vert followed by Hor and then Half. Thus

$$\text{(Half Hor) Vert} = \text{Half (Hor Vert)}$$

We are combining these operations together using a "followed by" rule and whenever we use the "followed by" operation we find that the associative rule holds.

Example 2

Here we have three operations called Splat, Splodge and Splinge. They operate on elements called A, B, C and D and we are told that

$$\text{Splat A} = \text{B}$$

$$\text{Splodge B} = \text{C}$$

$$\text{Splinge C} = \text{D}$$

The operations are combined according to a "followed by" rule so that Splinge Splodge B means do a Splodge and follow with a Splinge so that

$$\text{Splinge Splodge B} = \text{Splinge C} = \text{D}$$

If we examine Splinge (Splodge Splat A) and (Splinge Splodge) Splat A

we find that Splinge (Splodge Splat A) = Splinge (Splodge B) = Splinge C = D

and also that (Splinge Splodge) Splat A = Splinge Splodge B = Splinge C = D

Whatever we mean by these operations, if we combine them by a "followed by" rule then they will be associative:

$$\text{Splinge (Splodge Splat)} = \text{(Splinge Splodge) Splat}$$

Example 3

   Let Splat be the operation "add1"

   Let Splodge be the operation "times by 2"

   Let Splinge be the operation "subtract 3"

Here we will denote a binary operation on Splinge, Splat, Splodge by o.

   o will be the binary operation "followed by"

thus   Splodge o Splat x = Splodge (x+1) = 2(x+1)

whereas   Splat o Splodge x = Splat (2x) = 2x+1

Let us test the associative rule;

   Splinge o (Splodge o Splat x) = (Splodge o Splat x) - 3

$$= 2(\text{Splat } x) - 3$$

$$= 2(x+1) - 3 = 2x-1$$

   (Splinge o Splodge) o Splat x = (Splinge o Splodge) (x+1)

$$= \text{Splinge } (2(x+1))$$

$$= 2(x+1) - 3 = 2x-1$$

Finite Groups

Finite groups are groups with a finite number of elements. A finite group can be represented by a Cayley Table that shows the result of the group operation on any two elements.
(Arthur Cayley.. 1821-1895 was a senior wrangler at Cambridge at the age of 21 and discovered the theory of matrix algebra)

In a Cayley table, the identity element is represented by e.
The number of elements in a group is called the **order** of the group and is represented by the symbol $|G|$.

We now give a more formal definition of a group:

Definition of a Group

A group is a set of elements (i.e. things that we can label and talk about) which we will denote by G = {e, a, b, c, ...} with a binary operation or rule of combination o, such that every pair of elements in G determines a unique element a o b and

[1] G is closed under o

[2] o is associative

[3] there is a left identity e such that e o a = a $\forall$ a $\in$ G

[4] each element a has a left inverse a^{-1} such that $a^{-1}oa = e$

(Note: The symbol o for the group operation or rule of combination is often suppressed or replaced by a dot, thus instead of aob, we often find it more convenient to write just ab or sometimes a.b)

Example 4   The non-zero rational numbers form a group under x

Proof

[1] $\dfrac{a}{b} \times \dfrac{c}{d} = \dfrac{ac}{bd}$ which is a non zero rational, therefore, the set is closed under x

[2] $\dfrac{a}{b} \times \left(\dfrac{cxe}{d\ f}\right) = \left(\dfrac{axc}{b\ d}\right) \times \dfrac{e}{f}$,   both sides being $\dfrac{ace}{bdf}$

∴ the associative rule holds for these rational numbers

[3]   there is an identity $\dfrac{1}{1}$ since $\dfrac{1}{1} \times \dfrac{a}{b} = \dfrac{a}{b} \times \dfrac{1}{1} = \dfrac{a}{b}$

[4]   the inverse of $\dfrac{a}{b}$ is $\dfrac{b}{a}$

The four requirements for a group are therefore satisfied.

■

Example 5 Addition in clock arithmetic, modulo 5

The elements of the group are in the set of integers { **0, 1, 2, 3, 4** }
The operations are described by the following Cayley table:

+	0	1	2	3	4
0	0	1	2	3	4
1	1	2	3	4	0
2	2	3	4	0	1
3	3	4	0	1	2
4	4	0	1	2	3

[1] the set is closed because all the entries in the table belong to the set

[2]
The operation is associative, since if **a, b and c** are elements of the set
and ordinary addition gives **a+b+c = x + 5n** where $0 \le x \le 4$
Then we will have both

$$(a+b) + c = x+5n \quad \text{and} \quad a+ (b+c) = x + 5n$$

so that in modulo 5 arithmetic **(a+b) + c = x** and **a + (b+c) = x**

[3] **0** is the identity

[4] the inverse elements can be written

$$-1 = 4, \ -2 = 3, \ -3 = 2 \ \text{and} \ -4 = 1$$

Example 6

Multiplication in clock arithmetic modulo 5
The elements of the group are the integers { **1, 2, 3, 4** }

The operations are defined by the following Cayley table:

x	1	2	3	4
1	1	2	3	4
2	2	4	1	3
3	3	1	4	2
4	4	3	2	1

[1]   the set is closed under x (mod 5) since all the entries in the table are members of the set

[2]   if **ax(bxc) = y + 5n** where $1 \le y \le 4$ then **(axb)xc = y + 5n** and therefore, using mod 5 multiplication **ax(bxc) = (axb)xc = y** and we conclude that the operation x (mod 5) is associative

[3]   the identity element is **1**

[4]   inverses are $1^{-1} = 1$, $2^{-1} = 3$, $3^{-1} = 2$ and $4^{-1} = 4$

==

We now put our discussion on a more formal basis with proofs of introductory theorems.

Theorem 1   e is unique, there is only one identity

Proof   Suppose that both e and f are left identities

Let e^{-1} be the inverse of e

Then $e^{-1}e = e$ using rule 4, as e is a left identity

Also $e^{-1}e = f$ using rule 4, as f is a left identity

Therefore   e = f

================================

Theorem 2   Every left inverse is also a right inverse

Proof $ea^{-1} = a^{-1}$ e is the left identity

$\therefore$ $(a^{-1}a)a^{-1} = a^{-1}$ left inverse

$\therefore$ $a^{-1}(aa^{-1}) = a^{-1}$ associative rule

Let b be the left inverse of a^{-1}

Then $ba^{-1}(aa^{-1}) = ba^{-1}$

∴ $e(aa^{-1}) = e$

∴ $aa^{-1} = e$

Therefore a left inverse is also a right inverse thus if $a^{-1}a = e$ then $aa^{-1} = e$

=====================

Theorem 3   The left identity is also a right identity

Proof $ae = a(a^{-1}a)$

                $= (aa^{-1})a$ associative rule 2

                $= ea$ by theorem 2

                $= a$

========================

Theorem 4 a^{-1} is unique

Proof   **Let ba = e**

    Then   **$(ba)a^{-1} = ea^{-1}$**

                $b(aa^{-1}) = ea^{-1}$ associative rule

                $be = a^{-1}$ $aa^{-1} = e$ by Theorem 2

                $b = a^{-1}$ $be = b$ by Theorem 3

===========================

Exercise 1

[1]   Prove that $(a^{-1})^{-1} = a$

[2] Prove that $e^{-1} = e$

[3] Prove that $(ab)^{-1} = b^{-1}a^{-1}$

[4] Prove that $(abc)^{-1} = c^{-1}b^{-1}a^{-1}$

Theorem 5 Each element of the group occurs just once in each row and each column of the Cayley table.

Proof Suppose that $ax = z$ and the $ay = z$

Then $a^{-1}(ax) = a^{-1}z$ and $a^{-1}(ay) = a^{-1}z$

Now use the associative law with $a^{-1}a = e$ to get

$$ex = a^{-1}z \quad \text{and} \quad ey = a^{-1}z$$

therefore $x = y$ since they are both equal to $a^{-1}z$

This shows that each row of the Cayley table of a finite group is a permutation of the list of elements in the group.
In a similar way, we can show that each column must a permutation of the elements of the group.

========================

The group of order 1

The smallest group has only one element.
The Cayley table for the group of order 1:

$$
\begin{array}{c|c}
\cdot & e \\
\hline
e & e
\end{array}
$$

The single element e serves as its own inverse: $e^{-1} = e$

Euclid's zero dimensional point has this as its "symmetry group".. see later.

The group of order 2

Let e be the identity element and f be the other element, then we must have
The Cayley table for the group as

$$
\begin{array}{c|cc}
. & e & f \\
\hline
e & e & f \\
f & f &
\end{array}
$$

Using Theorem 5, the missing element must be e.
The group of order 2 is therefore

$$
\begin{array}{c|cc}
. & e & f \\
\hline
e & e & f \\
f & f & e
\end{array}
$$

Euclid's one dimensional line has this as its symmetry group.

A ————————— **B**

The two operations that leave the line looking the same are

e = leave the line alone, the "stay put" operation

f = flip the line over to get   **B** ————————— **A**

=================================

The group of order 3

Let the three elements be e, f, g with, as usual, e the identity, then we have
the group table:

$$
\begin{array}{c|ccc}
. & e & f & g \\
\hline
e & e & f & g \\
f & f & & \\
g & g & &
\end{array}
$$

the second row of the table must be f g e since f e g would result in two g
elements in the last column which would contradict theorem 5.

The group of order 3 is therefore

$$
\begin{array}{c|ccc}
\cdot & e & f & g \\
\hline
e & e & f & g \\
f & f & g & e \\
g & g & e & f \\
\end{array}
$$

Note that in this case we have $g = f^2$ and $fg = f^3 = e$

This is called the cyclic group of order 3, generated by f since we could write out the table as follows:

$$
\begin{array}{c|ccc}
\cdot & e & f & f^2 \\
\hline
e & e & f & f^2 \\
f & f & f^2 & e \\
f^2 & f^2 & e & f \\
\end{array}
$$

This group describes the rotational symmetries of the equilateral triangle.

Let f be a 120 degree rotation about its centre, then $g=f^2$ would be a 240 degree rotation.

Cyclic Groups

A cyclic group is a group that can be completely generated by powers of one of its elements.
The only group of order 2 is the cyclic group {e, f } with $f^2 = e$

The only group of order 3 is also a cyclic group {e, f, f^2 } with $f^3 = e$. But note also, that, using the notation of the previous paragraph, we could also describe the group of order 3 by {e, g, g^2 } with $g^3 = e$ writing g^2 wherever we find f in the original table.

The cyclic group of order 4

Clearly there is a cyclic group for any order we choose.
The cyclic group of order 4 has the following table:

.	e	f	f^2	f^3
e	e	f	f^2	f^3
f	f	f^2	f^3	e
f^2	f^2	f^3	e	f
f^3	f^3	e	f	f^2

but there is another group of order 4 which is not a cyclic group. It is called the Klein 4 group after the mathematician Felix Klein(1849-1925).

The Klein 4 group
The general form of the table for a group of order 4 is as follows:

.	e	f	g	h
e	e	f	g	h
f	f	x		
g	g			
h	h			

where the element x can be determined by trial and error:

Case 1   x = e gives two possibilities

	e	f	g	h
e	e	f	g	h
f	f	e	h	g
g	g	h	e	f
h	h	g	f	e

this is Klein's 4 group

or

	e	f	g	h
e	e	f	g	h
f	f	e	h	g
g	g	h	f	e
h	h	g	e	f

now in this case we observe that
$f = g^2$, $h = fg = g^3$ and $g^4 = e$
so that we could describe the group
as $\{e, g^2, g, g^3\}$, the cyclic group
generated by g.

alternatively, we may note that $f = h^2$, $g = hf = h^3$, $h^4 = e$ so that we could describe the group as $\{e, h^2, h^3, h\}$, the cyclic group generated by h.

The cyclic group generated by g, $\{e, g, g^2, g^3\}$ with $g^4 = e$ is also generated by g^3 since $\{e, g^3, g^6, g^9\} = \{e, g^3, g^2, g\}$.

Case 2   x = g gives

	e	f	g	h
e	e	f	g	h
f	f	g	h	e
g	g	h	e	f
h	h	e	f	g

Again, note that $g = f^2$, $h = fg = f^3$ and $f^4 = e$ so that this is a cyclic group $\{e, f, f^2, f^3\}$.
or we may note that $g = h^2$, $h = hg = h^3$ and $h^4 = e$ so that we could also regard this group as the cyclic group generated by h.

Case 3   x = h gives

	e	f	g	h
e	e	f	g	h
f	f	h	e	g
g	g	e	h	f
h	h	g	f	e

and this too turns out to be the cyclic group:

Since $h = f^2$, $g = hf = f^3$ and $f^4 = e$ we can describe this group as $\{e, f, f^3, f^2\}$ which is a cyclic group generated by f

Alternatively, we note that $h = g^2$, $f = gh = g^3$ and $g^4 = e$ so that we could also describe this group as $\{e, g^3, g, g^2\}$ which is a cyclic group generated by g.

Thus we come to the conclusion that a group of order 4 is either cyclic or it is

```
  | e f g h
--+---------
e | e f g h          Klein's 4 group
f | f e h g
g | g h e f
h | h g f e
```

Exercise 2

[1] Complete this Cayley table for a group or order 5:

```
  | a  b  c  d  e
--+----------------
a | b  c
b |       e  a
c |          b
d | e
e | a  b  c  d
```

Show that this is a cyclic group.

[2] Complete this Cayley table for a group of order 6:

```
  | a  b  c  d  e  f
--+-------------------
a |       f
b |
c |          a  f
d |    d
e |          f  a
f |       a     d  b
```

Symmetry and groups

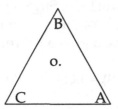

This equilateral triangle has 'rotational symmetry'. If we rotate by 120 degrees about the centre o (here called rot1), then the triangle maps onto itself:

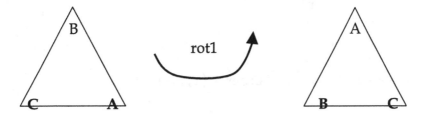

A rotation of 240 degrees (called rot2), also leaves the shape unchanged but rearranges the vertices.

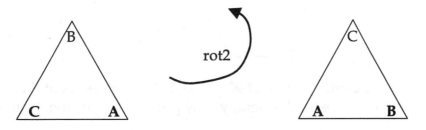

It is natural to write,   rot1(ABC) = CAB

and   rot2(ABC) = BCA

We can deduce that   rot1(BCA) = ABC by changing the labels and thus we can combine the transformations:

$$\text{rot1.rot2(ABC)} = \text{rot1(BCA)} = \text{ABC}$$

Thus rot1.rot2 is the 'stay put' or identity transformation. rot1 and rot2 are inverses of each other: rot1.rot2 = rot2.rot1 = e. Since we are combining these transformations using a 'followed by' operation, we know that we have an associative operation. Therefore we have a group { e, rot1, rot2 }

$$rot1.rot1(ABC) = rot1(CAB) = BCA = rot2(ABC)$$

$$rot2.rot2(ABC) = rot2(BCA) = CAB = rot1(ABC)$$

If we abbreviate rot1 to r1 and rot2 to r2 we have the following group table:

	e	r1	r2
e	e	r1	r2
r1	r1	r2	e
r2	r2	e	r1

which represents a cyclic group of order 3 { e, r1, r1^2 }

Rotations of a square

Let r1, r2 and r3 represent rotations of 90, 180 and 270 degrees about the centre of the square with e for the identity 'stay put' transformation, then we have:

e(ABCD) = ABCD

r1(ABCD) = DABC

r2(ABCD) = CDAB

and   r3(ABCD) = BCDA

We can combine these transformations using the 'followed by' operation so that, for example:

$$r1.r2(ABCD) = r1(CDAB) = BCDA = r3(ABCD)$$

$$r3.r1(ABCD) = r3(DABC) = ABCD = e(ABCD)$$

Since we have used the 'followed by' rule of combination, we know that these transformations are associative. e is the identity, r1 and r3 are inverses and r2 is self inverse. Therefore we have a group of transformations with Cayley table:

e	e	r1	r2	r3
r1	r1	r2	r3	e
r2	r2	r3	e	r1
r3	r3	e	r1	r2

which is a cyclic group generated by r1.

Reflections of a Rectangle

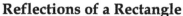

This rectangle can be reflected in the x axis to give: rx(ABCD) = DCBA

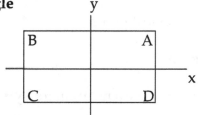

A reflection in the y axis gives ry(ABCD) = BADC

ro will be a reflection through the origin (i.e a half turn): ro(ABCD) = CDAB

As usual, the identity is represented by e: e(ABCD) = ABCD

Exercise 3: Evaluate the following:

[1] rx.rx(ABCD)	[2] rx.ry(ABCD)	[3] rx.ro(ABCD)
[4] ry.rx(ABCD)	[5] ry.ry(ABCD)	[6] ry.ro(ABCD)
[7] ro.rx(ABCD)	[8] ro.ry(ABCD)	[9] ro.ro(ABCD)

Now complete the group table for these transformations of the rectangle:

	e	rx	ry	ro
e	e	rx	ry	ro
rx	rx			
ry	ry			
ro	ro			

Show that this is a Klein 4 group

A Notation for Symmetries

In order to fix ideas we introduce a notation for symmetry transformations of a plane figure.
We take as our example object a regular hexagon ABCDEF centred on the origin as shown in the following figure:

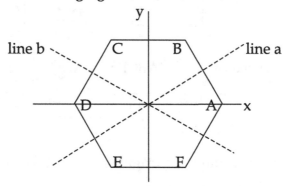

We suppose that the hexagon is a rigid flat object that can be reflected (which involves turning the hexagon over) or could be rotated, leaving outline of the hexagon in the same place but with the labeled vertices in different places. A symmetry is one of these transformations that leaves the outline of the figure in the same position.

Initially we start with the vertices in the positions shown and we describe each transformation by giving the vertices that land on positions A,B,C,D,E and F in this order. The hexagon has six axes of reflectional symmetry:

A reflection in the x axis is denoted by rx: rx(ABCDEF) = AFEDCB

A reflection in the y axis is denoted by ry: ry(ABCDEF) = DCBAFE

A reflection in the line a is denoted by ra: ra(ABCDEF) = BAFEDC

A reflection in the line b is denoted by rb: rb(ABCDEF) = FEDCBA

A reflection in the line BE is denoted by rbe: rbe(ABCDEF) = CBAFED

A reflection in CF is denoted by rcf: rcf(ABCDEF) = EDCBAF

We also have six rotations:

rot60(ABCDEF) = FABCDE

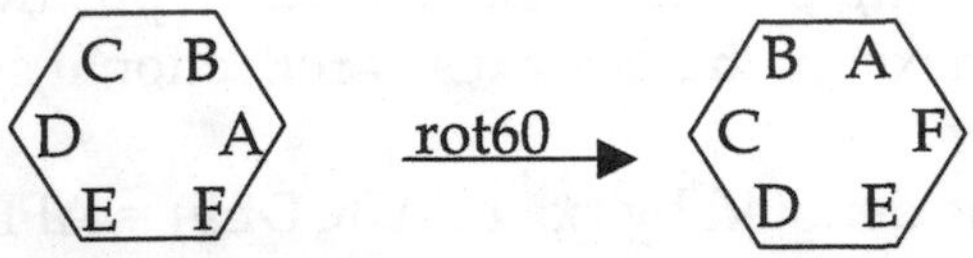

rot120(ABCDEF) = EFABCD

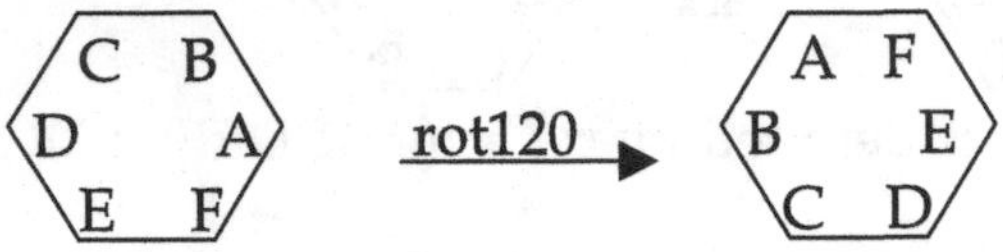

rot180(ABCDEF) = DEFABC

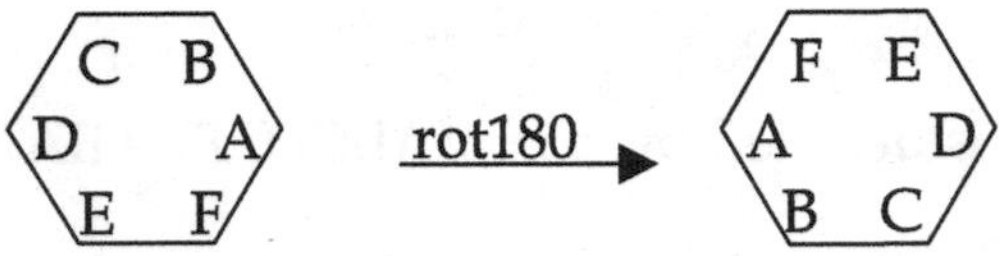

rot240(ABCDEF) = CDEFAB

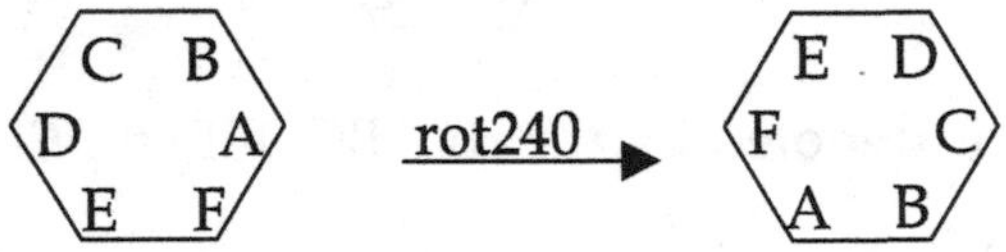

rot300(ABCDEF) = BCDEFA

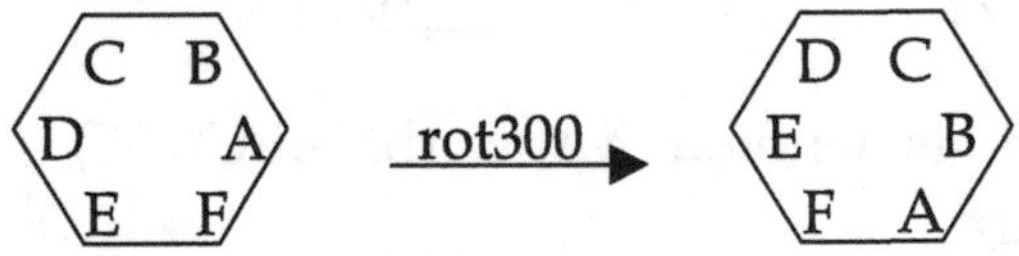

rot360 is the same as the identity,

e(ABCDEF) = ABCDEF

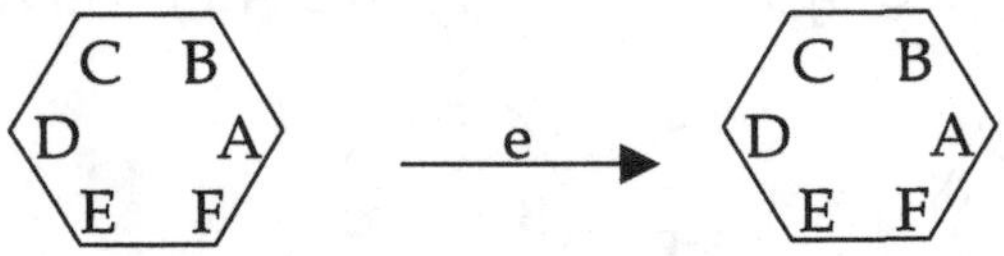

Each of these transformations preserves the outline of the figure but rearranges the vertices according to a particular symmetry.

If we examine the notation that we have used, we see that it is not important what letter or label we write on the hexagon. The important thing is the position in space before and after the transformation, for example, the following all illustrate a reflection in the y axis:

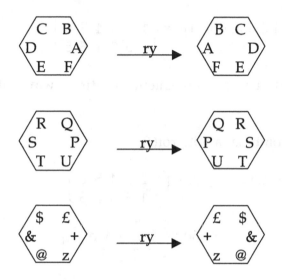

When we write
$$ry(ABCDEF) = DCBAFE$$

A is in position 1   A goes to position 4

In all of the above, whatever was in position 1 goes to position 4
Whatever was in position 2 goes to position 3
The C in position 3 goes to position 2...and so on

If we label the positions in space: (1,2,3,4,5,6)

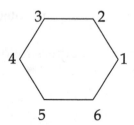

then we can describe the transformation ry by saying

the label at position 1 goes to position 4
the label at position 2 goes to position 3
the label at position 3 goes to position 2
the label at position 4 goes to position 1
the label at position 5 goes to position 6
the label at position 6 goes to position 5

which suggests a natural notation $ry = \begin{pmatrix} 1 & 2 & 3 & 4 & 5 & 6 \\ 4 & 3 & 2 & 1 & 6 & 5 \end{pmatrix}$

This notation gives us a convenient routine way of combining transformations.

Suppose we need to combine ra with rot60.

Using this notation we have $ra = \begin{pmatrix} 1 & 2 & 3 & 4 & 5 & 6 \\ 2 & 1 & 6 & 5 & 4 & 3 \end{pmatrix}$

$$rot60 = \begin{pmatrix} 1 & 2 & 3 & 4 & 5 & 6 \\ 2 & 3 & 4 & 5 & 6 & 1 \end{pmatrix}$$

remember that ra.rot60 means rot60 is performed first, and followed by ra. Giving

rot60 =   1 2 3 4 5 6   rot60
          2 3 4 5 6 1

now follow with   ra =   1 6 5 4 3 2   ra

thus $ra.rot60 = \begin{pmatrix} 1 & 2 & 3 & 4 & 5 & 6 \\ 1 & 6 & 5 & 4 & 3 & 2 \end{pmatrix}$

which is rx.   Thus   ra.rot60 = rx

A rotation of 60 degrees followed by the reflection in line a is a reflection in the x axis.

Exercise 4:   On the symmetries of the regular hexagon.

[1] Show that rot60.ra = $\begin{pmatrix} 1\ 2\ 3\ 4\ 5\ 6 \\ 3\ 2\ 1\ 6\ 5\ 4 \end{pmatrix}$ = rbe

[2] Show that rcf = $\begin{pmatrix} 1\ 2\ 3\ 4\ 5\ 6 \\ 5\ 4\ 3\ 2\ 1\ 6 \end{pmatrix}$ and   rb = $\begin{pmatrix} 1\ 2\ 3\ 4\ 5\ 6 \\ 6\ 5\ 4\ 3\ 2\ 1 \end{pmatrix}$

and hence show that (i)   rcf.rb = rot300   (ii) rb.rcf = rot60

[3] Complete the following group table:

	e	rot60	rot120	rot180	rot240	rot300
e	e	rot60	rot120	rot180	rot240	rot300
rot 60	rot60					
rot120	rot120					
rot180	rot180					
rot240	rot240					
rot300	rot300					

[4] Check the following transformations;

rx = $\begin{pmatrix} 1\ 2\ 3\ 4\ 5\ 6 \\ 1\ 6\ 5\ 4\ 3\ 2 \end{pmatrix}$ rot60 = $\begin{pmatrix} 1\ 2\ 3\ 4\ 5\ 6 \\ 2\ 3\ 4\ 5\ 6\ 1 \end{pmatrix}$

ry = $\begin{pmatrix} 1\ 2\ 3\ 4\ 5\ 6 \\ 4\ 3\ 2\ 1\ 6\ 5 \end{pmatrix}$ rot120 = $\begin{pmatrix} 1\ 2\ 3\ 4\ 5\ 6 \\ 3\ 4\ 5\ 6\ 1\ 2 \end{pmatrix}$

ra = $\begin{pmatrix} 1\ 2\ 3\ 4\ 5\ 6 \\ 2\ 1\ 6\ 5\ 4\ 3 \end{pmatrix}$ rot180 = $\begin{pmatrix} 1\ 2\ 3\ 4\ 5\ 6 \\ 4\ 5\ 6\ 1\ 2\ 3 \end{pmatrix}$

rb = $\begin{pmatrix} 1\ 2\ 3\ 4\ 5\ 6 \\ 6\ 5\ 4\ 3\ 2\ 1 \end{pmatrix}$ rot240 = $\begin{pmatrix} 1\ 2\ 3\ 4\ 5\ 6 \\ 5\ 6\ 1\ 2\ 3\ 4 \end{pmatrix}$

rbe = $\begin{pmatrix} 1\ 2\ 3\ 4\ 5\ 6 \\ 3\ 2\ 1\ 6\ 5\ 4 \end{pmatrix}$ rot300 = $\begin{pmatrix} 1\ 2\ 3\ 4\ 5\ 6 \\ 6\ 1\ 2\ 3\ 4\ 5 \end{pmatrix}$

rcf = $\begin{pmatrix} 1\ 2\ 3\ 4\ 5\ 6 \\ 5\ 4\ 3\ 2\ 1\ 6 \end{pmatrix}$ e = $\begin{pmatrix} 1\ 2\ 3\ 4\ 5\ 6 \\ 1\ 2\ 3\ 4\ 5\ 6 \end{pmatrix}$

Two reflections always commute and give a rotation.
Use these results to evaluate:

rx.rx =	rx.ry =	rx.ra =	rx.rb =	rx.rbe =	rx.rcf =
ry.rx =	ry.ry =	ry.ra =	ry.rb =	ry.rbe =	ry.rcf =
ra.rx =	ra.ry =	ra.ra =	ra.rb =	ra.rbe =	ra.rcf =

etc…

Two reflections always give a rotation through twice the angle between the
to mirror lines.
These symmetries of the hexagon form a group of order 12.

Exercise 5: Construct the group table for this group of transformations.

==================================

Cycles and Permutations

We introduce a useful shorthand notation and remark too that each
transformation is simply a permutation of the position numbers 1,2,3,4,5,6.

Take $rx = \begin{pmatrix} 1 & 2 & 3 & 4 & 5 & 6 \\ 1 & 6 & 5 & 4 & 3 & 2 \end{pmatrix}$ and note that $1 \to 1, 2 \to 6 \to 2, 3 \to 5 \to 3, 4 \to 4$

these are called the cycles that make up rx and we can write rx in terms of
these cycles as

$$rx = (1 \to 1)(2 \to 6 \to 2)(3 \to 5 \to 3)(4 \to 4)$$

$(1 \to 1)$ is called a one cycle, $(2 \to 6 \to 2)$ is called a two cycle.
The 60 degree rotation rot60 = $(1 \to 2 \to 3 \to 4 \to 5 \to 6 \to 1)$ and is a six cycle.
If we assume the cycles to be 'cyclic', meaning that $(2 \to 6)$ shows that 2 goes
to 6 and 6 goes to 2, then we have a useful shorthand:

$$rx = (1\ 1)(2\ 6)(3\ 5)(4\ 4)$$

This now means that 1 goes to 1, 2 goes to 6 and 6 goes to 2, 3 goes to 5 and 5 goes to 3, lastly 4 goes to 4.

We can omit the one cycles since a one cycle means that the position number does not move. Thus if we do not see a position number in any of the cycles it is understood that the position does not to move. If we do this, then we have

$$rx = (2\ 6)(3\ 5)$$

Meaning 1→1 (understood) 2→6 and 6→2, 3→5 and 5→3, 4→4 (understood).

Example:
$$ry = \begin{pmatrix} 1\ 2\ 3\ 4\ 5\ 6 \\ 4\ 3\ 2\ 1\ 6\ 5 \end{pmatrix}$$

Using cycles we have ry = (14)(23)(56)

Combining the transformations

First of all, remember to evaluate from the right hand of the expression:

$$ry.rx = (1\ 4)(2\ 3)(5\ 6)\ (2\ 6)(3\ 5)$$

$$= \begin{pmatrix} 1\ \ 2\ 3\ 4\ 5\ 6 \\ 1\ 6\ 5\ 4\ 3\ 2 \\ 4\ 5\ 6\ 1\ 2\ 3 \end{pmatrix} = \begin{pmatrix} 1\ 2\ 3\ 4\ 5\ 6 \\ 4\ 5\ 6\ 1\ 2\ 3 \end{pmatrix} = (1\ 4)(2\ 5)(3\ 6) = \text{rot180}$$

Example: Evaluate rb.rot60 and rot60.rb using permutation cycles

Solution

$$rb = \begin{pmatrix} 1\ 2\ 3\ 4\ 5\ 6 \\ 6\ 5\ 4\ 3\ 2\ 1 \end{pmatrix} = (1\ 6)(2\ 5)(3\ 4)$$

$$rot\ 60 = \begin{pmatrix} 1\ 2\ 3\ 4\ 5\ 6 \\ 2\ 3\ 4\ 5\ 6\ 1 \end{pmatrix} = (1\ 2\ 3\ 4\ 5\ 6)$$

[1]   rb.rot60 = (1 6)(2 5)(3 4) (1 2 3 4 5 6)

$$= \begin{pmatrix} 1\ 2\ 3\ 4\ 5\ 6 \\ 2\ 3\ 4\ 5\ 6\ 1 \\ 5\ 4\ 3\ 2\ 1\ 6 \end{pmatrix} = (1\ 5)(2\ 4) = \text{rcf}$$

[2]   rot60.rb = (1 2 3 4 5 6) (1 6)(2 5)(3 4)

$$= \begin{pmatrix} 1 & 2 & 3 & 4 & 5 & 6 \\ 6 & 5 & 4 & 3 & 2 & 1 \\ 1 & 6 & 5 & 4 & 3 & 2 \end{pmatrix} = (2\ 6)(3\ 5) = rx$$

Exercise 6

Show that

[1]   rot120 = (1 3 5)(2 4 6)

[2]   rot240 = (1 5 3)(2 6 4)

[3]   rot300 = (1 6 5 4 3 2)

[4]   rx.rot120 = rcf

[5]   ra.rot180 = rcf

[6]   rb.rot60   = rcf

[7]   rot60.rbe = ry

[8]   **Symmetries of the Equilateral Triangle**

Define the following transformations of an equilateral triangle:

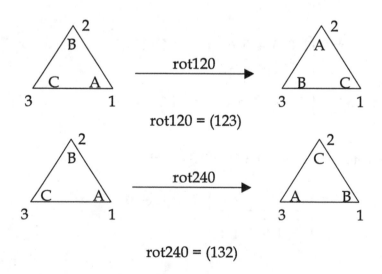

rot120 = (123)

rot240 = (132)

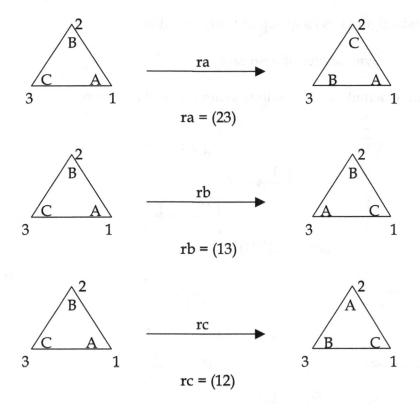

ra = (23)

rb = (13)

rc = (12)

Now complete this group table for the symmetries of the equilateral triangle:

	e	rot120	rot240	ra	rb	rc
e	e	rot120	rot240	ra	rb	rc
rot120	rot120					
rot240	rot240					
ra	ra					
rb	rb					
rc	rc					

For example   rot120.ra = (123)(23) = $\begin{pmatrix} 1\,2\,3 \\ 2\,1\,3 \end{pmatrix}$ = (12) = rc

ra.rot120 = (23)(123) = $\begin{pmatrix} 1\,2\,3 \\ 3\,2\,1 \end{pmatrix}$ = (13) = rb

Every symmetrical figure has its group of transformations.

[9]   **Symmetries of a Square**

Given these definitions for various symmetries of a square:

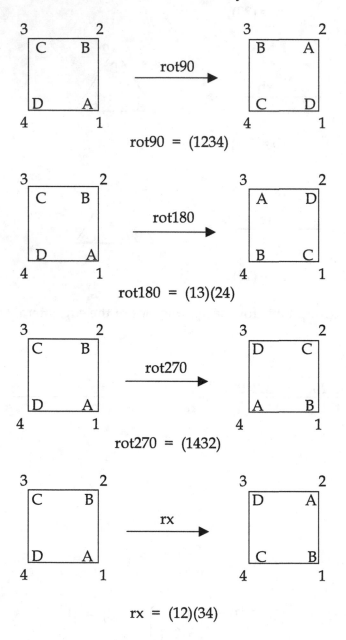

rot90 = (1234)

rot180 = (13)(24)

rot270 = (1432)

rx = (12)(34)

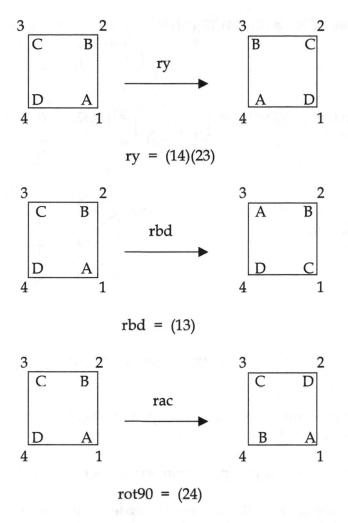

$$ry = (14)(23)$$

$$rbd = (13)$$

$$rot90 = (24)$$

Complete this group table for the order 8 group of symmetries of a square:

	e	rot90	rot180	rot270	rx	ry	rbd	rac
e	e							
rot90	rot90							
rot180	rot180							
rot270	rot270							
rx	rx							
ry	ry							
rbd	rbd							
rac	rac							

Method 1   rot180.rot180 = (13)(24)(13)(24) = $\begin{pmatrix} 1 & 2 & 3 & 4 \\ 3 & 4 & 1 & 2 \\ 1 & 2 & 3 & 4 \end{pmatrix}$ = e

ry.rbd = (14)(23)(24) = $\begin{pmatrix} 1 & 2 & 3 & 4 \\ 4 & 1 & 2 & 3 \end{pmatrix}$ = (1432) = rot270

or method 2,

alternatively,
work from a diagram:

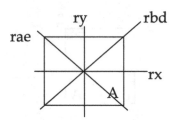

Exercise 7

[1]   Show that the symmetries of a regular pentagon form a group
of order 10.

[2]   If p is a prime number, show that the symmetries of a regular p-agon
form   a group of order 2p.

[3]   What is the order of the group of symmetries of a regular octagon?

[4]   Write down the group table for the symmetries of an isosceles triangle
that is **not** equilateral.

[5]   Write down the group table for the symmetries of a rhombus. What
group is it?

========================

Lagrange's Theorem for Finite Groups

The order of a subgroup of G divides the order of the group G.

Proof

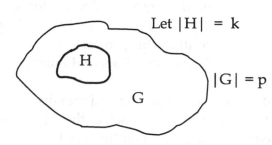

Let $|H| = k$

Let H be a subgroup of G

$|G| = p$

Suppose that $a \in G$ and $a \notin H$
Suppose that $h \in H$
aH is called a left coset and consists of all elements a.h where $h \in H$
We will show that
    [1] aH has k distinct elements
    [2] H and aH are disjoint (i.e. have no common members) and
    [3] if $a \notin H$, $b \notin H$ and $b \notin aH$ then aH and bH are also disjoint.
This then means that G can be completely divided up (partitioned) into
cosets like aH (one of them being eH = H) which means that the number of
elements in G can be divided up into lots of k elements, i.e. k divides into p.

The proofs

[1]   The elements of aH are all distinct hence, aH has k elements
Proof
    if $h, g \in H$ and $ah = ag$
    then $a^{-1}ah = a^{-1}ag$
    $\therefore$ $eh = eg$
    $\therefore$ $h = g$
    this means that aH has k elements

[2]   If $a \notin H$ then H and aH are disjoint
Proof
    ah is a member of the coset aH
    If ah is also a member of H then
        $ahh^{-1} \in H$ because H is a group and $h^{-1} \in H$
        so that $a \in H$
    which contradicts $a \notin H$
    therefore any member ah of aH cannot belong to H if a is not in H
    Therefore aH and H are disjoint and each have k members

[3]   If a∉H, b∉H and b∉aH then aH and bH are disjoint
Proof

Suppose that x∈aH and x∈bH

then   x=ah and x=bg for some h and g in the subgroup H

∴   bg = ah so that b = ahg⁻¹ and thus b ∈ aH

contradicting b∉aH

Therefore bH and aH must be disjoint

This means that G is partitioned into disjoint cosets aH, bH, cH... together with the subset H. Since each of these sets has k elements, the number of elements in G must be a multiple of k .

Therefore the order of the subset H divides the order of G.

■

=====================================

Theorem p: If p is prime then any group of order p is cyclic.

Proof

Suppose that G is a group of order p, where p is prime

Let a∈G where a ≠ e

Then there is a subgroup of G generated by a {e, a, a², a³, ...}

But by Lagrange's theorem, the order of this subgroup must divide p.

Therefore the order of the subgroup generated by a must be p since p is prime.

Therefore this subgroup must be G itself and so G is cyclic

==============================

There are, of course, many other theorems concerning the order of a group G. We state three more theorems but without proof:

Theorem p2:   If p is prime then there are 2 groups of order p^2

Theorem pq1: If p and q are primes and p does not divide q-1 then there is one group of order pq.

Theorem pq2: If p and q are prime and p divides q-1 then there are 2 groups of order pq

==

The classification of finite groups is a serious area of research in abstract algebra and group theory has applications in many areas of scientific research including crystallography and quantum theory.

Here is a list of the number of groups there are for orders 1 up to 32. Where the above theorems apply the name has been given.

Order of group	number of groups	theorem (if relevant)	Order of group	number of groups	theorem (if relevant)
1	1	{e}	17	1	p
2	1	p	18	5	?
3	1	p	19	1	p
4	2	p2	20	5	?
5	1	p	21	2	pq2
6	2	pq2	22	2	pq2
7	1	p	23	1	p
8	5	?	24	15	?
9	2	p2	25	2	p2
10	2	pq2	26	2	pq2
11	1	p	27	5	?
12	5	?	28	4	?
13	1	p	29	1	p
14	2	pq2	30	4	?
15	1	pq1	31	1	p
16	14	?	32	51	?

William Rowan Hamilton was born in Dublin in 1805 and was professor of mathematics at Trinity College, Dublin. Hamilton knew that multiplication of complex numbers could be used to represent rotations in the x-y plane and he wanted to extend this idea into three dimensions in order to represent rotations of a rigid body in three dimensional space. This led him to invent a new kinds of "numbers" called quaternions. A quaternion has four components, a real part here denoted by **a** and three other components, the **i**, **j**, and **k** components to represent three space dimensions.
Thus a quaternion q looks like $q = a + bi + cj + dk$

The rules for manipulating quaternions are

$$i^2 = j^2 = k^2 = -1 \qquad ij = -ji = k$$
$$jk = -kj = i$$
$$ki = -ik = j$$

plus the laws of arithmetic

Exercise 8

[1]   Show that (i) $2i \times 3j - 3j \times 2i = 12k$

           (ii) $(1+2i+3j+4k)(1+2i+3j+4k) = -28 + 4i + 6j + 8k$

[2] Show that the quaternions 1, -1, i, j, k, -i, -j, -k form an order 8 group under multiplication, by completing the following table:

The quaternion group

	1	-1	i	j	k	-i	-j	-k
1	1	-1	i	j	k	-i	-j	-k
-1	-1							
i	i							
j	j							
k	k							
-I	-i							
-j	-j							
-k	-k							

==

Solutions

Exercise 1

[1] $(a^{-1})^{-1}.a^{-1} = e$ $\therefore (a^{-1})^{-1}.a^{-1}.a = e.a$ $\therefore (a^{-1})^{-1}.e = a$ $\therefore (a^{-1})^{-1} = a$

[2]   Let $e^{-1} = x$ then $e.e^{-1} = e.x$ $\therefore e = x \therefore x = e \therefore e^{-1} = e$

[3] $(ab)^{-1}.(ab) = e$
   $\therefore (ab)^{-1}.(ab).b^{-1} = e.b^{-1} \therefore (ab)^{-1}.a = b^{-1} \therefore (ab)^{-1}.a.a^{-1} = b^{-1}.a^{-1} \therefore (ab)^{-1} = b^{-1}.a^{-1}$

[4] $(c^{-1}b^{-1}a^{-1}).(abc) = c^{-1}b^{-1}a^{-1}.abc = c^{-1}b^{-1}ebc = c^{-1}b^{-1}bc = c^{-1}c = e$
   therefore $(c^{-1}b^{-1}a^{-1})$ is the unique inverse of (abc), using theorem 1.

Exercise 2
(Filling in Cayley tables is very like solving a Sudoku)

[1]

	a	b	c	d	e
a	b	c	d	e	a
b	c	d	e	a	b
c	d	e	a	b	c
d	e	a	b	c	d
e	a	b	c	d	e

This the cyclic group
$\{e, a, a^2, a^3, a^4\}$

[2]

	a	b	c	d	e	f
a	e	a	f	c	b	d
b	a	b	c	d	e	f
c	d	c	b	a	f	e
d	f	d	e	b	c	a
e	b	e	d	f	a	c
f	c	f	a	e	d	b

Exercise 3

[1]	ABCD	[2]	CDAB	[3]	BADC
[4]	CDAB	[5]	ABCD	[6]	DCBA
[7]	BADC	[8]	DCBA	[9]	ABCD

	e	rx	ry	ro
e	e	rx	ry	ro
rx	rx	e	ro	ry
ry	ry	ro	e	rx
ro	ro	ry	rx	e

which has
the same
structure as

	e	f	g	h
e	e	f	g	h
f	f	e	h	g
g	g	h	e	f
h	h	g	f	e

Exercise 4

[3]

	e	rot60	rot120	rot180	rot240	rot300
e	e	rot60	rot120	rot180	rot240	rot300
rot60	rot60	rot120	rot180	rot240	rot300	e
rot120	rot120	rot180	rot240	rot300	e	rot60
rot180	rot180	rot240	rot300	e	rot60	rot120
rot240	rot240	rot300	e	rot60	rot120	rot180
rot300	rot300	e	rot60	rot120	rot180	rot240

[4]

	rx	ry	ra	rb	rbe	rcf
rx	e	rot180	rot60	rot300	rot120	rot240
ry	rot180	e	rot120	rot240	rot60	rot300
ra	rot60	rot240	e	rot120	rot300	rot180
rb	rot300	rot120	rot240	e	rot180	rot600
rbe	rot120	rot300	rot60	rot180	e	rot240
rcf	rot240	rot60	rot180	rot300	rot240	e

Exercise 5
(for convenience, the rotations are designated by the angles only)

	e	rx	ry	ra	rb	rbe	rcf	rot60	rot120	rot180	rot240	rot300
e	e	rx	ry	ra	rb	rbe	rcf	60	120	180	240	300
rx	rx	e	180	60	300	120	240	rb	rcf	ry	rbe	ra
ry	ry	180	e	120	240	60	300	rbe	ra	rx	rb	rcf
ra	ra	60	240	e	120	300	180	rx	rb	rcf	ry	rbe
rb	rb	300	120	240	e	180	60	rcf	ry	rbe	ra	rx
rbe	rbe	120	300	60	180	e	240	ra	rx	rb	rcf	ry
rcf	rcf	240	60	180	300	120	e	ry	rbe	ra	rx	rb
rot60	60	ra	rcf	rbe	rx	ry	rb	120	180	240	300	e
rot120	120	rbe	rb	ry	ra	rcf	rx	180	240	300	e	60
rot180	180	ry	rx	rcf	rbe	rb	ra	240	300	e	60	120
rot240	240	rcf	ra	rb	ry	rx	rbe	300	e	60	120	180
rot300	300	rb	rbe	rx	rcf	ra	ry	e	60	120	180	240

Exercise 6

[8] ### Symmetries of the Equilateral triangle

	e	120	240	ra	rb	rc
e	e	120	240	ra	rb	rc
120	120	240	e	rc	ra	rb
240	240	e	120	rb	rc	ra
ra	ra	rb	rc	e	120	240
rb	rb	rc	ra	240	e	120
rc	rc	ra	rb	120	240	e

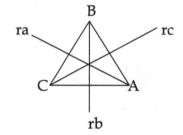

[9]   **Symmetries of a Square**

	e	90	180	270	rx	ry	rbd	rac
e	e	90	180	270	rx	ry	rbd	rac
90	90	180	270	e	rbd	rac	ry	rx
180	180	270	e	90	ry	rx	rac	rbd
270	270	e	90	180	rac	rbd	rx	ry
rx	rx	rac	ry	rbd	e	180	270	90
ry	ry	rbd	rx	rac	180	e	90	270
rbd	rbd	rx	rac	ry	90	270	e	180
rac	ra	ry	rbd	rx	270	90	180	e

Exercise 7

[1]   There are 5 axes of reflection,
      one through each vertex and
      five rotations of 0, 72, 144, 216,
      and 288 degrees giving a group
      of order 10.

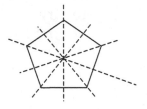

[2]   As in qu. 1, there are p axes of reflection and p rotations of $(360n)/p$
      ($n=1\ldots p-1$) giving a group of order 2p.

[3]   16

[4]

	e	r
e	e	r
r	r	e

where r is a reflection in the axis of symmetry.

[5]

	e	ra	rb	180
e	e	ra	rb	180
ra	ra	e	180	rb
rb	rb	180	e	ra
180	180	rb	ra	e

ra and rb are reflections
and 180 is the half turn

This is Klein's 4 group

Exercise 8

Qu 2 The Quaternion Group

	1	-1	i	j	k	-i	-j	-k
1	1	-1	i	j	k	-i	-j	-k
-1	-1	1	-i	-j	-k	i	j	k
i	i	-i	-1	k	-j	1	-k	-j
j	j	-j	-k	-1	i	k	1	-i
k	k	-k	j	-i	-1	-j	i	1
-i	-i	i`	1	-k	j	-1	k	-j
-j	-j	j	k	1	-i	k	-1	i
-k	-k	k	-j	i	1	j	-i	-1

CHAPTER 14

Induction

Predicates, Propositions and Sets

If I say that "the number N is a prime number" you will not be able to say if I am telling the truth or not until someone specifies the value of N. Such a statement that refers to unknown value is called a **predicate**. Once a value is assigned to the unknown number N then we should be able to say definitely that the statement is now either true or false. The statement has now become a **proposition** and has a truth value, true or false, depending on the value that is assigned to the unknown number N.

It is natural to write such statements in a more symbolic form, for example I might write P(n) to stand for the statement " n is a prime number".

Thus **P(n)** would be the predicate **" n is a prime number"**.
Once I have assigned a value to the variable n then the predicate becomes a proposition and I must be able to decide the truth value for the proposition. The proposition
$$P(5) \text{ has a truth value "true"}$$
whereas the proposition
$$P(6) \text{ has a truth value "false"}$$

Here are some other examples of predicates:

A(x):   x is bigger than 3

B(x):   x is more than 5 years old

C(n):   n(n+1)(n+2) is divisible by 4

The proposition A(3) has a truth value "false". The proposition B(me) is true, the proposition C(2k) is true for any integer k.

For any such predicate, it is assumed that we know what things, objects or kinds of number we are talking about. For predicate A(x), we could be talking about the set of natural numbers

N={1, 2, 3, 4,.....} or it may be the case that we are talking about the rational numbers (numbers of the form m/n where m and are integers) or, it may be the case that we were considering all real numbers.

The set of objects, things, numbers that we are assumed to be talking about is called the **Universal Set** for the predicate.

A Universal Set for predicate A(x) could be the set of natural numbers N={1, 2, 3, 4,.....}.
A Universal set for predicate B(x) might be all people on earth.

Examples of other Universal Sets are

The set of positive or negative integers

$$Z= \{0, \pm1, \pm 2, \pm 3, \pm4,.....\}$$

The set of real numbers R

All students enrolled at a given college

All cars registered in U.K.

We will use the symbol U to denote the Universal Set for those objects under discussion.

Truth sets

Any predicate P(x) with its associated Universal Set U defines a set called the truth set for P(x) being the set of all values x in the universal set that make the corresponding proposition P(x) true. If we denote this set by T then we would write
$$T = \{ x \mid P(x) \text{ is true)}\}$$

Assuming that the universal set is the set of all real numbers we have the following examples:

Predicate   truth set

(i)   2x + 1 = 3 $T = \{ 1 \}$

(ii) $x^2 + 1 = 3$ $T = \{ +2, -2 \}$

(iii) $2x^2 + 1 = 3$ $T = \{ +1, -1 \}$

(iv) $x^2 + 3 = 1$ $T = \{ \}$

Notation:

     $x \, \varepsilon \, U$ x is a member of the set U

     ϕ the symbol for the empty set { }

If we were to change the universal set then the truth set for a particular predicate may also change, for example, if we were to allow x to range over the set of all complex numbers, (see chapter 24), then the truth set for the predicate

     $x^2 + 3 = 1$ is now $T = \{ i\sqrt{2}, -i\sqrt{2} \}$

Similarly, for real numbers, the statement

     $x^3 + x = x^2 + 1$ has a truth set $T = \{ 1 \}$

Whereas for the field of complex numbers

     $x^3 + x = x^2 + 1$ has a truth set $T = \{ 1, i, -i \}$

The Induction Party

A group of friends meet at a party and greet each other by clinking glasses. As the party goes on the partygoers make more and more clinks with their glasses.
Prove that the number of partygoers that have made an odd numbers of clinks, is always even.

Proof

     Suppose that the number of odd clinkers, at some stage, is even.
     Suppose two friends then clink glasses.

[1]   If they were both odd clinkers then they will both become even clinkers, so the number of odd clinkers will go down by two. The number of odd clinkers therefore stays even.

[2]   If the two friends were both even clinkers, then they will both become odd clinkers, so the number of odd clinkers will increase by two. The number of odd clinkers therefore stays even.

[3]   If one friend is an odd clinker and the other is an even clinker, then the odd clinker becomes even and the even clinker becomes odd. The number of even clinkers therefore stays the same.

If the number of odd clinkers is even, therefore, the number of odd clinkers stays even.
But initially, the number of odd clinkers is zero, which is even.
Therefore the number of odd clinkers remains even all through the party.

The Principle of Mathematical Induction

We are concerned here, with predicates of the form P(n) where our universal set is the set of natural numbers N={1, 2, 3, 4,.....}.
Let T be the truth set for the proposition P(n) then the principle of mathematical induction states that:

If   (i) $n \varepsilon T \Rightarrow n+1 \varepsilon T$

and   (ii) $1 \varepsilon T$ then the truth set for P(n) is {1, 2, 3, 4,.....}.

In words, this means that

        If   (i) whenever P(n) is true,
                    we can show that P(n+1) must also be true
        and
                (ii) P(1) is true

        then P(n) must be true for all numbers 1,2,3,4,....

To illustrate the principle, consider a row of dominoes standing on end, close enough so that if any one of the dominoes falls over to the right, then it will knock over the next one.

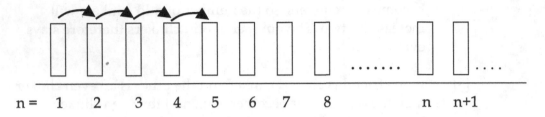

We ask the question "Will the dominoes fall over?" and the answer to the question might be "If you push the first one over then they will all fall over."

More symbolically we could express this as:

  If fallsover(n) means that fallsover(n+1)

  and if fallsover(1)

  then they all fall over.

Note: most questions on induction involve n = 1, 2, 3, 4, ….. but induction is often used starting with n = 0 and sometimes starting with some other value of n that is greater than 1.
The principle of induction can sometimes be used for negative values of n but the proposition P(n) would have to make sense for n < 0 in this case.

In the case of the induction party, we could say P(n) = "the number of odd clinkers out of n clinks is even" and we showed that if two friends add an extra clink, then the number of odd clinkers remains even, so P(n) is true implies that P(n+1) is true. If T is the truth set for P(n) then this means that n $\in$ T implies that (n+1) $\in$ T. But 0 $\in$ T and therefore we conclude that P(n) is always true.

In the following example we will start off the induction at n=0.

Example 1.

Prove that $n^3 - n$ is always divisible by 6

Proof

Let T be the truth set for the proposition

Suppose $\quad$ n ε T

Now $\qquad$ $(n+1)^3 - (n+1)$

$\qquad\qquad = n^3 + 3n^2 + 3n + 1 - n - 1$

$\qquad\qquad = n^3 - n + 3n(n + 1)$

now $n(n + 1)$ is always even so $3n(n + 1)$ must be divisible by 6 so if $n^3 - n$ is divisible by 6 then the whole expression must be divisible by 6.

$\therefore$ n ε T $\quad\Rightarrow\quad$ n+1 ε T

But 0 ε T because $0^3 - 0$ is divisible by 6

Therefore, by the principle of mathematical induction, $n^3 - n$ is divisible by 6 for all $n \geq 0$.

Note : We could start the induction at any integer, for example, at n=-10, for, then $n^3 - n = -1000 + 10 = -900$ which is divisible by 6. This would then prove that $n^3 - n$ is divisible by 6 for all $n \geq -10$.

In fact, if $n^3 - n = 6k$, then $(-n)^3 - (-n) = -6k$ so $n^3 - n$ is divisible by 6 for all integer values of n.

Example 2.

Prove that an n x n grid of squares has 2n(n+1) links joining adjacent points.

For example, a 3x3 grid
has 2x3(3+1) = 24 links

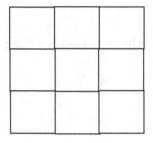

Proof

Suppose that T is the truth set for this proposition and that n ε T., that is, for some particular integer n, the n x n grid has 2n(n+1) links.

Now consider thet (n+1) x (n+1) square drawn below:

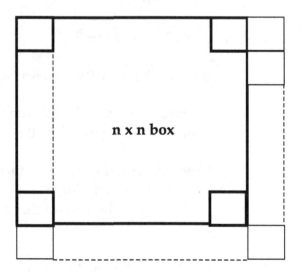

n x n box

We have an extra n+2 horizontal links down the side and an extra n horizontal links along the base, making altogether 2n+2 extra horizontal links.

Similarly, there are an extra 2n+2 extra vertical links.
Therefore, in total, there are 4n+4 extra links.

Assuming that the nxn square has 2n(n+1) links, this means that the (n+1)x(n+1) square will have 2n(n+1) + 4n + 4 links.

Now

$$2n(n+1) + 4n + 4$$

$$= 2n(n+1) + 4(n+1)$$

$$= 2(n+1)(n + 2)$$

$$= 2K(K + 1) \quad \text{where K is equal to n+1}$$

That is, the KxK square will have 2K(K+1) links, where K=n+1

Hence we have shown that $n\varepsilon T \Rightarrow n{+}1\varepsilon T$

Now $1\varepsilon T$ since a 1x1 square has 2x1(1+1) = 4 links.

Therefore, by the principle of mathematical induction, an nxn square will have 2n(n+1) links for any value of n≥1.

Exercise 1

I wish to build a wooden trellis fence on the top of a brick wall for my honeysuckle to climb over. The trellis will be 4 ft high and formed by one foot squares.

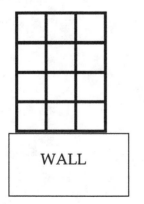

This 3ft long trellis needs 31 feet of wood.

The length of wood needed for this 3 ft trellis is
$$L(3) = 31$$

Prove **by induction,** that

$$L(n) = 9n+4$$

Summation of series

The sum of n terms of a series is denoted by S_n thus

$$S_n = U_1 + U_2 + U_3 + U_4 + \ldots + U_{n-1} + U_n$$

Using sigma notation this would be written

$$S_n = \sum_{r=1}^{r=n} U_r$$

The sigma notation says "add all the numbers using r=1 up to r=n".
Where the limits of the summation are clear, it is not necessary to put in the limits, thus we could write

$$S_n = \sum U_r$$

Exercise 2   **Using** $\sum$

Calculate

1. $$\sum_{r=1}^{r=10} r$$

2. $$\sum_{21}^{30} r$$

3. $$\sum_{1001}^{1010} r$$

4. $$\sum_{1}^{5} (2r-1)$$

5.
$$\sum_{1}^{6} (2r-1)$$

6.
$$\sum_{1}^{10} (2r)$$

7.
$$\sum_{1}^{10} (r-1)$$

8.
$$\sum_{n=1}^{n=5} \left(\sum_{1}^{n} (2r-1) \right)$$

9.
$$\sum_{r=1}^{5} r(r+1)$$

10.
$$\sum_{r=1}^{6} r(r+1)$$

Induction is a useful tool for summation of series providing you know the answer!

Example 3

Show that

$$1\times2 + 2\times3 + 3\times4 + 4\times5 + \dots\dots + n(n+1) = \frac{n(n+1)(n+2)}{3}$$

Proof

Suppose that the truth set for the exercise is T and that some integer n is in this truth set, that is, $n \varepsilon T$.

Then
$$\sum_{r=1}^{r=n} r(r+1) = \frac{n(n+1)(n+2)}{3}$$

$\therefore$
$$\sum_{r=1}^{r=n} r(r+1) + (n+1)(n+2) = \frac{n(n+1)(n+2)}{3} + (n+1)(n+2)$$

$\therefore$
$$\sum_{r=1}^{r=n+1} r(r+1) = \frac{n(n+1)(n+2)}{3} + (n+1)(n+2)$$

$$= \frac{(n+1)(n+2)[\,n+3]}{3}$$

$\therefore$
$$\sum_{1}^{K} r(r+1) = \frac{K(K+1)(K+2)}{3} \quad \text{where } K=n+1$$

This shows that $(n+1)\ \varepsilon\ T$

Thus $n\ \varepsilon\ T$ implies that $(n+1)\ \varepsilon\ T$
But $1\ \varepsilon\ T$ since

$$\sum_{1}^{1} r(r+1) = 1 \times 2 = \frac{1(1+1)(1+2)}{3}$$

$\therefore$ by the principle of mathematical induction, $n\ \varepsilon\ T$ for all **n≥1**

■

Exercise 3   **On Induction**

1. Prove that $2 \times 7^n + 4$ is always divisible by 6.

2. Prove that $6^n + 4$ is a multiple of 5.

3. Prove that $7^{2n+1} + 11^{2n+1}$ is always a multiple of 6.

4. Prove that $10^{2n+1} + 50^{2n+1}$ is a multiple of 6.

5. Prove that $45^{2n+1} + 55^{2n+1}$ is a multiple of 100.

6.
$$1+2+3+4+ \ldots +n = \frac{n(n+1)}{2}$$

7.
$$1^2+2^2+3^2+4^2+ \ldots +n^2 = \frac{n}{6}(n+1)(2n+1)$$

8.
$$1^3+2^3+3^3+4^3+ \ldots +n^3 = \frac{n^2(n+1)^2}{4}$$

9.
$$1^4+2^4+3^4+4^4+ \ldots +n^4 = \frac{n(n+1)(2n+1)(3n^2+3n+1)}{30}$$

10.
$$1^5+2^5+3^5+4^5+ \ldots +n^5 = \frac{n^2(n+1)^2(2n^2+2n+1)}{12}$$

11.
$$1 \times 2 + 2 \times 3 + 3 \times 4 + \ldots + n(n+1) = \frac{n(n+1)(n+2)}{3}$$

12.
$$1 \times 2 \times 3 + 2 \times 3 \times 4 + 3 \times 4 \times 5 + \ldots + n(n+1)(n+2) = \frac{n(n+1)(n+2)(n+3)}{4}$$

13.
$$1+3+5+7+9+ \ldots (2n-1) = n^2$$

===============================

Example 4.

Prove that

$$1^6+2^6+3^6+4^6+ \ldots +n^6 = \frac{n(n+1)(2n+1)(3n^4+6n^3-3n+1)}{42}$$

Proof

Let the truth set for the proposition be T
Suppose n ε T for some particular integer n.

i.e. $\quad \displaystyle\sum_1^n r^6 = \frac{n(n+1)(2n+1)(3n^4+6n^3-3n+1)}{42}$

Then

$$\sum_1^{n+1} r^6 = \frac{n(n+1)(2n+1)(3n^4+6n^3-3n+1)}{42} + (n+1)^6$$

$$= \frac{(n+1)(\, n(2n+1)(3n^4+6n^3-3n+1) + 42\,(n+1)^5\,)}{42}$$

$$= \frac{(n+1)(\, 6n^6 + 15n^5 +6n^4 - 6n^3 - n^2 + n + 42n^5 + 210n^4 + 420n^3 + 420n^2 + 210n + 42\,)}{42}$$

$$= \frac{(n+1)(n+2)(\, 6n^5 + 45n^4 +126n^3 + 162n^2 + 95n + 21\,)}{42}$$

$$= \frac{(n+1)(n+2)(2n+3)(\, 3n^4 + 18n^3 + 36n^2 + 27n + 7)}{42}$$

$$= \frac{(n+1)(n+2)(2n+3)(\, 3[n^4 + 4n^3 + 6n^2 + 4n + 1] + 6[n3 + 3n2 + 3n + 1] - 3[n+1] + 1)}{42}$$

$$= \frac{(n+1)(n+2)(2n+3)(\, 3[n + 1]^4 + 6[n + 1]^3 - 3[n+1] + 1)}{42}$$

$$= \frac{K(K+1)(2K+1)(\, 3[\, K^4 + 6K^3 - 3K + 1)}{42} \qquad \text{where } K = n+1$$

Therefore we have shown that $\quad n\,\varepsilon\,T \quad \Rightarrow\ n+1\,\varepsilon\,T$

Now 1 ε T since $\quad 1^6 = \dfrac{1(1+1)(2+1)(\, 3\mathrm{x}1^4 + 6\mathrm{x}1^3 - 3\mathrm{x}1 + 1)}{42}$

Therefore, by the principle of mathematical induction,

$$\sum_{1}^{n} r^6 = \frac{n(n+1)(2n+1)(3n^4+6n^3-3n+1)}{42} \qquad \text{for all } n \geq 1$$

The Binomial Theorem

At around 300 years B.C. the Greek mathematician **Euclid** discovered that the squared bracket $(1+x)^2$ could be expanded as

$$(1+x)^2 = 1 + 2x + x^2$$

The Persian mathematician **Omar Khayyam** (1048..1131) discovered that higher powers could also be expanded:

$$(1+x)^3 = 1 + 3x + 3x^2 + x^3$$

$$(1+x)^4 = 1 + 4x + 6x^2 + 4x^3 + x^4$$

The Chinese mathematician **Yang Hui** (1238..1298) presented the following diagram for the coefficents in these expansions:

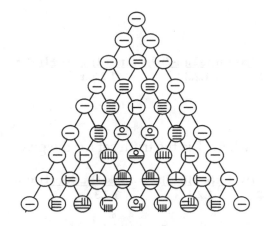

The symbols used are counting board numerals (see page 37). Note the use of zero in row 1, 5, 10, 10, 5, 1 and also in the next row: 1, 6, 15, 20, 15, 6, 1

This is a Chinese representation of what we now call Pascal's triangle, described by the French mathematician **Blaise Pascal** (1623..1662)

Pascal's triangle

$$1$$
$$1 \; 2 \; 1$$
$$1 \; 3 \; 3 \; 1$$
$$1 \; 4 \; 6 \; 4 \; 1$$
$$1 \; 5 \; 10 \; 10 \; 5 \; 1$$
$$1 \; 6 \; 15 \; 20 \; 15 \; 6 \; 1$$

Pascal's triangle gives us the coefficients of powers of x in the expansion of $(1+x)^n$ where n is a positive integer.

The binomial expansion can be written

$$(1+x)^n = 1 + nx + \frac{n(n-1)}{1.2}x^2 + \frac{n(n-1)(n-2)}{1.2.3}x^3 + \ldots \frac{n(n-1)\ldots(n-r+1)}{1.2.3\ldots r} x^r + \ldots + nx^{(n-1)} + x^n$$

In 1665, **Sir Isaac Newton (1643..1727)** proved that if , $-1 < x < +1$, then this formula can also be used when n is a negative number or a rational number in which case the expansion in powers of x in a infinite series of terms.

Here, we use induction to prove the binomial theorem for a positive integral index:

To prove that

$$(1+x)^n = 1 + nx + \frac{n(n-1)}{1.2}x^2 + \frac{n(n-1)(n-2)}{1.2.3}x^3 + \ldots \frac{n(n-1)\ldots(n-r+1)}{1.2.3..r} x^r + \ldots + nx^{(n-1)} + x^n$$

Proof

Suppose that $k \in T$ where T is the truth set for the given statement i.e.

$$(1+x)^k = 1 + kx + \frac{k(k-1)}{1.2}x^2 + \frac{k(k-1)(k-2)}{1.2.3}x^3 + \ldots \frac{k(k-1)\ldots(k-r+1)}{1.2.3\ldots r} x^r + \ldots + kx^{(k-1)} + x^k$$

then, multiplying both side by $(1+x)$, we have

$(1+x)^{k+1}$

$$= 1 + kx + \frac{k(k-1)x^2}{1.2} + \frac{k(k-1)(k-2)x^3}{1.2.3} + ... + \frac{k(k-1)...(k-r+1)x^r}{1.2.3...r} + ... + kx^{(k-1)} + x^k$$

$$+ x + kx^2 + \frac{k(k-1)x^3}{1.2} + ... + \frac{k(k-1)...(k-r)x^r}{1.2.3...(r-1)} + ... + kx^k + x^{(k+1)}$$

$$= 1 + (1+k)x + \frac{k[k-1+2]x^2}{1.2} + \frac{k(k-1)(k-2+3)x^3}{1.2.3} + ... + \frac{k(k-1)...(k-r)(k-r+1+r)x^r}{1.2.3...r} + ... + (1+k)x^k + x^{(k+1)}$$

$$= 1 + (1+k)x + \frac{(k+1)kx^2}{1.2} + \frac{(k+1)k(k-1)x^3}{1.2.3} + ... + \frac{(k+1)k(k-1)...(k-r)x^r}{1.2.3...r} + ... + (1+k)x^k + x^{(k+1)}$$

by transferring the last bracket (k+1) to the front in each of the numerators,

$$= 1 + Nx + \frac{N(N-1)x^2}{1.2} + \frac{N(N-1)(N-2)x^3}{1.2.3} + ... + \frac{N(N-1)(N-2)...(N-r+1)x^r}{1.2.3...r} + ... + Nx^{N-1} + x^N$$

where N=k+1. Thus, if $k \in T$ then $k+1 \in T$. Clearly, $0 \in T$ and $1 \in T$ thus, by the principal of mathematical induction, the result is proved.

Practice exercise. Using the above binomial expansion,

1. Prove that $(1+2x)^3 = 1+6x+12x^2+8x^3$

2. Prove that $(1+2x)^4 = 1+8x+24x^2+32x^3+16x^4$

3. Prove that $(1+2x)^5 = 1+10x+40x^2+80x^3+80x^4+32x^5$

Induction and recursion (for the computer programmer)

It is interesting to compare mathematical induction with the programming method called recursion in which a function "calls itself" on the assumption that the function call will return the correct value.

Mathematical induction verifies the result say for N=1 and shows that if the result is true for N then it must also be true for N+1. Proof for all N then follows.

Recursion assumes that the result for N can be deduced from the result for N-1 and provides the result for N=1.

We illustrate with a few examples using the C++ and Pascal programming languages.

Example 1
       Let sigma(n) represent the sum of the first n odd numbers.

Prove that sigma(n) = n^2

 Proof (by mathematical induction)

 Suppose that sigma(N) = N^2 for some particular integer N.

Then
       sigma(N+1) = N^2 + (2N+1) since the next odd number will be (2N+1)

   $\therefore$ sigma(N+1) = N^2 + 2N+1

                 = $(N+1)^2$

   $\therefore$ if sigma(n) = n^2 is correct when n = N

   then it is also correct when   n = N+1

   but sigma(1) = 1 = 1^2 so sigma(n) = n^2 is correct when n=1

Therefore, by the principle of mathematical induction,
the formula sigma(n) = n^2 is correct for all n $\geq$ 1

Example 2

```
// C routine to calculate the sum of the first n odd numbers
int sigma(int n)
{
        if (n = =1)
                return 1;
        else
                return(2n-1 + sigma(n-1));
}
```

Complete program

```
//program to display the sum of the first n odd numbers

int main()
{
    int n;
    cout<<"input number n/n";
    cin>>n;
    cout<<"1+...+"<<2*n-1<<"="<<sigma(n)<<"/n";
}
```

```
// C routine to calculate the sum of the first n odd numbers
int sigma(int n)
{
        if (n = =1)
                return 1;
        else
                return(2n-1 + sigma(n-1));
}
```

Example 3

(* Pascal routine for calculating sigma(n) *)

```pascal
function sigma(n:integer):integer
begin
   if(n=1)
          sigma:=1
   else
          sigma:=2*n-1 + sigma(n-1);
end;
```

Complete program

```pascal
program oddnumbers;
{program to display the sum of n odd integers}
uses crt;
var
   n:integer;

function sigma(n:integer):integer
begin
   if(n=1)
          sigma:=1
   else
          sigma:=2*n-1 + sigma(n-1);
end;

begin
   writeln('input n');
   readln(n);
   writeln('1+..+',2*n-1,'=',sigma(n));
end;
```

==============================

Answers

Exercise 1.   Suppose that $L(n) = 9n+4$ for one particular value n
Then $L(n+1) = 9n+4 + 9 = 9(n+1)+4$

$\therefore n\varepsilon T \Rightarrow (n+1)\ \varepsilon\ T$

But $L(1) = 13 = 9.1+4\ \therefore 1\varepsilon T$
Therefore, by the principal of induction $L(n) = 9n+4$ for all $n\geq 1$

Exercise 2

1. 55, 2. 234, 3. 10055, 4. 25, 5. 36, 6. 1010, 7. 45, 8. 55, 9. 69, 10. 111

Exercise 3   Short hints on questions 1 to 5

[1] $F(n+1) = 2\times7^{n+1} + 4 = 7(2\times7^n + 4) - 24 = 7.F(n) - 24$
$\therefore$ if $F(n)$ is divisible by 6 then so is $F(n+1)$. $F(0) = 6$ so therefore $F(n)$ is divisible by 6 for all $n\geq 0$.

[2] $F(n+1) = 6^{n+1} + 4 = 6(6^n + 4) - 20 = 6.F(n) - 20$
$\therefore$ if $F(n)$ is divisible by 5 then so is $F(n+1)$. Now $F(0) = 5$ so therefore $F(n)$ is divisible by 5 for all $n\geq 0$.

[3] $F(n+1) = 7^{2(n+1)+1} + 11^{2(n+1)+1} = 7^2(7^{2n+1} + 11^{2n+1}) + 11^{2n+1}(11^2 - 7^2)$

$$= 7^2.F(n) + 11^{2n+1}(18\times4)$$

$\therefore$ if $F(n)$ is divisible by 18 then so is $F(n+1)$. Now $F(0) = 7+11 = 18$ so therefore $F(n)$ is divisible by 18 for all $n\geq 0$.

[4] $F(n+1) = 10^{2(n+1)+1} + 50^{2(n+1)+1} = 10^2(10^{2n+1} + 50^{2n+1}) + 50^{2n+1}(50^2 - 10^2)$

$$= 10^2.F(n) + 50^{2n+1}(60\times40)$$

$\therefore$ if $F(n)$ is divisible by 60 then so is $F(n+1)$. Now $F(0) = 10+50 = 60$ so therefore $F(n)$ is divisible by 18 for all $n\geq 0$.

[5] $F(n+1) = 45^{2(n+1)+1} + 55^{2(n+1)+1} = 45^2(45^{2n+1} + 55^{2n+1}) + 55^{2n+1}(55^2 - 45^2)$

$$= 45^2 . F(n) + 55^{2n+1}(100 \times 10)$$

$\therefore$ if $F(n)$ is divisible by 100 then so is $F(n+1)$. Now $F(0) = 45+55 = 100$ so therefore $F(n)$ is divisible by 100 for all **n≥0**.

==========================

CHAPTER 15

Probability

"As I was going to St. Ives,
I met a man with seven wives,
Every wife had seven sacks and
Every sack had seven cats,
Every cat had seven kits.
Kits, cats sacks and wives,
How many were going to St. Ives"

(From Mother Goose)

Most people, at one time or another, take an interest in probability and chance. At the sub-atomic level, according to quantum theory, the evolution of the universe is governed by probabilities and random events, but for the average person it is probably more important to know what chance there is of winning the National Lottery or what chance Cloudy Lane has of winning the Grand National or whether it will rain tomorrow.
If you ask what is the chance of throwing a six with a fair dice, you would expect to get a reply of one in six and in sixty throws you would expect to get around 10 sixes.

There are two common views that we can adopt when dealing with probability and chance.

First view: **Trials, outcomes and events**

We carry out an **experiment** (or **trial**):
(Throw a dice and record the number)
We decide what are the equally likely **outcomes**...
(1,2,3,4,5,6)
We count the number of ways for the **event** to be successful:
(number of ways to throw a six = 1)
We divide the number of ways for a successful outcome by the total number of outcomes:

$$P(six) = 1/6$$

Second view: Frequency

Throw the dice a large number of times and record the **frequency** of scoring six then

$$P(six) = \frac{\textbf{number of sixes}}{\textbf{total number of throws}}$$

Example 1 If you draw a card from a standard pack, what is the probability of getting a spade?

Solution   number of successful outcomes = 13

Total number of outcomes = 52

$$P(spade) = \frac{13}{52} = \frac{1}{4}$$

Example 2   What is the probability that it will rain tomorrow?

Solution: We cannot predict what the weather will be tomorrow, but we can use the frequency definition of probability to calculate the chance that it will rain. Suppose that "tomorrow" is 2nd April. We look up the weather records for the past 50 years and find that it rained on 18 of those days. We therefore estimate that the probability of it raining tomorrow is

$$P(rain) = 18/50 = 0.36$$

=====================================

Most of the examples we consider here will conform to the first view described above and to this purpose we describe a number of formal terms:

Jargon: **Possibility spaces and events**

We introduce some of the common terms used in the study of probability by means of further examples:

Example 3

I have two dice. One is red and one is blue.

If I throw the dice, what is the chance that the red dice scores more than the blue?

Solution;

The equally likely **outcomes** are

(1,1), (1,2), (1,3), (1,4), (1,5), (1,6),
(2,1), (2,2), (2,3), (2,4), (2,5), (2,6),
(3,1), (3,2), (3,3), (3,4), (3,5), (3,6),
(4,1),.(4,2), (4,3), (4,4), (4,5), (4,6),
(5,1), (5,2), (5,3), (5,4), (5,5), (5,6),
(6,1), (6,2), (6,3), (6,4), (6,5), (6,6)

We can represent these equally likely outcomes or possibilities in a diagram of the **possibility space E:**

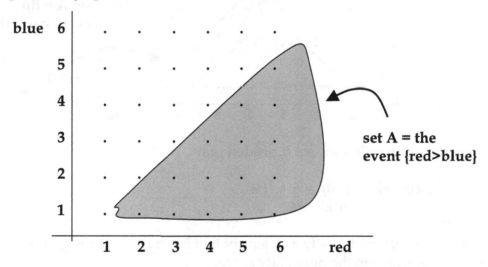

The **event** red>blue is the subset of the **possibility space** labelled set A.

$$P(\text{red>blue}) = \frac{n(A)}{n(E)} = \frac{15}{36} = \frac{5}{12}$$

Example 4 What is the probability that we score 7?

The possible scores are 2,3,4,5,6,7,8,9,10,11,12 but these are not equally likely.
P(score =7) is not 1/11.
The most likely score is in fact 7 and the chance of getting a score of 1 is zero.
The chance of getting a score of 12, with a double six is only one in 36.
The equally likely outcomes that make up the possibility space are the same
as in example 3:

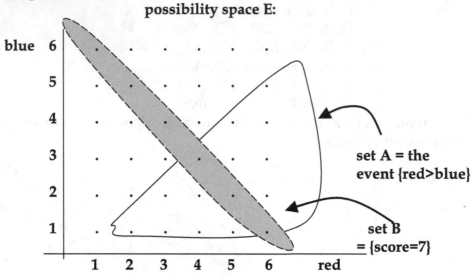

possibility space E:

set A = the
event {red>blue}

set B
= {score=7}

The event {**score=7**} is the set B, shaded gray

$$P(score=7) = \frac{n(B)}{n(E)} = \frac{6}{36} = 1/6$$

The event score = 7 is really a statement that has a truth value "yes" or "no"
for each outcome in the possibility space.
Thus we could regard the event {score=7} as the truth set for the statement
"score=7".
To emphasize this distinction we may sometimes write the lower case for the
statement and use upper case for the event which is the truth set for the
statement.
Thus

$$b = \text{"score=7"}$$

$$P(b) = n(B)/n(E) = 1/6$$

Compound Events

Example 5   What is the probability that we score 7 with the red > blue?

Solution

This is a **compound event,** given by the conjunction of the statements

a = "score=7" and b = "red>blue"

using the diagram for example 4;

The truth set for **a∧b** is the intersection **A∩B**

Thus $P(a∧b) = \dfrac{n(A∩B)}{n(E)} = \dfrac{3}{36}$

Example 6   What is the chance that
                 (i)   we score >6 with a double
                 (ii)   we score either a double or >6

Solution

In this example, A is the truth set for the statement a = "total>6"
and B is the truth set for the statement b = "we score a double", with the
same score on each dice.
Set A is coloured light gray and set B is darker

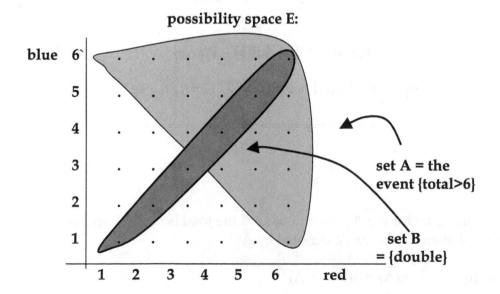

possibility space E:

set A = the
event {total>6}

set B
= {double}

$$P(a) = n(A)/36 = 21/36 = 7/12$$

$$P(b) = n(B)/36 = 6/36 = 1/6$$

The event **a and b** is represented by the set $A \cap B$.

$$P(a \text{ and } b) = n(A \cap B)/36 = 3/36 = 1/12$$

The event **a or b** is represented by the set $A \cup B$

$$P(a \text{ or } b) = n(A \cup B)/36 = 24/36 = 2/3$$

Thus, $A \cap B$ is the truth set for the statement $a \wedge b$. $P(a \wedge b) = 1/12$

and $A \cup B$ is the truth set for the statement $a \vee b$. $P(a \vee b) = 2/3$

From set algebra, we know that

$$n(A \cup B) \;=\; n(A) + n(B) - n(A \cap B)$$

dividing by $n(E)$ we have

$$\frac{n(A \cup B)}{n(E)} \;=\; \frac{n(A)}{n(E)} + \frac{n(B)}{n(E)} - \frac{n(A \cap B)}{n(E)}$$

therefore

$$P(a \vee b) \;=\; P(a) + P(b) - P(a \wedge b) \qquad\qquad \text{Law 1}$$

For this example, we should have $2/3 = 7/12 + 1/6 - 1/12$

$$============================$$

Negation

The truth set for the statement ¬**a**, i.e. " the total is not >6" will be the complement of the set A, denoted by $\overline{A}$

Since $n(\overline{A}) = n(E) - n(A)$

Dividing by **n(E)** gives

$$\frac{n(\overline{A})}{n(E)} = \frac{n(E)}{n(E)} - \frac{n(A)}{n(E)}$$

giving $\qquad P(\neg a) = 1 - P(a)$ $\qquad\qquad\qquad$ **Law 2**

The probability of not getting more than 6 is then, $1-7/12 = 5/12$

========================

Exclusive events

Two events are **mutually exclusive** if they cannot both be true for the same outcome.

Example 7: Illustrate the events **a = "score>10"** and **b = "score<4"**

Solution

$n(A \cap B) = 0$ and thus $P(a \wedge b) = 0$

Therefore, in **Law 1,** we have $P(a \vee b) = P(a) + P(b) - 0$

For mutually exclusive events $P(a \wedge b) = 0$

$$\text{and} \quad P(a \vee b) = P(a) + P(b) \qquad \text{Law 3}$$

================================

Exercise 1

[1]   Prove that

$$P(a) = P(a \wedge b) + P(a \wedge \neg b) \qquad \text{Law 4}$$

[2]   Two dice are thrown and we record the scores.
       Let **a="score<6" b="we get a 3"**

   [a]   show that $P(a \wedge b) = 4/36$, $P(\neg a) = 26/36$ and $P(\neg b) = 25/36$

   [b]   show that $P \neg(a \wedge b) = P(\neg a \vee \neg b)$

   [c]   show that $P \neg(a \vee b) = P(\neg a \wedge \neg b)$

[3]   Given $P(a) = 4/12$, $P(b) = 6/12$ and $P(a \wedge b) = 2/12$

       find $P(\neg a \wedge \neg b)$

Conditional Probability

Example 8

What is the probability of getting a total greater than 6 if we get the same number of each dice?

Solution
Here, we are told that we score a double and therefore we have to restrict ourselves to looking in the set B.

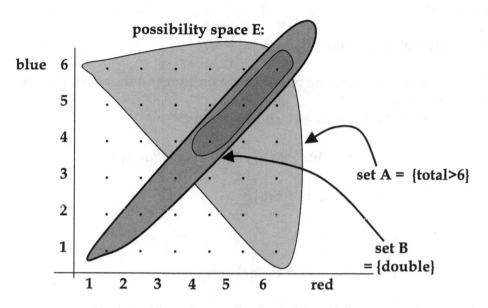

possibility space E:

set A = {total>6}

set B = {double}

Given that we are only looking at set B, then the event {total>6} is now represented by the darkest set which is the intersection of A an B. We only have a successful outcome in the set A∩B.

Given **B**, then, the probability of **A** is $\dfrac{n(A \cap B)}{n(B)}$

we write $\quad P(a\,|\,b) = \dfrac{n(A \cap B)}{n(B)} = \dfrac{n(A \cap B)/n(E)}{n(B)/n(E)} = \dfrac{P(a \wedge b)}{P(b)}$

Therefore, the probability of event a, given that event b occurs is

$$P(a\,|\,b) = \dfrac{P(a \wedge b)}{P(b)} \qquad\qquad \textbf{Law 5}$$

===============================

This is called the conditional probability of **a** given **b**.
In our example,

$$P(a\,|\,b) = \dfrac{P(a \wedge b)}{P(b)} = \dfrac{3/36}{6/36} = \tfrac{1}{2}$$

Exercise 2

[1]   In example 8, show that **P(b | a) = 1/7**

[2]   The red dice and the blue dice are thrown and the two scores are
       recorded.
       Let **a = "total score>6" and b = "red<3"**
       Find

       [a]   P(a | b), P(b | a) , P(a∨b | a), P(a∧b | a)

       [b]   P(¬a | b) + P(a | b)

       [c]   P(¬a | b)

       [d]   P(a∧¬b) + P(¬a∧b) + P(a∧b)

[3]   Given P(a) = 3/12, P(b) = 4/12 and P(a∧b) 2/12
       Find

       [a]   P(a | b)

       [b]   P(b | a)

       [c]   P(a∨b)

       [d]   P(¬a∧¬b)

[4]   A card is drawn from a pack. Its value is recorded and then it is
       replaced. A second card is then drawn and it value noted.

       Let **a = "same denomination on both cards",
          b = "second card is a court card"**

       Find P(a | b) and P(b | a)
       ========================
We so often use Law 5 in its multiplied out form that we can justify this form
as a law in its own right:

$$P(a∧b) = P(a | b) \times P(b) \qquad\qquad \textbf{(Law 6)}$$

Example 9

Two cards are drawn from a pack of 52.
What is the probability that they are both spades?

Solution

A tree diagram gives a representation of the possibility space.

There are 52x51 outcomes in the possibility space so clearly, we cannot draw a complete representation, however☺, we draw cards and record the suit

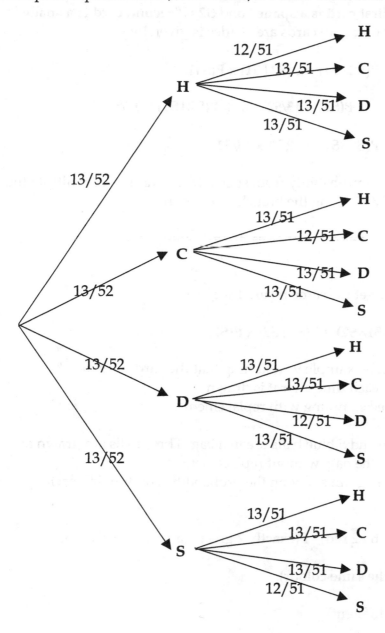

The first column of probabilities are the probabilities of drawing a particular suit and are all 13/52 since there are 13 cards in each suit.

The second column of probabilities are **conditional probabilities**, given that the suit in the first column of cards has been drawn.

Now $P(a \wedge b) = P(a \mid b) \times P(b)$ (Law 6)

If we denote **S1 = "first card is a spade"** and **S2 = "second card is a spade"** then the probability that both cards are spades is given by

$$P(S2 \wedge S1) = P(S2 \mid S1) \times P(S1)$$

Now   **P(S1) = 13/52   and   P(S2 | S1) = 12/51**

therefore $P(S1 \wedge S2) = 12/51 \times 13/52$

To find the compound probability from such a tree diagram we multiply the conditional probabilities along the branches of the tree.

============================

Exercise 3

[1]   Show that we get the same result if we use

$$P(S1 \wedge S2) = P(S1 \mid S2) \times P(S2)$$

[2]   Follow through example 9, assuming that the card is replaced into the pack each time after it is drawn.
(This is termed sampling with replacement!)

[3]   2 red, 3 green and 5 blue balls are in a bag. Three balls are drawn at random from the bag, without replacement.
Draw a tree diagram showing the probabilities of drawing each colour.
Find
[a]   P(exactly 2 red), P(exactly 2 green) and P(exactly 2 blue)

[b]   P(all the same colour)

[c]   P(all different)

Example 10

Draw three cards from the pack and record the suit. What is the probability that we get three spades?

Solution

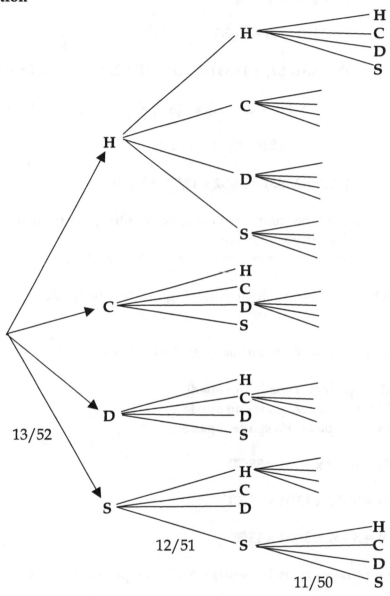

Let **S1, S2 and S3** represent the events **S1="first card is a spade"**, **S2="second card is a spade"** and **S3="third card is a spade"**, then the incomplete probability tree illustrates

$$P(S1) = 13/52$$

$$P(S2\,|\,S1) = 12/51$$

$$P(S3\,|\,S1{\land}S2) = 11/50$$

Now
$$P(S3{\land}S2{\land}S1) = P(S3\,|\,S2{\land}S1) \times P(S2{\land}S1) \qquad \textbf{Law 6}$$

$$= P(S3\,|\,S1{\land}S2) \times P(S2\,|\,S1) \times P(S1) \qquad \textbf{Law 6}$$

$$= 11/50 \times 12/51 \times 13/52$$

or,
$$P(S1{\land}S2{\land}S3) = 13/52 \times 12/51 \times 11/50$$

To find the probability of the compound event we multiply the probabilities along the branches of the probability tree.

===================================

Example 11   What is the probability of drawing exactly one spade?

Solution

We have to combine three mutually exclusive events:

$$E1 = \textbf{spade}{\land}\textbf{no spade}{\land}\textbf{no spade}$$
$$E2 = \textbf{no spade}{\land}\textbf{spade}{\land}\textbf{no spade}$$
$$E3 = \textbf{no spade}{\land}\textbf{no spade}{\land}\textbf{spade}$$

$$P(E1) = 13/52 \times 39/51 \times 38/50$$

$$P(E2) = 39/52 \times 13/51 \times 38/50$$

$$P(E3) = 39/52 \times 38/51 \times 13/50$$

Since these are mutually exclusive events we add the probabilities, hence we have

$$P(\textbf{just one spade}) = \frac{3 \times 13 \times 39 \times 38}{52 \times 51 \times 50} = \frac{741}{1700}$$

Exercise 4

[1]   Three cards are drawn from the pack, without replacement:
        What is the probability of drawing exactly two diamonds?

[2]   What is the probability of drawing three spades if the card is replaced
        after each draw?

[3]   What is the probability of drawing exactly one spade if the card is
        replaced after each draw?

[4]   Two dice are thrown. Let **a** = "score > 3" and **b** = "score a double".

        Find   P(a), P(b), P(a∧b), P(a | b).

===============================

Independent Events

Two events **a** and **b** are independent if the probability of **a** is the same
whether **b** happens or not, thus **a** and **b** are independent events if

$$P(a \mid b) = P(a)$$

If **a** and **b** are independent then

$$P(a \mid b) = P(a)$$

thus $\dfrac{P(a \wedge b)}{P(b)} = P(a)$ ∴   P(a∧b) = P(a).P(b)

=================

Dividing by **P(a)** we also have

        $\dfrac{P(a \wedge b)}{P(a)} = P(b)$ or $\dfrac{P(b \wedge a)}{P(a)} = P(b)$

i.e. $P(b \mid a) = P(b)$

Which means that **if a and b are independent then b and a are also
independent.**

Thus, if **a** and **b** are independent events then

$$P(a \mid b) = P(a), \quad P(b \mid a) = P(b) \quad \text{and} \quad P(a \wedge b) = P(a).P(b) \qquad \text{Law 7}$$

==============================

Example 12

We have two dice, a red one and a blue one. We throw the dice and record the number on each dice.

Let   **event 1 = "we score more than 2 on the red dice"**

      event 2 = "we score more than 3 on the blue dice"

The possible results for the experiment are illustrated by this diagram:

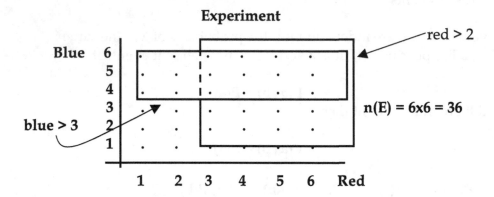

Whatever number shows on the blue dice, there are just 4 ways for the red score to be greater than 2.
$$P(\text{event 1}) = 4/6$$

Whatever number shows on the red dice, there are just 3 ways for the blue score to be greater than 3.
$$P(\text{event 2}) = 3/6$$

{ red>2 } and { blue>3 } are independent events.

$$P(\text{event 1 and event 2}) = 4/6 \times 3/6 = 12/36$$

Exercise 5

[1]   An experiment consists of drawing a card from a pack and recording the suit, and then tossing a dice and recording the number.

Let **a** = "red card", **b** = "score>4"

Draw the truth sets for **a** and **b** on the following diagram:

```
C  |   .   .   .   .   .   .
   |
H  |   .   .   .   .   .   .
   |
D  |   .   .   .   .   .   .
   |
S  |   .   .   .   .   .   .
   |_____
       1   2   3   4   5   6
```

Show that $P(a \mid b) = P(a)$

$P(b \mid a) = P(b)$

$P(a \wedge b) = P(a) \times P(b)$

[2]   If $P(a \mid b) = P(a)$

Prove that $P(a \mid \neg b) = P(a)$

i.e. if **a** and **b** are independent then **a** and **¬b** are also independent.

(Hint: Use $P(\neg b) = 1 - P(b)$ with $P(a) = P(a \wedge b) + P(a \wedge \neg b)$)

Prove also, that $P(\neg a \mid b) = P(\neg a)$

Three independent events

Suppose that the three events **a, b** and **c** are independent of each other

Thus **a and b** are independent and **a and c** are independent
then we shall assume that **a** is independent of the event **b∧c**
then, by definition

$$P(a) = P(a \mid b \wedge c) = \frac{P(a \wedge (b \wedge c))}{P(b \wedge c)} = \frac{P(a \wedge b \wedge c)}{P(b).P(c)}$$

therefore $P(a \wedge b \wedge c) = P(a).P(b).P(c)$

In general, the multiplying rule can be extended to any number of
independent events, if **a, b, c, d, e**.....are independent of each other then

$$P(a \wedge b \wedge c \wedge d...) = P(a).P(b).P(c).P(d)... \qquad \text{Law 8}$$

Example 13

Fred carries out three independent experiments:

Experiment 1: Draw a card from a pack and note its colour, for which

$$P(red) = \tfrac{1}{2} \text{ and } P(black) = \tfrac{1}{2}$$

Experiment 2: Toss a coin and note **head or tail**, for which

$$P(head) = \tfrac{1}{2} \text{ and } P(tail) = \tfrac{1}{2}$$

Experiment 3: Choose a number at random from {1,2,3}, for which

$$P(1) = 1/3, \; P(2) = 1/3 \text{ and } P(3) = 1/3$$

The results of these experiments are independent of each other. The
probability of a head does not depend on whether Fred draws a red or black
card and his choice of a number is completely random.

The possibility space for the three experiments can be represented by a tree diagram:

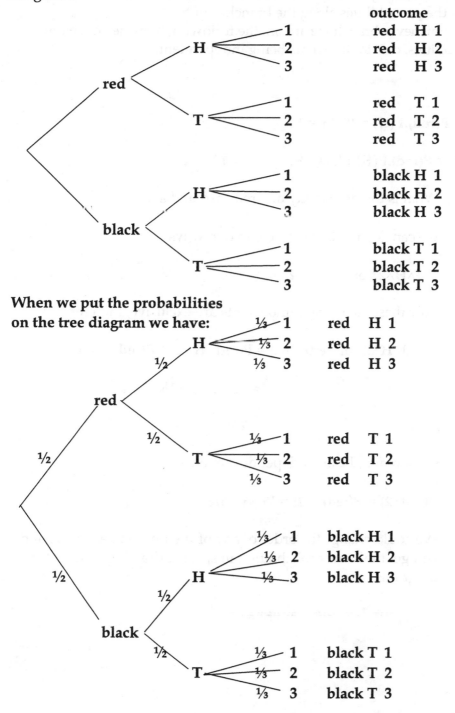

When we put the probabilities
on the tree diagram we have:

When we put the probabilities for each independent event on the tree diagram, the probability for any compound event can be found by multiplying the probabilities along the branches of the tree:
Notice that whichever branch we are on, the following branches have the same probabilities, due to the events being independent.

P(red) = ½

P(red∧H) = P(red).P(H) = ½ .½ = ¼

P(red∧H∧3) = P(red).P(H).P(3) = ½ .½ .⅓ = 1/12

Example 14 What is the probability of a red card and a 2?

Solution 1: we can get a **red∧2** outcome in two ways

        red∧H∧2 or **red∧T∧2**

        now these two compound events are exclusive, therefore

        P(red∧H∧2 or red∧T∧2) = P(red∧H∧2) + P(red∧T∧2)

                              = ½ .½ .⅓ + ½ .½ .⅓

                              = 1/6

Solution 2: since **red** and **2** are independent events,

        P(red∧2) = P(red).P(2) = ½ .⅓ = 1/6

        We could change the order of any of the three experiments and still get the same probabilities, in spite of drawing different tree diagrams.

================================

Exercise 6

[1]   In Fred's three experiments, find the probability of

   [a]   spinning a head and choosing a three

   [b]   choosing a black card and the number 1

[2]

   Toss a coin three times.
   Draw the appropriate tree diagram.
   [a] Given that the first spin is a head, what is the chance of getting two
      tails?
   [b] Given that the second spin is a head, what is the probability of
      getting two tails?
   [c] What is the chance of getting two heads and spinning a head with
      the second toss?
   [d] What is the chance of getting two heads and spinning a tail with
   the second toss?

The Multiplying Principle

You are setting out on a journey from a town called A to a town called C.
The only way to get to C is to pass through a town called B but there are
three roads from A to B and two roads from B to C.
How many ways are there of completing your journey?

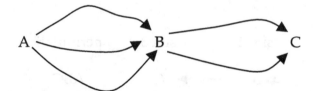

There are three ways of completing the first part of your journey and
whichever way you choose, there are then two ways of completing it.

There are then, 3x2 = 6 ways of going from A to C.

This is an example of the **multiplying principle.**
If there are M ways of completing one action and then N ways of completing
another independent action then there are MxN ways of completing both.
P.S.   The answer to the Mother Goose rhyme is 1 as all the others were
going the other way.

Permutations

A permutation is an arrangement, of some or all of a set of distinguishable objects. Many problems involve permutations of sets of objects and the multiplication principle is a useful tool for their solution:

Example 15

Three cards are marked with the letters A, B and C.
How many arrangements are there if these cards are placed in a row?

Solution

Suppose that we label the position for the cards as slot 1, slot 2 and slot 3:

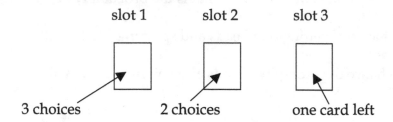

There are 3 ways of choosing which card to place in slot 1.
Having chosen card 1, there will be 2 ways of choosing which card to put in slot 2 and having placed the first two cards, there is only 1 left for slot 3.
Therefore, using the multiplying principle, there are 3x2x1 = 6 ways of placing the cards in the three slots. The solution can be illustrated using a tree diagram:

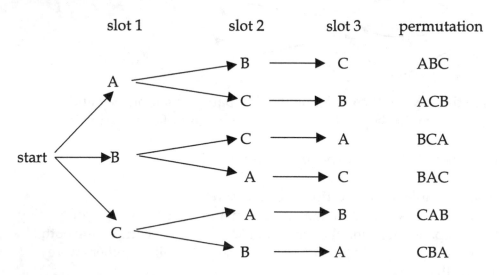

Example 16

A fair coin is tossed three times.

[a] How many different outcomes are there?

[b] What is the probability of getting three heads?

Solution

[a] Mathematically, there is no difference between tossing one coin three times and recording each result and tossing three different coins once each and recording the three results.

Suppose that the three results are called toss 1, toss 2 and toss 3.

Each toss could be either Heads (H) or tails (T).

toss 1	toss 2	toss 3
☐	☐	☐
H or T	H or T	H or T
2 choices	2 choices	2 choices

There are 2 possibilities for toss 1. and after toss 1, there are 2 possibilities for toss 2. After toss 2, there are 2 possibilities for toss 3.

Using the multiplying principle, we have 2x2x2 = 8 outcomes.

The tree diagram for this exercise is as follows;

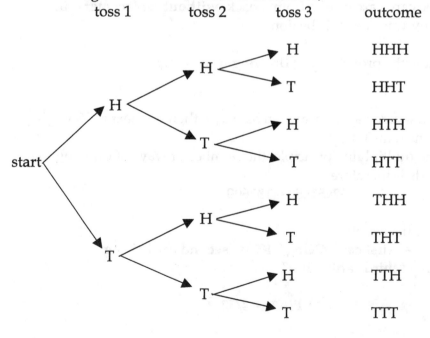

The last column represents the possibility space for the experiment and there are 8 outcomes.

[b]   There are 8 equally likely outcomes and only one of these is **HHH,** therefore

$$P(HHH) = 1/8$$

Alternatively, we may argue that the three tosses are independent

with $P(H) = \frac{1}{2}$ at any stage, therefore

$$P(HHH) = P(H) \times P(H) \times P(H) = \frac{1}{2} \times \frac{1}{2} \times \frac{1}{2} = 1/8$$

==============================

With/without replacement

When the outcome of a previous action can be repeated as in example 16, where we could toss a head as many times as we like, if we are lucky, then we refer to sampling or arranging **with replacement**. Otherwise, we sample **without replacement.**

Example 17

[a] Three cards are drawn from a pack, **without replacement**. In how many ways can this be done?

[b] What is the probability of drawing three kings?

Solution

[a] The first card can be drawn in 52 ways, then the next in 51 ways and the third in 50 ways.
Using the multiplying principle, the number of ways of drawing the three cards is therefore

$$52 \times 51 \times 50 = 132600$$

[b] Define the events

K1 = "first card= King", K2 = "second card=King"
and   K3 = "third card=King"

Then $P(K1) = 4/52,$

$$P(K2 \mid K1) = 3/51$$

$$P(K3 \mid K2 \wedge K1) = 2/50$$

$\therefore$ $P(K3 \wedge K2 \wedge K1) = P(K3 \mid K2 \wedge K1) \times P(K2 \wedge K1)$

$$= P(K3 \mid K2 \wedge K1) \times P(K2 \mid K1) \times P(K1)$$

$$= 2/50 \times 3/51 \times 4/52$$

Alternatively we can argue as follows: The first can be a King in 4 ways, then the next card can be a king in 3 ways and the third card can then be a King in 2 ways
Therefore, using the multiplying principle, the three cards can all be Kings in **4x3x2** ways
The probability of drawing three kings is therefore

$$\frac{4 \times 3 \times 2}{52 \times 51 \times 50}$$

Example 18

[a] Three cards are drawn from a pack, **with replacement**.
   In how many ways can this be done?

[b] What is the probability of drawing three kings?

Solution

[a] The first card can be drawn in 52 ways, then the next in 52 ways and the third in 52 ways.
Using the multiplying principle, the number of ways of drawing the three cards is therefore
$$52 \times 52 \times 52 = 140608$$

[b] Define the events **K1** = "first card= King",
K2 = "second card=King" and **K3** = "third card=King"

Then $P(K1) = 4/52,$
                    $P(K2) = 4/52$
                    $P(K3) = 4/52$

Thus **K1, K2 and K3** are independent events

$\therefore$ $\quad$ P(K1∧K2∧K3) = P(K1)xP(K2)xP(K3)

$$= 4/52 \times 4/52 \times 4/52$$

Alternatively we can argue as follows:
The first can be a King in 4 ways, then the next card can be a King in 4 ways and the third card can be a King in 4 ways
Therefore, using the multiplying principle, the three cards can all be Kings in **4x4x4** ways
The probability of drawing three kings is therefore

$$\frac{4 \times 4 \times 4}{52x52x52}$$

Example 19
How many arrangement are there of three letters
taken from the set A, B, B, C, C, C, D, D, D, D.?

Solution
Here, we suppose that the repeated letters are indistinguishable.
We have to divide the selections into different cases.

Case 1 $\qquad$ Three letters the same:
$\qquad$ selections are CCC, DDD $\qquad$ **2 arrangements**

Case 2 $\qquad$ Two letters the same:
$\qquad$ selections are ABB, ACC, ADD,
$\qquad\qquad\qquad$ BCC, BDD,
$\qquad\qquad\qquad$ CBB, CDD,
$\qquad\qquad\qquad$ DBB, DCC $\qquad$ 9 ways

$\qquad$ For each selection there are 3 arrangements, e.g. ABB, BAB, BBA
$\qquad$ Using the multiplying principle,
$\qquad$ there are **3x9 = 27 arrangements** with two letters the same

Case 3 All letters different : there are 4 selections BCD, CDA, DAB, ABC.
$\qquad$ For each selection there are 3x2x1 = 6 arrangements
$\qquad$ Using the multiplying principle,
$\qquad$ there are **4x6 = 24 arrangements** with all letters different.

$\qquad$ The total number of arrangements is therefore 2+27+24 = 53

Example 20

How many three digit numbers are there?

Solution

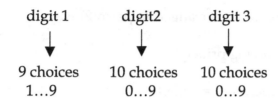

digit 1   digit2   digit 3

9 choices   10 choices   10 choices
1...9   0...9   0...9

Using the multiplying principle, there are 9x10x10 = 900 three digit numbers.

=========================

Exercise 7

[1]   How many arrangements are there of the letters FLOUR?

[2]   How many arrangements are there of the letters FLOOR?

[3]   How many three digit numbers are there that do not have an 8?

[4]   How many arrangements are there of the letters ABC if repetitions are allowed?
How many of these have just two letters the same?

[5]   I throw a dice three times. What is the probability of scoring
[a] no sixes
[b] exactly one six
[c] exactly two sixes
[d] exactly three sixes

==============================

Arranging in a row

If I have two cards ☐ A ☐ and ☐ B ☐ then there are two possible arrangements or

permutations   ☐ A ☐ B ☐   or   ☐ B ☐ A ☐

There are two permutations for two cards and we write

$$\text{perm}(2) = 2\text{x}1 = 2$$

With three cards, $\boxed{A}\;\boxed{B}\boxed{C}$, we have 3 ways of starting the arrangement,

and then two cards left that can be arranged in perm(2) ways.

Therefore, using the multiplying principle,

$$\text{perm}(3) = 3\text{xperm}(2) = 3\text{x}2\text{x}1$$

With four cards $\boxed{A}\;\boxed{B}\boxed{C}\boxed{D}$ there are 4 ways of starting the arrangement

and, whichever card we start with, there will be perm(3) ways of arranging the rest. Therefore, again using the multiplying principle:

$$\text{perm}(4) = 4\text{xperm}(3) = 4\text{x}3\text{x}2\text{x}1$$

We can continue this argument to give

$$\text{perm}(n) = n\text{x}(n\text{-}1)\text{x}(n\text{-}2)\text{x}..3\text{x}2\text{x}1$$

The expression on the right is called **factorial n** and written $\lfloor n \rfloor$ in early

books but nowadays is usually written **n!** and called **n shriek**
Therefore the number of arrangement of **n** different objects in a row is:

$$\textbf{perm(n) = n!}$$

Exercise 8

Using the multiplication principle, show that the number of arrangements of r things taken from n different objects is

$$^{n}P_{r} = \frac{n!}{(n\text{-}r)!}$$

Arranging round a Table

Suppose that we seat n people around a table and we want to know how many seating plans there could be.

Let the number of arrangements be called **table(n)**.

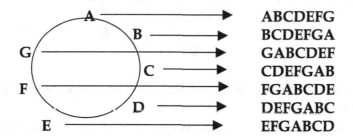

For each clockwise arrangement around the table, we can construct **n** different arrangements in a row because we can choose any of the **n** table positions to start the row.

Therefore, the number of arrangements of the **n** persons in a row, from left to right, is **n times** the number of clockwise arrangements round the table. Therefore

$$\text{perm(n)} = \text{n} \times \text{table(n)}$$

Thus

$$\text{table(n)} = \frac{\text{perm(n)}}{\text{n}} = \frac{\text{n!}}{\text{n}} = \text{(n-1)!}$$

$$\text{table(n)} = \text{(n-1)!}$$

Threading beads on a string

Suppose that the beads **A,B,C,D,E,F,G and H** are to be threaded on a circular necklace that has no start and no end, in other words, there is no visible clasp to indicate the ends of the necklace.

How many different patterns can we form?

It is a similar problem to that of seating people round the table, however, the two arrangements shown in the next figure will be the same when the beads

ABCDEFH are threaded on the necklace because the necklace on the right is the necklace on the left, but turned over.

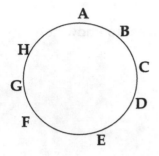 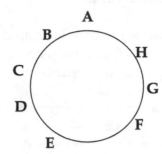

Arrangements of beads on a string that are the same. The left hand picture is the same as the right and picture, but viewed from behind.

If we call the number of arrangements of **n** beads on a ring **ring(n)** then this shows that

$$\text{table(n)} = 2 \times \text{ring(n)}$$

therefore

$$\text{ring(n)} = \tfrac{1}{2}\,\text{table(n)} = \frac{(n-1)!}{2}$$

================================

Example 21

Five friends sit down for a meal together. If they sit at random round the table, what is the probability that Fred will sit next to his wife?

Solution

[1] The total number of arrangements is **table(5) = 4! = 24**

Fred can sit next to his wife in two ways e.g. FW or WF

Therefore, the number of ways in which Fred can sit next to his wife is

$$2 \times \text{table(4)} = 2 \times 3! = 12$$

The probability is then **12/24**

Therefore Fred has a fifty-fifty chance of sitting next to his wife.

================================

alternatively:

[2] Let the three friends plus Fred's wife sit at the table:

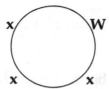

When Fred arrives he sees there are 4 places where he could sit and in two of those, he is next to his wife.
If he chooses a place at random then he has a 2 in 4 chance of sitting next to his wife.

=============================

Exercise 9

[1]   Freda has 6 beads. Two are Red and the other four are Green, Blue, Orange and Yellow. She randomly threads the beads on a string to make a necklace. What is the probability that the two Red beads will be together?

      [a]   If there is a clasp that keeps the ends apart

      [b]   If the beads can slide all the way around the string.

[2]   If **n** people sit a random places round a table show that the probability that two friends do not sit together is $\dfrac{(n-3)}{(n-1)}$

=====================================

Permutations with Indistinguishable Objects

Example 22: How many arrangements are there of the letters **ABBBCDE** ?

Solution:

We start off by assuming that the three letters B are in fact different objects, say **B1 , B2, B3.**

Then we will have 7 different objects that can be arranged in

perm(7) = 7x6x5x4x3x2x1 ways

Now we could arrange these 7 objects in two stages:

Action 1   arrange the objects with identical letters **B,** which is the number we are trying to find, say **N** ways.

Action 2   whatever the outcome of action 1, change the three letters **B** into **B1 , B2, B3** and arrange these letters in their perm(3) ways to get arrangements of the seven different objects:

for example
Action 1 might produce **A B C B B D E**

and then Action 2 produces

A B1 C B2 B3 D E   A B1 C B3 B2 D E   A B2 C B1 B3 D E
A B2 C B3 B1 D E   A B3 C B1 B2 D E   A B3 C B2 B1 D E

Whatever the outcome of action 1, action 2 results in six different arrangements of the seven objects.
Using the multiplying principle, we have

$$\textbf{perm(7)} \ = \ \textbf{N x perm(3)}$$

therefore the result we are looking for is **N** = $\dfrac{\textbf{perm(7)}}{\textbf{perm(3)}}$ = $\dfrac{\textbf{7x6x5x4x3x2x1}}{\textbf{3x2x1}}$

$$= \ \textbf{840}$$

=======================

More general cases

Suppose that we have n objects with p objects alike of one kind, q objects alike of another kind and r objects alike of a third kind.
How many arrangements are there?

Solution

Suppose that these objects can be arranged in **N ways**

Now **n different** objects can be arranged in **perm(n)** ways and we can construct these arrangements using four actions:

Action 1   make an arrangement with **p alike, q alike and r alike.** There are **N** ways of doing this

Action 2   mark the **p** objects and arrange them. There are **perm(p) ways** of doing this

Action 3   mark the **q** objects and arrange them. There are **perm(q) ways** of doing this.

Action 4   mark the **r** objects and arrange them. There are **perm(r) ways** of doing this

After these four independent actions we have an arrangement of **n** different objects and we know that there are **perm(n)** ways of doing this.

Therefore, using the multiplication principle we have

$$\textbf{perm(n) = N x perm(p) x perm(q) x perm(r)}$$

therefore

$$N = \frac{n!}{p!\ q!\ r!}$$

This result of course, extends to any number of groups of indistinguishable objects.

===============================

Example 23

How many arrangements are there of the letters **ABBCCCDDDD**

Solution

> There are 10 letters
> 2 Bs, 3 Cs, 4 Ds.

> Answer $\dfrac{10!}{2!\ 3!\ 4!}$

Exercise 10

[1]   How many arrangements are there of the letters

    [a]   AARDVARK

    [b]   MISSISSIPPI

[2]   How many six digit numbers can be formed from the digits 1,2,2,3,3,3?

[3]   How many five digit numbers can be formed from the digits 1,2,2,3,3,3?

Selections and Combinations

Fred is captain of the college darts team and has to select a team of 5 players to play a rival college. There are eleven players who want to play.
Captain Fred wants to know how many different teams could be selected.
He hits on a simple solution:
Fred writes out the eleven names and places a tick against five selected names and a cross against the others for example:

Alf Bob Chris Dave Ethel Fred Guy Hal Ian Joe Karol
 x   √   √   x   x   √   √   x   x   √   x

The number of different teams Fred can choose is the same as the number of ways of arranging 5 ticks and 6 crosses, which we know is

$$\dfrac{11!}{5!\ 6!}$$

===================

The General result

The number of ways of choosing **r** objects from **n** given objects is the same as the number of ways of arranging **r ticks and (n-r) crosses** and is therefore

$$\frac{n!}{r!\,(n\text{-}r)!}$$

We write $^nC_r = \dfrac{n!}{r!\,(n\text{-}r)!}$ for **"n choose r"**

Note that each selection of **r** things is a rejection of **(n-r)** things, therefore

$$^nC_r = {}^nC_{n\text{-}r}$$

===

Example 24

How many different ways are there of choosing 3 cards from a pack of 52?

Solution

$$52 \text{ choose } 3 = {}^{52}C_3 = \frac{52!}{3! \times 49!} = \frac{52 \times 51 \times 50 \times 49 \times \ldots \times 3 \times 2 \times 1}{3 \times 2 \times 1 \times 49 \times 48 \times 47 \times \ldots \times 3 \times 2 \times 1}$$

$$= \frac{52 \times 51 \times 50}{3 \times 2 \times 1}$$

$$= 22100$$

===================================

Example 25

In the National Lottery, 6 numbers are selected from the numbers 1 to 49. One selection will win.
What are my chances of winning??

Solution

$$49 \text{ choose } 6 = {}^{49}C_6 = \frac{49!}{6! \, 44!} = \frac{49.48.47.46.45.44}{6 . 5 . 4 . 3 . 2 . 1}$$

$$= 13983816$$

My chances of winning are about 1 in 14 million.

==============================

Example 26

In how many ways can three letters be chosen from
ABBCCCDDDD ?
If three letters are chosen at random, what is the probability that they will be the same?

Solution

[a]   3 letters the same   CCC, DDD   2 ways

[b]   2 letters the same: 3 ways for the double
then 3 ways for the 3rd letter: 3x3 = 9 ways

[c]   all different: ${}^4C_3 = 4$ ways

To find the probability that the three letters are the same, we note that there is only 1 chance of getting CCC, but there are 4 chances of getting DDD giving a total of 5 chances that we choose the same three letters.

The total number of equally likely choices is ${}^{10}C_3 = \dfrac{10.9.8}{3.2.1} = 120$

Therefore

$$P(\text{same letters}) = 5/120 = 1/24$$

=====================================

Exercise 11

[1]   Fred decides to go to the gym 3 times per week. How many different timetables can he make up for the three visits?

[2]   Freda has a clock with 12 numbers on the dial. She decided to paint four of the numbers red.
In how many different ways can she do this?

[3]   A football team of 11 players is to be selected from a rugby side of 15 players. How many different football teams could be chosen?
If the selection is completely random, what is the chance that two friends will be selected?

[4]   How many hands of six cards can be dealt from a pack of 52 cards?
What is the chance of being dealt four aces?

[5]   [a] Ten players are to be divided into a group of 6 and a group of 4.
In how many ways can this be done?
What is the chance that two friends end up in the same side?

[b] Ten players are to be divided into 2 teams of 5 players each.
In how many ways can this be done?
What is the chance that two friends end up in the same side?

[6]   Six players are to be divided into 3 teams of 2 each. In how many ways can this be done?
What is the chance that two friends are in the same team?

[7]   [a] In how many ways can 12 players be divided into 4 teams of 3 players each ?
What is the chance that two friends end up in the same side?

[b] In how many ways can 12 players be divided into 3 teams of 4 players each?
What is the chance that two friends end up in the same side?

=================================

Binomial Probability

Example 26

If you throw a dice five times, what is the probability of getting

0 sixes, 1 six, 2 sixes, 3 sixes, 4 sixes or 5 sixes?

Solution

To fix ideas, let us try for 4 sixes.
This can be done in five different ways:

6666x, 666x6, 66x66, 6x666 or x6666

The probability of a six in any toss is **1/6** and the probability of not getting a six is **5/6.**

Since the five tosses are independent of each other, we find the probability of any of these five results by multiplying the independent probabilities.

The probability of each one of these results is therefore

$$\frac{1.1.1.1.5}{6\,6\,6\,6\,6}$$

The five ways of getting the four sixes are also **mutually exclusive,** any two of them cannot happen together. Therefore, we get the total probability of four sixes by adding.

Therefore

$$P(4 \text{ sixes}) = 5 \times \left(\frac{1}{6}\right)^4 \left(\frac{5}{6}\right)$$

Suppose that we now look at getting 3 sixes
First we check how many ways this can be done, e.g **6xx66**

The number of ways of arranging **6 x x 6 6** is $^5C_3$, because it is the same as choosing which 3 throws are to be sixes

Each of these $^5C_3$ ways has the same probability $\left(\frac{1}{6}\right)^3 \left(\frac{5}{6}\right)^2$

Therefore

$$P(3 \text{ sixes}) = {}^5C_3 \left(\frac{1}{6}\right)^3 \left(\frac{5}{6}\right)^2$$

Using the same reasoning, we have:

$$P(0 \text{ sixes}) = {}^5C_0 \left(\frac{1}{6}\right)^0 \left(\frac{5}{6}\right)^5$$

$$P(1 \text{ six}) = {}^5C_1 \left(\frac{1}{6}\right)^1 \left(\frac{5}{6}\right)^4$$

$$P(2 \text{ sixes}) = {}^5C_2 \left(\frac{1}{6}\right)^2 \left(\frac{5}{6}\right)^3$$

$$P(3 \text{ sixes}) = {}^5C_3 \left(\frac{1}{6}\right)^3 \left(\frac{5}{6}\right)^2$$

$$P(4 \text{ sixes}) = {}^5C_4 \left(\frac{1}{6}\right)^4 \left(\frac{5}{6}\right)^1$$

$$P(5 \text{ sixes}) = {}^5C_5 \left(\frac{1}{6}\right)^5 \left(\frac{5}{6}\right)^0$$

Exercise 12

[1]   In 6 tosses of a fair dice,
        write down the probabilities of getting **0, 1, 2, 3, 4, 5 and 6** sixes

[2]   6 cards are drawn from the pack with the card being replaced after
        each   draw.
        What is the chance of getting **4 aces**?

[3]   A coin is tossed ten times.
        What is the probability of getting exactly **6 Heads**?

==================================

Pascal's Triangle

$$1$$
$$1 \quad 2 \quad 1$$
$$1 \quad 3 \quad 3 \quad 1$$
$$1 \quad 4 \quad 6 \quad 4 \quad 1$$
$$1 \quad 5 \quad 10 \quad 10 \quad 5 \quad 1$$

gives the coefficients of x^r in the expansions of $(1+x)^n$,

for example:

$$(1+x)^5 = 1 + 5x + 10x^2 + 10x^3 + 5x^4 + x^5$$

The coefficients of x in the expansion are also the values of

$$^5C_0 \quad ^5C_1 \quad ^5C_2 \quad ^5C_3 \quad ^5C_4 \quad ^5C_5$$

For example, if we wanted to find the x^3 term in the expansion

$$(1+x)^5 = (1+x)(1+x)(1+x)(1+x)(1+x)$$

we observe that each bracket contributes either a **1** or an **x** to the final expansion. There are $^5C_3$ ways of choosing which three brackets should contribute an **x** in order to make a term x^3.

Therefore there are $^5C_3 \ x^3$ terms in the expansion.

Note that the number of ways of choosing zero objects from n objects is one, an so we write $^5C_0 = 1$.

Similarly, the number of ways of choosing n objects from n objects is also 1 so we write $^5C_5 = 1$.

This also gives us a reason for defining $\lfloor 0 = 0! = 1$

so that $^nC_r = \dfrac{n!}{r! \ (n-r)!}$ also works for

$$^nC_0 = \dfrac{n!}{0! \ (n-0)!} = 1 \qquad \text{and} \qquad ^nC_n = \dfrac{n!}{n! \ (n-n)!} = 1$$

Using these findings, we can therefore write

$$(1+x)^5 = {}^5C_0\,x^0 + {}^5C_1 x^1 + {}^5C_2 x^2 + {}^5C_3 x^3 + {}^5C_4 x^4 + {}^5C_5 x^5$$

================================

Note that the sums of the numbers in the rows of Pascal's Triangle are all powers of 2, for example,

$$1 + 2 + 1 = 4$$

$$1 + 3 + 3 + 1 = 8$$

$$1 + 4 + 6 + 4 + 1 = 16$$

If we put **x=1** in the expansion we find 2^n = **sum of coefficients**

thus, for example, we have

$${}^5C_0 + {}^5C_1 + {}^5C_2 + {}^5C_3 + {}^5C_4 + {}^5C_5 = 2^5$$

================================

The probabilities of getting **r** sixes in **5** throws of the dice are connected with the coefficients in the Binomial expansion of $\left(\dfrac{5}{6} + \dfrac{1.t}{6}\right)^5$

This expansion is called the probability generating function for the experiment.

The coefficient of t^r in the expansion is the probability of getting **r** sixes in five throws of the dice.

$$\left(\frac{5}{6}+\frac{1}{6}t\right)^5 = \left(\frac{5}{6}\right)^5 + 5\left(\frac{5}{6}\right)^4\left(\frac{1}{6}\right)t + 10\left(\frac{5}{6}\right)^3\left(\frac{1}{6}\right)^2 t^2 + 10\left(\frac{5}{6}\right)^2\left(\frac{1}{6}\right)^3 t^3 + 5\left(\frac{5}{6}\right)\left(\frac{1}{6}\right)^4 t^4 + \left(\frac{5}{6}\right)^5 t^5$$

For example, the probability of throwing three sixes is given by:

$$P(3 \text{ sixes}) = \text{coefficient of } t^3 = 10\left(\frac{5}{6}\right)^2\left(\frac{1}{6}\right)^3$$

Exercise 13

[1]   During April there is a 75% chance that it will rain on any particular day.

What is the probability that it will rain on 4 of the days in the last week of April?

Show that it is worth betting that it will rain on more than four days in the last week.

[2] **Swop or No Swop**

You are in a games show and the quizmaster shows you three doors. Behind one of the doors is the star prize but behind the other two doors there is nothing. The quizmaster knows where the star prize is. You pick one of the doors and then the quizmaster opens one of the other two, revealing nothing.
He then asks you if you want to swop your first choice with the other closed door.

To swop or not to swop, that is the question?

===================================

Answers

Exercise 1

[1] Let the truth set for a be A
Let the truth set for b be B

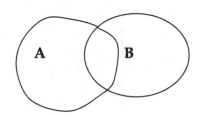

Then

$$A = A \cap B + A \cap \overline{B}$$

and since the two sets on the right hand side are disjoint,

$$n(A) = n(A \cap B) + n(A \cap \overline{B})$$

∴ $$\frac{n(A)}{n(E)} = \frac{n(A \cap B)}{n(E)} + \frac{n(A \cap B)}{n(E)}$$

∴ $$P(a) = P(a \wedge b) + P(a \wedge \neg b)$$

[2] throw two dice and record both numbers

Let **a** = "total score<6"

Let **b** = "we get a 3"

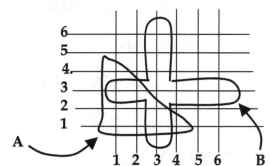

Counting points in the Venn diagram:

$$n(A) = 10, \ n(B) = 11, \ n(A \cap B) = 4, \ n(E) = 36$$

Therefore $P(a) = 10/36$ and $P(b) = 11/36$

We can now reconstruct the Venn diagram for this possibility space showing the numbers of elements in each of its regions:

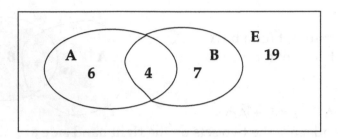

[a] $P(a \wedge b)$ = $\dfrac{n(A \cap B)}{n(E)}$ = 4/36

     $P(\neg a)$ = 1- $P(A)$ = 1 – 10/36 = 26/36

     $P(\neg b)$ = 1 – $P(b)$ = 1- 11/36 = 25/36

[b] $P \neg(a \wedge b)$ = 1 – $P(a \wedge b)$ = 1 – 4/36 = 32/36

     $P(\neg a \vee \neg b)$ = $P(\neg a)$ + $P(\neg b)$ – $P(\neg a \wedge \neg b)$

           = 26/36 + 25/36 – 19/36

           = 32/36

or, better is

$$\overline{A \cap B} = \overline{A} \cup \overline{B} \qquad \text{(De Morgan)}$$

$$\frac{n(\overline{A \cap B})}{n(E)} = \frac{n(\overline{A} \cup \overline{B})}{n(E)}$$

hence $P\neg(a \wedge b)$ = $P(\neg a \vee \neg b)$

[c] $P\neg(a \lor b) = 1 - P(a \lor b) = 1 - 17/36 = 19/36$

$P(\neg a \land \neg b) = 19/36$ from the Venn diagram

Probably better is:

$$\overline{A \cup B} = \overline{A} \cap \overline{B} \qquad \textbf{(De Morgan)}$$

$$\frac{n(\overline{A \cup B})}{n(E)} = \frac{n(\overline{A} \cap \overline{B})}{n(E)}$$

hence $P\neg(a \lor b) = P(\neg a \land \neg b)$

================================

[3]   Given $P(a) = 4/12$, $P(b) = 6/12$ and $P(a \land b) = 2/12$,

we can deduce that

$$P(a \lor b) = P(a) + P(b) - P(a \land b) = 4/12 + 6/12 - 2/12 = 8/12$$

$\therefore$ $P\neg(a \lor b) = 1 - 8/12 = 4/12$

$\therefore$ $P\neg a \land \neg b) = P\neg(a \lor b) = 4/12$

================================

Exercise 2

[1] $P(b \mid a) = \dfrac{P(b \land a)}{P(a)} = \dfrac{3/36}{21/36} = 1/7$

     Or

$$P(b \mid a) = \frac{n(B \cap A)}{n(a)} = \frac{3}{21} = 1/7$$

[2]　　The red and blue dice are thrown and the scores are recorded.

Let **a** = "total score>6"

Let **b** = "red < 3"

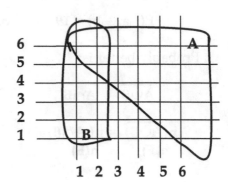

[a]

$$P(a\,|\,b) = 3/12 \ = \ 1/4$$

$$P(b\,|\,a) = 3/21 \ = \ 1/7$$

$$P(a{\wedge}b\,|\,a) = 3/21 \ = \ 1/7$$

[b]　　$P(\neg a\,|\,b) + P(a\,|\,b) \ = \ \dfrac{P(\neg a{\wedge}b)}{P(b)} + \dfrac{P(a{\wedge}b)}{P(b)}$

$$= \ \frac{P(\neg a{\wedge}b) \ + \ P(a{\wedge}b)}{P(b)}$$

$$= \ \frac{P(b)}{P(b)} \qquad \text{Law 3}$$

$$= \ 1$$

[c]　　　　$P(\neg a\,|\,b) \ = \ 1 - P(a\,|\,b)$　　　　**from part [b]**

$$= \ 1 - 1/4 = \ 3/4$$

[d] $$A \cap \overline{B} + \overline{A} \cap B + A \cap B = A \cup B$$

But these are disjoint set with no common members

$$\therefore \quad n(A \cap \overline{B}) + n(\overline{A} \cap B) + n(A \cap B) = n(A \cup B)$$

$$\therefore \quad \frac{n(A \cap \overline{B})}{n(E)} + \frac{n(\overline{A} \cap B)}{n(E)} + \frac{n(A \cap B)}{n(E)} = \frac{n(A \cup B)}{n(E)}$$

$$\therefore \quad P(a \wedge \neg b) + P(\neg a \wedge b) + P(a \wedge b) = P(a \vee b)$$

$$= P(a) + P(b) - P(a \wedge b) = 21/36 + 12/36 - 3/36 = 30/36 = 5/6$$

========================

[3] [a] $P(a \mid b) = \frac{1}{2}$

[b] $P(b \mid a) = 2/3$

[c] $P(a \vee b) = 5/12$

[d] $P(\neg a \wedge \neg b) = 7/12$

==========================

[4]   Two cards are drawn from the pack with replacement.

   Let   **a = "same denomination on both cards"**

           b = " second card is a court card"

   Find **P(a | b) and P(b | a)**

   solution

   We may choose a 13x13 possibility space:

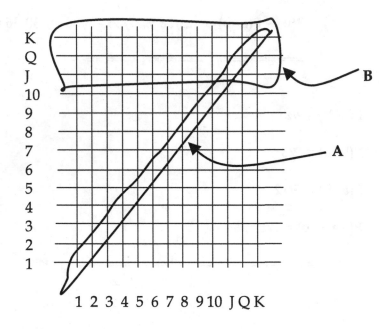

   n(A) = 13, n(B) = 3x13, n(A∩B) = 3

   P(a | b) = 3/ 3x13 = 1/13

   P(b | a) = 3/13

   ==

Exercise 3

[1]   Two cards are drawn from the pack, without replacement. Discuss the probability of getting two spades using

$$P(S1 \wedge S2) = P(S1 \mid S2) \times P(S2)$$

Solution

The probability space has **52x51** equally likely outcomes

The event **S2** consists of **HS, CS, DS and SS**
With respective numbers **13x13 for HS, 13x13 for CS, 13x13 for DS and 13x12 for SS**

Giving the total numbers of outcomes in **S2** as

$$n(S2) = 13x13+13x13+13x13=13x12 = 13x51$$

therefore $P(S1 \mid S2) = \dfrac{n(S1 \cap S2)}{n(S2)} = \dfrac{13x12}{13x51} = 12/51$

also $P(S2) = \dfrac{13x51}{52x51} = 13/52$

$\therefore P(S1 \wedge S2) = P(S1 \mid S2) \times P(S2)$

$$= 12/51 \times 13/52 = \dfrac{12x13}{51x52}$$

=========================

[2]   With replacement,

$$P(S1 \wedge S2) = P(S1) \times P(S2)$$

$$= \dfrac{13x13}{52x52}$$

=============================

[3]   A bag contains 2 red, 3 green and 5 blue balls.
        Three balls are drawn, without replacement.
        Tree diagram:

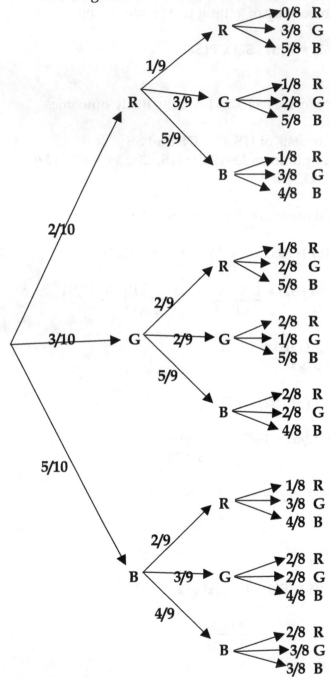

P(2 red)   = 3xP(RRG)+3xP(RRB) = (3x6+3x10)/720   = 48/720
P(2 green) = 3xP(GGR)+3xP(GGB) = (3x12+3x30)/720 = 126/720
P(2 blue)   = 3xP(BBR)+3xP(BBG)   = (3x40+3x60)/720 = 300/720
P(same colour) = P(RRR)+P(GGG)+P(BBB) = (0+6+60)/720 = 66/720
P(all different) = 6xP(RGB)   = 6x30/720 = 180/720

Exercise 4

[1]   n(E)   = 52x51x50

P(xDD) = 39/52 x 13/51 x 12/50

P(DxD) = 13/52 x 39/51 x 12/50

P(DDx) = 13/52 x 12/51 x 39/50

P(2 diamonds) = 3x $\frac{39 \times 13 \times 12}{52 \times 51 \times 50}$ = $\frac{117}{850}$

[2]   1/64

[3]   with replacement,

P(one spade) = P(Sxx)+P(xSx)+P(xxS)

= 3x ¼ x ¾ x ¾ = 27/64

[4]   Two dice are thrown.
a = "score>3"

P(a) = 1 – P(score<=3) = 1 – P({1,1} or {1,2} or {2,1})

= 1 – 3/36 = 33/36

b = "score a double"

P(b) = 6/36

P(a∧b) = P({2,2} ∨ {3,3} ∨ {4,4} ∨ {5,5} ∨{6,6}) = 5/36

P(a|b) = P(a∧b)/P(b) = 5/6

Exercise 5

[1]   Draw a card from the pack and then toss the dice.

     Let **a = "red card",   b = "score > 4"**

     Possibility space:

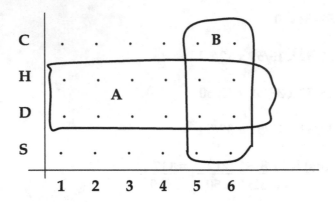

$P(a \mid b) = \dfrac{n(A \cap B)}{n(B)} \quad = \quad 4/8 = 1/2$

$P(a) \quad = 12/24 \quad = \frac{1}{2}$

Therefore $P(a \mid b) = P(a)$

                ======================

$P(b \mid a) = \dfrac{n(A \cap B)}{n(A)} = 4/12 = 1/3$

$P(b) \quad = 8/24 \quad = 1/3$

Therefore $P(b \mid a) = P(b)$

                ======================

$P(a \wedge b) = \dfrac{n(A \cap B)}{n(E)} = 4/24 = 1/6 = P(a) \times P(b)$

                ======================

[2]   **Given** $P(a\,|\,b) = P(a)$ **prove** that $P(a\,|\,\neg b) = P(a)$

Proof

$$P(a \wedge b) = P(a) \times P(b) \qquad \text{(Given)}$$

but $P(a) = P(a \wedge b) + P(a \wedge \neg b)$

∴ $P(a) = P(a) \times P(b) + P(a \wedge \neg b)$

∴ $P(a \wedge \neg b) = P(a)\,\{\,1 - P(b)\,\}$

$$= P(a) \times P(\neg b)$$

∴ $\dfrac{P(a \wedge \neg b)}{P(\neg b)} = P(a)$

therefore $P(a\,|\,\neg b) = P(a)$

=================================

To prove that $P(\neg a\,|\,b) = P(\neg a)$

Proof

$$P(b) = P(b \wedge a) + P(b \wedge \neg a)$$

$$= P(b) \times P(a) + P(b \wedge \neg a)$$

∴ $P(b \wedge \neg a) = P(b)\,\{\,1 - P(a)\,\}$

$P(\neg a \wedge b) = P(b) \times P(\neg a)$

∴ $\dfrac{P(\neg a \wedge b)}{P(b)} = P(\neg a)$

Therefore $P(\neg a\,|\,b) = P(\neg a)$

Thus, if **a** is independent of **b**

then **a** will be independent of ¬**b**

and ¬**a** will be independent of **b**

and further still, since $P(\neg a) = P(\neg a \wedge b) + P(\neg a \wedge \neg b)$

we have $P(\neg a \wedge \neg b) = P(\neg a) - P(\neg a \wedge b)$

$$= P(\neg a) - P(\neg a) \times P(b) \quad \text{from the last result}$$

$$= P(\neg a)\{1 - P(b)\}$$

$$= P(\neg a) \times P(\neg b)$$

so that ¬**a** is also independent of ¬**b**

===========================

Exercise 6

[1]   Fred draws a card and notes **red or black,** then he spins a coin and
       notes **H or T,** and the picks a number and notes **1,2 or 3.**

       [a] $P(H \wedge 3) = \frac{1}{2} \times \frac{1}{3} = \frac{1}{6}$

       [b] $P(B \wedge 1) = \frac{1}{2} \times \frac{1}{3} = \frac{1}{6}$

[2]   [a] $P(TT \mid H1) = \frac{1}{4}$

       [b] $P(T1 \wedge T3 \mid H2) = \frac{1}{4}$

       [c] $P(H1 \wedge H2 \wedge T3) + P(T1 \wedge H2 \wedge H3) = 1/8 + 1/8 = \frac{1}{4}$

       [d] $P(H1 \wedge T2 \wedge H3) = 1/8$

The tree diagram for three independent tosses of a fair coin.

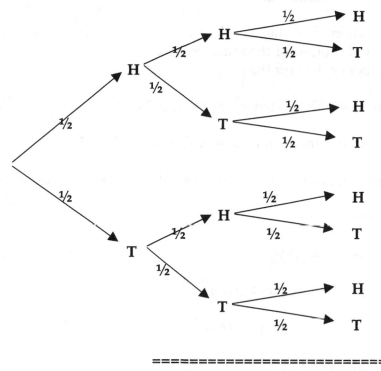

===========================

Exercise 7

[1]   Arrangements of **FLOUR**:

$\underline{|5} = 120$

[2]   Arrangements of **FLOOR**:

½ $\underline{|5} = 60$

[3]   Three digit numbers without number 8:

8	x	9	x	9	
					= 648

1...9   0...9   0...9
  -1   -1   -1
  =8   =9   = 9

[4]   arrange the letters **ABC**, repetitions allowed

$$3 \times 3 \times 3 = 27$$

With just two letters the same,
      number of choices for the same letter ..3
      number of choices for the other letter ..2

number of choices with two letters the same is $3 \times 2 = 6$

for each choice, the number of arrangements is ½ $\lfloor 3 \rfloor$ = 3

answer: number of arrangements with two letters the same = $6 \times 3 = 18$

[5]   I throw a dice three times:

    [a]   **P(no sixes)**   = $(5/6)^3$

    [b]   **P(one six)**   = $3 \times (5/6)^2 \times (1/6)$

    [c]   **P(two sixes)**   = $3 \times (5/6)^1 \times (1/6)^2$

    [d]   **P(three sixes)** = $(1/6)^3$

===============================

Exercise 8

Arranging r things from n

Position   1   2   3   4   ....r

Number of choices   n   (n-1) (n-2) (n-3) ....(n-r+1)

Number of arrangements = $\dfrac{n \times (n-1) \times (n-2) \times \ldots (n-r+1) \times (n-r) \times (n-r-1) \times \ldots \times 3 \times 2 \times 1}{(n-r) \times (n-r-1) \times \ldots \times 3 \times 2 \times 1}$

$$= {}^n P_r = \frac{n!}{(n-r)!}$$

=======================

Exercise 9

[1]   Freda has 6 beads to make a necklace:

2 red, one green, blue, orange and a yellow.

[a]

The total number of ways of threading the beads is **6!**
(we assume that Freda threads the beads on one end an that they do not fall off the other end)

With the red beads together there are **2 x 5!** arrangements since the red beads can be together in two ways, e.g. **r1.r2 or r2.r1**

∴   **P(red beads together) =** $\dfrac{2 \times 5!}{6!}$ **= 1/3**

[b]

The total number of ways of threading the beads is

ring(6) = ½ . 5!

The number of ways ending up with the red beads together is

2 x ring(5) = 2 x ½ 4!

∴   **P(red beads together) =** $\dfrac{2 \times ½ \, 4!}{½ \, 5!}$ **= 2/5**

[2]   n people sit at random places round a table.
       What is the probability that two friends do not sit together?

Solution

The friends can sit together in two ways: e.g. **AB or BA**

The numbers of ways that they can sit together round the table is
therefore
$$2 \text{xtable}(n-1) = 2 \times (n-2)!$$

The total numbers of ways for the **n** people to sit at the table is

$$\text{table(n)} = (n-1)!$$

$$\therefore \text{P(they sit together)} = 2 \times \frac{(n-2)!}{(n-1)!} = \frac{2}{(n-1)}$$

therefore, **P(not together)** $= 1 - \frac{2}{(n-1)} = \frac{(n-3)}{(n-1)}$

===================================

Exercise 10

[1] [a] $\frac{8!}{3! \, 2!} = 3360$ [b] $\frac{11!}{4! \, 4! \, 2!} = 34650$

[2] $\frac{6!}{3! \, 2!} = 60$

[3]   60

==============================

Exercise 11

[1] $^7C_3 = \dfrac{7.6.5}{1.2.3} = 35$

[2] $^{12}C_4 = \dfrac{12.11.10.9}{1.2.3.4} = 495$

[3] $^{15}C_{11} = \dfrac{15.14.13.12}{1.2.3.4} = 1365$

Teams with the two friends can be selected in $^{13}C_9$ ways

P(both selected) $= {}^{13}C_9/{}^{15}C_{11} = \dfrac{13.12.11.10}{15.14.13.12} = \dfrac{11}{21}$

[4] Number of 6 card hands $= {}^{52}C_6 = 20358520$

P(4 aces) $= \dfrac{{}^{48}C_2}{{}^{52}C_6} = \dfrac{3}{54145}$

[5] [a] Ten players into groups of 6 and 5

$^{10}C_6 = 210$

The two friends can be on the same side in

$^8C_4 + {}^8C_2 = 98$ ways

P(on same side) $= \dfrac{98}{210} = 7/15$

[b] Ten players into two groups of 5

$\frac{1}{2} \, {}^{10}C_5 = 126$

Two friends are on the same side in $^8C_3 = 56$ ways

P(on same side) $= \dfrac{56}{126} = 4/9$

[6]   Six players are divided into 3 teams of 2 each:

$$^6C_2 \times {}^4C_2 \div 3! = 15 \text{ ways}$$

With two friends on the same team:

$$4C2 \times 2C2 \div 2 = 3 \text{ ways}$$

P(two friends together) = 3/15 = 1/5

[7]

[a]   Twelve players divided into 4 teams of 3 each:

$$^{12}C_3 \times {}^9C_3 \times {}^6C_3 \div 4! = 15400 \text{ ways}$$

With two friends on the same side:

$$^{10}C_1 \times {}^9C_3 \times {}^6C_3 \div 4! \quad \text{Ways}$$

$$\textbf{P(two friends together)} = \frac{^{10}C_1 \times {}^9C_3 \times {}^6C_3 \div 4!}{^{12}C_3 \times {}^9C_3 \times {}^6C_3 \div 4!} = 1/22$$

[b]   Twelve players divided into 3 teams of 4 each:

$$^{12}C_4 \times {}^8C_4 \div 3! = 5775 \text{ ways}$$

With two friends on the same side:

$$^{10}C_2 \times {}^8C_4 \div 3! \text{ ways}$$

$$\textbf{P(two friends together)} = \frac{^{10}C_2 \times {}^8C_4 \div 3!}{^{12}C_4 \times {}^8C_4 \div 3!} = 1/11$$

=========================

Exercise 12

[1] Using $(1+x)^6 = 1 + 6x + 15x^2 + 20x^3 + 15x^4 + 6x^5 + x^6$

in six tosses of the dice, the probabilities of sixes are:

P(0 sixes) = $(5/6)^6$

P(1 six)   = $6.(5/6)^5.(1/6)$

P(2 sixes) = $15.(5/6)^4.(1/6)^2$

P(3 sixes) = $20.(5/6)^3.(1/6)^3$

P(4 sixes) = $15.(5/6)^2.(1/6)^4$

P(5 sixes) = $6.(5/6).(1/6)^5$

P(6 sixes) = $(1/6)^6$

[2] 6 cards are drawn from the pack with replacement after each draw.
The probability of getting four aces is

P(4 aces) = $^6C_4.(1/13)^4.(12/13)^2$ = $15.(1/13)^4.(12/13)^2$

[3] A coin is tossed ten times. The probability of getting exactly 6 heads
is:

P(6 heads) = $^{10}C_6.(1/2)^6.(1/2)^4$ = $15.(1/2)^{10}$ = 15/1024

===

Exercise13

The chance that it rains on any particular day in April is ¾ .
The chance that it will rain on four of the 7 days in the last week of
April is:

P(rains on four days) = $^{7}C_{4}.(\,¾\,)^{4}.(\,¼\,)^{3}$ = $\underline{2835}$
                                                   16384

The chance that it will rain on more than 4 days is

P(5 days) + P(6 days) + P(7 days)

$$= \,^{7}C_{5}.(\,¾\,)^{5}.(\,¼\,)^{2} \,+\, ^{7}C_{6}.(\,¾\,)^{6}.(\,¼\,) \,+\, (\,¾\,)^{7}$$

$$= \frac{21.3^{5} \,+\, 7.3^{6} \,+\, 3^{7}}{4^{7}} \;=\; \frac{51.3^{5}}{4^{7}} \;=\; 0.76 \;\;(2\text{ d.p.})$$

Therefore, there is roughly a 3 in 4 chance that it will rain on more
than four days.

==

"Swop" v "No Swop"

The case for No Swop

If you stick with your original door, then your chance of winning the
star prize will be 1/3.

The case for Swop

If you have chosen the star door, then you will lose the game.
If you have chosen the wrong door then when you swop, you will win
the game.
Therefore, if you decide to adopt a Swop policy your chances of
winning the star prize will be 2/3 because you have two ways of
choosing the wrong door in the first place.

The Judgement

The court finds in favour of **Swop** which gives a 2/3 chance of winning the star prize.

==

P.S.   If you spin a coin to decide whether to swop or not then your chances of winning will be reduced to 50:50

Index for Volume 1

================   =================